Mathematical
Optimization
Techniques

MATHEMATICAL OPTIMIZATION TECHNIQUES

Edited by Richard Bellman
The RAND Corporation

UNIVERSITY OF CALIFORNIA PRESS
Berkeley and Los Angeles, 1963

University of California Press
Berkeley and Los Angeles, California

Cambridge University Press
London, England

Preface

The papers collected in this volume were presented at the Symposium on Mathematical Optimization Techniques held in the Santa Monica Civic Auditorium, Santa Monica, California, on October 18–20, 1960.

The objective of the symposium was to bring together, for the purpose of mutual education, mathematicians, scientists, and engineers interested in modern optimization techniques. Some 250 persons attended. The techniques discussed included recent developments in linear, integer, convex, and dynamic programming as well as the variational processes surrounding optimal guidance, flight trajectories, statistical decisions, structural configurations, and adaptive control systems.

The symposium was sponsored jointly by the University of California, with assistance from the National Science Foundation, the Office of Naval Research, the National Aeronautics and Space Administration, and The RAND Corporation, through Air Force Project RAND.

Richard Bellman

Santa Monica

Committee for the Symposium

Richard Bellman: The RAND Corporation, Santa Monica
David Blackwell: University of California, Berkeley
Alexander Boldyreff: University of California, Los Angeles
William S. Jewell: Broadview Research Corporation, Burlingame
Robert Kalaba: The RAND Corporation, Santa Monica
Robert M. Oliver: University of California, Berkeley
Ronald W. Shephard: University of California, Berkeley

Welcoming Address

John D. Williams: The RAND Corporation, Santa Monica

Session Chairmen

Mario L. Juncosa: The RAND Corporation, Santa Monica
George B. Dantzig: University of California, Berkeley
Edward W. Barankin: University of California, Berkeley
Lotfi A. Zadeh: University of California, Berkeley
John M. Richardson: Hughes Research Laboratories, Malibu
Joseph P. LaSalle: Research Institute for Advanced Studies,
  Baltimore

Contents

PART FOUR: MODELS, AUTOMATION, AND CONTROL

Introduction

RICHARD BELLMAN

One of the self-imposed, and sometimes unappreciated, tasks of the mathematician is that of providing a choice of firm bases for the quantitative description of natural processes. Among the several motivations for these Herculean labors, the fact that they are interesting is certainly paramount. The physical world has been and continues to be the primary source of intriguing and significant mathematical problems. Although one would think that the armchair philosopher with his ability to conjure up countless infinities of universes could easily create arbitrarily many fascinating fictitious worlds, historically this has not been the case. The mathematician, from all appearances, needs the constant infusion of ideas from the outside. Without these stimuli the pure breed of axiomatics, as pure breeds are wont to do, becomes sterile and decadent.

The second motivation is pragmatic. The task, if successful, has many important ramifications. Predicted results, derived from mathematical models of physical phenomena, can be compared with experimental data obtained from a study of the actual physical phenomena, and thus used to test the validity of the fundamental assumptions of a physical theory. Mathematics can therefore, if not capable of constructing the actual universe by the critique of pure reason, at least play an essential role in demonstrating what hypotheses should not be put forth.

Finally, there is the perennial hope that with sufficient understanding of a physical system will come the ability to control it. Thus, celestial mechanics leads to improved calendars, or at least to theories capable of constructing improved calendars, to more accurate navigation, and to a very profitable and thriving trade in horoscopes; nuclear physics leads to industrial reactors and cancer cures; and so on.

In view of what has been said, it is perhaps natural to expect that purely descriptive studies would precede a theory of control processes.

Historically, this has indeed been the case to a great extent. Yet, even so, seventeenth-century theology led to the postulation of various economical principles of natural behavior that turned out to be of enormous mathematical and physical significance. We know now that there is no clear-cut line of demarcation between descriptive and control processes. It is to a great extent a matter of analytic convenience as to how we propose to derive the basic equations and to conceive of the various physical images of a particular equation. This "as if" quality of mathematics is one of the most powerful aspects of the scientific method.

One of the most interesting and important classes of optimization problems of contemporary technology is that connected with the determination of optimal trajectories for manned and unmanned flights. Initially, these questions can be formulated in terms of the classical calculus of variations. A fundamental quantity associated with the trajectory can be written as a functional,

$$J(x, y) = \int_0^T g(x, y) \, dt, \tag{1}$$

where $x = x(t)$ is the position vector and $y = y(t)$ is the control vector, connected by a vector differential equation,

$$\frac{dx}{dt} = h(x, y), \qquad x(0) = c, \tag{2}$$

and generally by some local and global constraints of the form

$$k_i(x, y) \le 0, \qquad i = 1, 2, \cdots, M,$$

$$\int_0^T r_j(x, y) \, dt \le a_j, \qquad j = 1, 2, \cdots, R. \tag{3}$$

Even if the constraints of (3) are not present, the problem of determining the analytic structure of the optimal trajectory and of the optimal control policy is a difficult one, and the question of computational solution is even more complex. When the constraints are present, we face the delicate juggling act of balancing a set of differential equations and differential inequalities involving state variables and Lagrange multipliers.

The first three chapters, by Miele, Dergarabedian and Ten Dyke, and Breakwell, present the approaches to these matters by means of the conventional calculus of variations. The fourth chapter, by Dreyfus, combines these techniques, the method of successive approximations, and the theory of dynamic programming to provide a new approach. Many further references will be found in these chapters.

Another important area of modern life is that concerned with the communication and interpretation of signals, and with their use in various multistage decision processes such as radar detection and equipment replacement. We find a large area of overlap between the by now classical theory of prediction and filtering of Wiener–Kolmogorov and the modern statistical theories of estimation of Wald, Girshick, Blackwell, and others. The fifth and sixth chapters, by Parzen and Kailath, are devoted to prediction and filtering; the seventh, by Middleton, is a comprehensive study of the formidable optimization problems encountered throughout the theory of communication. The eighth chapter, by Hall, discusses questions of the optimal allocation of effort in testing and experimentation. The last chapter of part two, by Derman, presents an application of dynamic programming to the study of a class of replacement processes.

The third part of the book consists of chapters devoted to geometric and combinatorial questions directly or indirectly connected with linear and nonlinear programming, and to applications of these theories. Linear programming is devoted to the study of ways and means of maximizing the linear form

$$L(x) = \sum_{i=1}^{N} a_i x_i \tag{4}$$

over all x_i subject to the constraints

$$\sum_{j=1}^{N} b_{ij} x_j \leq c_i, \qquad i = 1, 2, \cdots, M. \tag{5}$$

Geometrically, this forces us to examine the vertices of a simplex defined by the inequalities of (5). The chapter by Kruskal is devoted to an aspect of this, and the one by Tucker illustrates the applicability of the simplex method of Dantzig to the systematic exposition of a number of questions in the field of linear inequalities. Thus a method designed primarily as a computational tool turns out to be of fundamental theoretical significance. There is certainly a moral attached to this.

The theory of nonlinear programming is devoted to the study of the maximization of a general function of N variables,

$$F(x) = F(x_1, x_2, \cdots, x_N), \tag{6}$$

over all x_i subject to the constraints

$$G_i(x_1, x_2, \cdots, x_N) \leq 0, \qquad i = 1, 2, \cdots, M. \tag{7}$$

The chapter by Wolfe contains an expository account of some of the principal analytic and computational results concerning this ubiquitous problem.

In a slightly different vein are the chapters by Elfving and Prager. The first considers a problem related to that discussed by Hall, using the ideas of game theory, and the second considers some optimization problems arising in the design of structures.

The fourth part of the volume contains three chapters on automation and control and the use of digital computers. The first, by LaSalle, is a survey of the modern theory of control processes in the Soviet Union, where some of the leading mathematicians and engineers, such as Pontryagin and Letov, are devoting their energies to a determined attack on the theoretical aspects of control theory. The second chapter, by Kalman, shows how the functional-equation technique of dynamic programming can be established along the lines of Hamilton–Jacobi theory and the work of Carathéodory, and discusses some further results in the theory of optimal control. The last chapter, by Bellman, is devoted to a formulation of mathematical model making as an adaptive control process, and thus as a process that can in part be carried out by digital computers.

PART ONE:

AIRCRAFT, ROCKETS, AND GUIDANCE

Chapter 1

A Survey of the Problem of Optimizing Flight Paths of Aircraft and Missiles†

ANGELO MIELE

1. Introduction

This chapter reviews the problems associated with the optimization of aircraft and missile flight paths. From a physical point of view, these problems are of two types: problems of quasi-steady flight and problems of nonsteady flight. The quasi-steady approach, in which the inertia terms appearing in the dynamical equations are regarded as negligible, is of considerable interest along a large part of the flight path of an aircraft powered by air-breathing engines. On the other hand, the nonsteady approach is indispensable in the analysis of rocket-powered aircraft, guided missiles, skip vehicles, and hypervelocity gliders; it is also of interest in the study of the transient behavior of aircraft powered by air-breathing engines.

Regardless of the steadiness or nonsteadiness of the motion, the determination of optimum flight programs requires the study of functional forms that depend on the flight path in its entirety. Thus, the calculus of variations [1] is of primary importance in flight mechanics, even though there are certain simplified problems of quasi-steady flight in which it is by no means an indispensable tool. As a matter of fact, for these simplified problems, the optimization on an *integral* basis by the calculus of variations and the optimization on a *local* basis by the ordinary theory of maxima and minima yield identical results [2], [3], [4].

---

† This paper was presented also at the semiannual meeting of the American Rocket Society, Los Angeles, California, May 9–12, 1960.

However, since *all* optimum problems of the mechanics of flight can be handled by means of the calculus of variations, it follows that the most economical and general theory of the flight paths is a variational theory. The results relative to quasi-steady flight can be obtained as a particular case of those relative to nonsteady flight by letting the acceleration terms appearing in the equations of motion decrease, tending to zero in the limit.

Historical Sketch

Although the application of the calculus of variations to flight mechanics is quite recent, it is of interest to notice that Goddard [5] recognized that the calculus of variations is an important tool in the performance analysis of rockets in an early paper published about 40 years ago. Hamel [6], on the other hand, formulated the problem of the optimum burning program for vertical flight about 30 years ago.

Despite these sporadic attempts, however, the need for an entirely new approach to the problem of optimum aircraft performance was realized by the Germans only during World War II. Lippisch [7], designer of the Messerschmitt 163, investigated the most economic climb for rocket-powered aircraft and shed considerable light on a new class of problems of the mechanics of flight. In the years following World War II, the optimum climbing program of turbojet aircraft attracted considerable interest and was investigated in a highly simplified form by Lush [8] and Miele [9] using techniques other than the indirect methods of the calculus of variations.

A short time later, a rigorous variational formulation of the problem of the optimum flight paths became possible as a result of the work of Hestenes [10], Garfinkel [11], and Cicala [12], [13] on the formulations of Bolza, Lagrange, and Mayer; subsequently, a general theory of these problems was formulated by Breakwell [14], Leitmann [15], Fried [16], and Miele [17]. Incidentally, while the indirect methods of the calculus of variations are of fundamental importance in solving extremal problems, several other optimization techniques have been employed in recent years—more specifically, the theory of dynamic programming [18], [19], [20], the theory of linear integrals by Green's theorem [21], [22], and the gradient theory of optimum flight paths [23].

Since most of the recent developments are based on indirect variational methods and since the results of the quasi-steady theory can be obtained from the variational procedure, this chapter is organized as follows. First, the problems of Bolza, Mayer, and Lagrange are formulated; then, the following problems are reviewed: (a) quasi-steady flight over a flat earth; (b) nonsteady flight over a flat earth; and (c) nonsteady flight over a spherical earth.

2. Techniques of the Calculus of Variations

The calculus of variations is a branch of calculus that investigates minimal problems under more general conditions than those considered by the ordinary theory of maxima and minima. More specifically, the calculus of variations is concerned with the maxima and minima of *functional expressions* in which entire functions must be determined. Thus, the unknown in this case is not a discrete number of points, but rather the succession or the assembly of an infinite set of points—all those identifying a curve, a surface, or a hypersurface, depending upon the nature of the problem.

Applications of the calculus of variations occur in several fields of science and engineering—for instance, classical geometry, elasticity, aerolasticity, optics, fluid dynamics, and flight mechanics. Nevertheless, this branch of mathematics has thus far received little attention from engineers, the probable reason being that the applications described in almost every known textbook (the classical brachistochronic problem, the curve of minimum distance between two given points, the isoperimetric problem of the ancient Greeks, etc.) are either obsolete or susceptible to obvious answers. In the last 15 years, however, the calculus of variations has experienced a revival in engineering. Two fields of problems are mainly responsible for this: applied aerodynamics and the study of the optimum shapes of aircraft components; flight mechanics and the study of the optimum trajectories of aircraft and missiles.

The Problem of Bolza

The most general problems of the calculus of variations in one dimension are the problems of Bolza, Mayer, and Lagrange. Perhaps the simplest way to approach these problems is to study first the problem of Bolza, and then to derive the other two problems as particular cases. Theoretically, however, these three problems are equivalent, since it is known that any one of them can be transformed into another by a change of coordinates [1].

The problem of Bolza is now stated as follows: "Consider the class of functions

$$y_k(x), \qquad k = 1, \cdots, n, \tag{2.1}$$

satisfying the constraints

$$\phi_j(x, y_k, \dot{y}_k) = 0, \qquad j = 1, \cdots, p, \tag{2.2}$$

and involving

$$f = n - p > 0$$

degrees of freedom. Assume that these functions must be consistent with the end-conditions

$$\omega_r(x_i, y_{ki}, x_f, y_{kf}) = 0, \qquad r = 1, \cdots, s < 2n + 2, \tag{2.3}$$

where the subscripts i and f designate the initial and final point, respectively. Find that special set for which the functional form

$$\psi \equiv G(x_i, y_{ki}, x_f, y_{kf}) + \int_{x_i}^{x_f} H(x, y_k, \dot{y}_k)\, dx \tag{2.4}$$

is minimized."

For the particular case $H \equiv 0$, the problem of Bolza is reduced to the problem of Mayer. Furthermore, if $G \equiv 0$, the problem of Bolza is reduced to the problem of Lagrange.

Euler–Lagrange equations. The problem formulated above can be treated in a simple and elegant manner if a set of variable Lagrange multipliers

$$\lambda_j(x), \qquad j = 1, \cdots, p, \tag{2.5}$$

is introduced and if the following expression, called the *fundamental function* or *augmented function*, is formed:

$$F = H + \sum_{j=1}^{p} \lambda_j \phi_j. \tag{2.6}$$

It is known [1] that the *extremal arc*, the special curve extremizing ψ, must satisfy not only the set of equations (2.2) but also the following *Euler–Lagrange equations:*

$$\frac{d}{dx}\left(\frac{\partial F}{\partial \dot{y}_k}\right) - \frac{\partial F}{\partial y_k} = 0, \qquad k = 1, \cdots, n. \tag{2.7}$$

The system composed of the constraining equations and the Euler–Lagrange relations includes $n + p$ equations and unknowns; consequently, its solution yields the n dependent variables and the p Lagrange multipliers simultaneously.

The boundary conditions for this differential system are partly of the fixed end-point type and partly of the natural type. The latter must be determined from the *transversality condition*

$$dG + \left[\left(F - \sum_{k=1}^{n} \frac{\partial F}{\partial \dot{y}_k} \dot{y}_k \right) dx + \sum_{k=1}^{n} \frac{\partial F}{\partial \dot{y}_k} dy_k \right]_i^f = 0, \qquad (2.8)$$

which is to be satisfied identically for all systems of displacements consistent with the prescribed end-conditions.

Discontinuous solutions. There are problems of the calculus of variations that are characterized by discontinuous solutions, that is, solutions in which one or more of the derivatives $\dot{y}_k$ experience a jump at a finite number of points. These points are called *corner points*; the entire solution is still called the *extremal arc*, while each component piece is called a *subarc*.

When discontinuities occur, a mathematical criterion is needed to join the different pieces of the extremal arc. This criterion is supplied by the *Erdmann–Weierstrass corner conditions*, which are written as

$$\left(\frac{\partial F}{\partial \dot{y}_k} \right)_- = \left(\frac{\partial F}{\partial \dot{y}_k} \right)_+, \qquad (2.9)$$

$$\left(F - \sum_{k=1}^{n} \frac{\partial F}{\partial \dot{y}_k} \dot{y}_k \right)_- = \left(F - \sum_{k=1}^{n} \frac{\partial F}{\partial \dot{y}_k} \dot{y}_k \right)_+, \qquad (2.10)$$

where the negative sign denotes conditions immediately before a corner point and the positive sign denotes conditions immediately after such a point.

Incidentally, discontinuous solutions are of particular importance in engineering. In fact, while nature forbids discontinuities on a macroscopic scale, not infrequently the very process of idealization that is intrinsic to all engineering applications leads to a mathematical scheme that forces a discontinuity into the solution.

First integral. A mathematical consequence of the Euler equation is

$$\frac{d}{dx} \left(F - \sum_{k=1}^{n} \frac{\partial F}{\partial \dot{y}_k} \dot{y}_k \right) - \frac{\partial F}{\partial x} = 0. \qquad (2.11)$$

Consequently, for problems in which the augmented function is formally independent of x, the following first integral occurs:

$$-F + \sum_{k=1}^{n} \frac{\partial F}{\partial \dot{y}_k} \dot{y}_k = C, \qquad (2.12)$$

where C is an integration constant. For the case of a discontinuous solution this first integral is valid for each component subarc; furthermore, because of the corner conditions, the constant C has the same value for all the subarcs composing the extremal arc.

Legendre–Clebsch condition. After an extremal arc has been determined, it is necessary to investigate whether the function ψ attains a maximum or a minimum value. In this connection, the *necessary* condition due to Legendre and Clebsch is of considerable assistance. This condition states that the functional ψ attains a minimum if the following inequality is satisfied at all points of the extremal arc:

$$\sum_{k=1}^{n}\sum_{j=1}^{n}\frac{\partial^2 F}{\partial \dot{y}_k \partial \dot{y}_j}\delta\dot{y}_k\delta\dot{y}_j > 0, \tag{2.13}$$

for all systems of variations $\delta\dot{y}_k$ consistent with the constraining equations

$$\sum_{k=1}^{n}\frac{\partial \phi_j}{\partial \dot{y}_k}\delta\dot{y}_k = 0, \qquad j = 1, \cdots, p. \tag{2.14}$$

It is emphasized that condition (2.13) is only a *necessary condition*. The development of a complete sufficiency proof requires that several other conditions be met. For this, the reader is referred to the specialized literature on the subject [1].

The Problem of Mayer with Separated End-Conditions

An important subcase of the Mayer problem is that in which the end-conditions are separated. In this particular problem, the functional to be extremized takes the form

$$\psi \equiv [G(x, y_k)]_i^f, \tag{2.15}$$

while the end-conditions appear as

$$\omega_r(x_i, y_{ki}) = 0, \qquad r = 1, \cdots, q,$$
$$\omega_r(x_f, y_{kf}) = 0, \qquad r = q+1, \cdots, s.$$

It is worth mentioning that, in the general case, the transversality condition reduces to

$$\left[\left(\frac{\partial G}{\partial x} - \sum_{k=1}^{n}\frac{\partial F}{\partial \dot{y}_k}\dot{y}_k\right)dx + \sum_{k=1}^{n}\left(\frac{\partial G}{\partial y_k} + \frac{\partial F}{\partial \dot{y}_k}\right)dy_k\right]_i^f = 0, \tag{2.16}$$

and to

$$\left[\left(\frac{\partial G}{\partial x} - C\right)dx + \sum_{k=1}^{n}\left(\frac{\partial G}{\partial y_k} + \frac{\partial F}{\partial \dot{y}_k}\right)dy_k\right]_i^f = 0 \qquad (2.17)$$

if the fundamental function is formally independent of x.

3. Quasi-Steady Flight over a Flat Earth

Consider an aircraft operating over a flat earth, and assume that the inertia terms in the equations of motion are negligible. Denote by T the thrust, D the drag, L the lift, m the mass, g the acceleration of gravity, X the horizontal distance, h the altitude, V the velocity, γ the inclination of the velocity with respect to the horizon, and ϵ the inclination of the thrust with respect to the velocity. Assume that the drag function has the form

$$D = D(h, V, L)$$

and that the thrust and mass flow of fuel are functions of the following type:

$$T = T(h, V, \alpha),$$
$$\beta = \beta(h, V, \alpha),$$

where α is a variable controlling the engine performance and is called the engine-control parameter, the thrust-control parameter, or the power setting.

With these considerations in mind, we write the equations governing quasi-steady flight in a vertical plane as

$$\phi_1 \equiv \dot{X} - V\cos\gamma = 0, \qquad (3.1)$$
$$\phi_2 \equiv \dot{h} - V\sin\gamma = 0, \qquad (3.2)$$
$$\phi_3 \equiv T(h, V, \alpha)\cos\epsilon - D(h, V, L) - mg\sin\gamma = 0, \qquad (3.3)$$
$$\phi_4 \equiv T(h, V, \alpha)\sin\epsilon + L - mg\cos\gamma = 0, \qquad (3.4)$$
$$\phi_5 \equiv \dot{m} + \beta(h, V, \alpha) = 0, \qquad (3.5)$$

where the dot denotes a derivative with respect to time. These equations contain one independent variable, the time t, and eight dependent variables, $X, h, V, \gamma, m, L, \alpha, \epsilon$. Consequently, three degrees of freedom are left, as is logical in view of the possibility of controlling the time history of the lift, the thrust direction, and the thrust modulus.

Because of the characteristics of the engine, the thrust modulus cannot have any arbitrary value but only those values that are bounded by lower and upper limits. Assuming that the lower limit is ideally zero, we complete the equations (3.1) through (3.5) by the inequality

$$0 \leq T(h, V, \alpha) \leq T_{\max}(h, V),$$

which can be replaced by the constraints

$$\phi_6 \equiv T(h, V, \alpha) - \xi^2 = 0, \tag{3.6}$$

$$\phi_7 \equiv T_{\max}(h, V) - T(h, V, \alpha) - \eta^2 = 0, \tag{3.7}$$

where ξ and η are *real* variables.

Additional Constraints

In many engineering applications it is of interest to study particular solutions of the equations of motion—more specifically, those solutions that simultaneously satisfy either one or two *additional constraints* having the form

$$\phi_8 \equiv A(X, h, V, \gamma, m, L, \alpha, \epsilon) = 0, \tag{3.8}$$

$$\phi_9 \equiv B(X, h, V, \gamma, m, L, \alpha, \epsilon) = 0. \tag{3.9}$$

The effect of these additional constraints is to reduce the number of degrees of freedom of the problem and to modify the Euler–Lagrange equations. Consequently, the solution of the variational problem is altered.

The Mayer Problem

In the class of functions $X(t)$, $h(t)$, $V(t)$, $\gamma(t)$, $m(t)$, $L(t)$, $\alpha(t)$, $\epsilon(t)$, $\xi(t)$, $\eta(t)$, which are solutions of the system composed of (3.1) through (3.9), the Mayer problem seeks the particular set extremizing the difference ΔG between the final and the initial values of an arbitrarily specified function $G = G(X, h, m, t)$.

The Euler–Lagrange equations associated with this variational problem are written as follows:

$$\dot{\lambda}_1 = \lambda_8 \frac{\partial A}{\partial X} + \lambda_9 \frac{\partial B}{\partial X}, \tag{3.10}$$

$$\dot{\lambda}_2 = \lambda_3 \left(\frac{\partial T}{\partial h} \cos \epsilon - \frac{\partial D}{\partial h} \right) + \lambda_4 \frac{\partial T}{\partial h} \sin \epsilon + \lambda_5 \frac{\partial \beta}{\partial h} + \lambda_6 \frac{\partial T}{\partial h}$$

$$+ \lambda_7 \frac{\partial}{\partial h} (T_{\max} - T) + \lambda_8 \frac{\partial A}{\partial h} + \lambda_9 \frac{\partial B}{\partial h}, \tag{3.11}$$

$$0 = -\lambda_1 \cos \gamma - \lambda_2 \sin \gamma + \lambda_3 \left(\frac{\partial T}{\partial V} \cos \epsilon - \frac{\partial D}{\partial V} \right) + \lambda_4 \frac{\partial T}{\partial V} \sin \epsilon$$

$$+ \lambda_5 \frac{\partial \beta}{\partial V} + \lambda_6 \frac{\partial T}{\partial V} + \lambda_7 \frac{\partial}{\partial V} (T_{\max} - T) + \lambda_8 \frac{\partial A}{\partial V} + \lambda_9 \frac{\partial B}{\partial V}, \quad (3.12)$$

$$0 = V(\lambda_1 \sin \gamma - \lambda_2 \cos \gamma) + mg(-\lambda_3 \cos \gamma + \lambda_4 \sin \gamma) + \lambda_8 \frac{\partial A}{\partial \gamma}$$

$$+ \lambda_9 \frac{\partial B}{\partial \gamma}, \quad (3.13)$$

$$\lambda_5 = -g(\lambda_3 \sin \gamma + \lambda_4 \cos \gamma) + \lambda_8 \frac{\partial A}{\partial m} + \lambda_9 \frac{\partial B}{\partial m}, \quad (3.14)$$

$$0 = -\lambda_3 \frac{\partial D}{\partial L} + \lambda_4 + \lambda_8 \frac{\partial A}{\partial L} + \lambda_9 \frac{\partial B}{\partial L}, \quad (3.15)$$

$$0 = \frac{\partial T}{\partial \alpha} (\lambda_3 \cos \epsilon + \lambda_4 \sin \epsilon) + \lambda_5 \frac{\partial \beta}{\partial \alpha} + (\lambda_6 - \lambda_7) \frac{\partial T}{\partial \alpha} + \lambda_8 \frac{\partial A}{\partial \alpha}$$

$$+ \lambda_9 \frac{\partial B}{\partial \alpha}, \quad (3.16)$$

$$0 = T(-\lambda_3 \sin \epsilon + \lambda_4 \cos \epsilon) + \lambda_8 \frac{\partial A}{\partial \epsilon} + \lambda_9 \frac{\partial B}{\partial \epsilon}, \quad (3.17)$$

$$0 = \lambda_6 \xi, \quad (3.18)$$

$$0 = \lambda_7 \eta, \quad (3.19)$$

and admit the first integral

$$V(\lambda_1 \cos \gamma + \lambda_2 \sin \gamma) - \lambda_5 \beta = C,$$

where C is an integration constant. Furthermore, these equations must be solved for boundary conditions consistent with the transversality condition, which is rewritten here as

$$[dG + \lambda_1 \, dX + \lambda_2 \, dh + \lambda_5 \, dm - C \, dt]_i^f = 0.$$

Problems with Three Degrees of Freedom

If there are no additional constraints, that is, if the two functions A and B are identically zero, it is possible to obtain several general results by inspection of equations (3.1) through (3.19).

Concerning the optimization of the thrust direction, (3.15) and (3.17) yield the important result that

$$\epsilon = \arctan \frac{\partial D}{\partial L}, \qquad (3.20)$$

which, for flight at subsonic speeds with a low angle of attack and for a parabolic drag polar, leads to the following conclusion: The flight performance is extremized when the inclination of the thrust axis is equal to twice the downwash angle [14].

Concerning the optimization of the thrust modulus, (3.18) and (3.19) indicate that the extremal arc is discontinuous and is composed of sub-arcs of three kinds [17]:

a. $\xi = 0,$ $\lambda_7 = 0,$

b. $\eta = 0,$ $\lambda_6 = 0,$ (3.21)

c. $\lambda_6 = 0,$ $\lambda_7 = 0.$

Subarcs of type (a) are flown by coasting $(T=0)$; subarcs of type (b) are flown with maximum engine output $(T = T_{max})$; and subarcs of type (c) are flown with continuously varying thrust. The way in which these different subarcs must be combined depends on the nature of the function G and the boundary conditions of the problem. This problem is not analyzed here, because of space considerations, but must be solved with the combined use of the Euler equations, the corner conditions, the Legendre–Clebsch condition, and the Weierstrass condition (see, for instance, [15]).

Problems with One Degree of Freedom

By specifying the form of the functions A and B for particular cases, we can obtain a wide variety of engineering information on the nature of the optimum paths for quasi-steady flight.

Maximum range at a given altitude.   Consider the problem of maximizing the range $(G \equiv -X)$ for a given fuel weight, the flight time being free. If we assume that the trajectory is horizontal and the thrust and the velocity are parallel, the two additional constraints take the form

$$A \equiv \gamma = 0, \qquad (3.22)$$

$$B \equiv \epsilon = 0, \qquad (3.23)$$

and the number of degrees of freedom is reduced to one. After laborious manipulations, it is possible to eliminate the Lagrange multipliers and

to obtain the following result: The optimum path includes subarcs
along which $T = T_{max}$ and subarcs flown with variable thrust along
which

$$J\left(\frac{V/cT}{V} \quad \frac{T - D}{\alpha}\right) = 0, \tag{3.24}$$

where $c = \beta g/T$ is the specific fuel consumption and J is the Jacobian
determinant of the functions V/cT and $T - D$ with respect to the
velocity and the power setting. In an explicit form, (3.24) can be rewritten as

$$\begin{vmatrix} \dfrac{\partial}{\partial V}(V/cT) & \dfrac{\partial}{\partial \alpha}(V/cT) \\[2ex] \dfrac{\partial}{\partial V}(T - D) & \dfrac{\partial}{\partial \alpha}(T - D) \end{vmatrix} = 0.$$

This leads to

$$\frac{\partial}{\partial V}\left(\frac{cD}{V}\right) = 0, \tag{3.25}$$

if the specific fuel consumption is independent of the power setting.

Now, denote the zero-lift drag by D_0 and the induced drag by D_i. If
it is assumed that the drag polar is parabolic with constant coefficients
and that the specific fuel consumption is independent of the speed
(turbojet aircraft operating at low subsonic speeds), (3.25) leads to
the well-known result (see [7]) that $D_i/D_0 = 1/3$. This solution is modified considerably if compressibility effects are considered [4].

Maximum endurance at a given altitude.   A modification of the previous problem consists of maximizing the flight time ($G \equiv -t$), assuming
that $\gamma = \epsilon = 0$, that the fuel weight is given, and that the range is
free. The optimum path includes subarcs $T = T_{max}$ and subarcs along
which

$$J\left(\frac{1/cT}{V} \quad \frac{T - D}{\alpha}\right) = 0. \tag{3.26}$$

This expression reduces to

$$\frac{\partial}{\partial V}(cD) = 0, \tag{3.27}$$

if the specific fuel consumption is independent of the power setting. For the particular case in which the specific fuel consumption is independent of the flight speed and the drag polar is parabolic with constant coefficients, (3.27) leads to $D_i/D_0 = 1$ (see [24]).

Maximum range at a given power setting. Consider the problem of maximizing the range $(G \equiv -X)$ for a given fuel weight, the flight time being free. If we assume that the power setting is given and that the thrust is tangent to the flight path, the additional constraints take the form

$$A \equiv \alpha - \text{Const} = 0, \qquad (3.28)$$

$$B \equiv \epsilon = 0. \qquad (3.29)$$

If the inclination of the trajectory with respect to the horizon is such that $\cos \gamma \cong 1$ and $mg \sin \gamma \ll T$, the following optimizing condition is obtained [4]:

$$J\begin{pmatrix} V/cT & T-D \\ V & h \end{pmatrix} = 0. \qquad (3.30)$$

Numerical analyses indicate that, as the weight decreases as a result of the consumption of fuel, the flight altitude resulting from equation (3.30) increases continuously. The associated flight technique is called cruise-climb [25] and is characterized by a constant Mach number and a constant lift coefficient in the following cases:

a. A turbojet-powered aircraft operating at constant rotor speed in an isothermal stratosphere.

b. A turbojet-powered aircraft operating at a constant *corrected* rotor speed in an arbitrary atmosphere.

For the particular case of a turbojet aircraft flying at low subsonic speeds in the stratosphere [25], the optimum ratio of the induced drag to the zero-lift drag is $D_i/D_0 = 1/2$. Compressibility effects cause a substantial departure from this result [26].

Maximum endurance at a given power setting. A modification of the previous problem consists of maximizing the flight time $(G \equiv -t)$, assuming that the fuel weight is given and the range is free. Retaining all the foregoing maximum-range hypotheses, we may express the optimizing condition by

$$J\begin{pmatrix} 1/cT & T-D \\ V & h \end{pmatrix} = 0. \qquad (3.31)$$

In analogy with the maximum-range discussion, numerical analyses indicate that best endurance is obtained by operating along a cruise-climb trajectory. For the particular case of a turbojet aircraft flying at low subsonic speeds in the stratosphere, best endurance is obtained when $D_i/D_0 = 1$. Thus, the optimum operating altitude is the instantaneous ceiling of the aircraft [27].

Minimum time to climb. Consider the problem of minimizing the time employed in climbing from one altitude to another ($G \equiv t$), retaining constraints (3.28) and (3.29). If we neglect the variations in the weight of the aircraft due to the fuel consumption and consider the horizontal distance traveled by the aircraft as free, the optimizing condition is given by

$$J\left(\frac{V \sin \gamma}{V} \quad \frac{T - D - mg \sin \gamma}{\gamma} \quad \frac{L - mg \cos \gamma}{L}\right) = 0 \quad (3.32)$$

and implies that

$$\frac{\partial}{\partial V}(TV - DV) - mg \frac{\sin^2 \gamma}{\cos \gamma} \frac{\partial D}{\partial L} = 0. \quad (3.33)$$

An important particular case occurs when the induced drag is calculated by approximating the lift with the weight, that is,

$$D_i(h, V, L) \cong D_i(h, V, mg). \quad (3.34)$$

In this case, the optimizing condition reduces to subarcs along which $\cos \gamma = 0$ and subarcs along which

$$\frac{\partial}{\partial V}(TV - DV) = 0. \quad (3.35)$$

Equation (3.35) states that the fastest quasi-steady ascent occurs when the net power (difference between the available power and the power required to overcome the aerodynamic drag) is a maximum with respect to the velocity for a constant altitude.

Most economic climb. Under the hypotheses of the minimum time-to-climb discussion, the climbing technique for minimum fuel consumption ($G \equiv -m$) can be investigated. The optimizing condition is expressed by

$$J\left(\begin{array}{cccc} \dfrac{V \sin \gamma}{cT} & T - D - mg \sin \gamma & L - mg \cos \gamma \\ V & \gamma & L \end{array}\right) = 0 \qquad (3.36)$$

and implies that

$$\frac{\partial}{\partial V}\left(\frac{TV - DV}{cT}\right) - \frac{mg \sin^2 \gamma}{cT \cos \gamma} \frac{\partial D}{\partial L}\left[1 - \frac{\partial \log (cT)}{\partial \log V}\right] = 0.$$

If the induced drag is approximated as in equation (3.34), then the optimizing condition reduces to subarcs along which $\cos \gamma = 0$ and subarcs along which

$$\frac{\partial}{\partial V}\left(\frac{TV - DV}{cT}\right) = 0. \qquad (3.37)$$

Maximum range for a glider. The problem of maximizing the range $(G \equiv -X)$ of a glider $(T = 0)$ is now considered, assuming that the flight time is free. Simple manipulations yield the result that

$$\frac{\partial D}{\partial V} = 0.$$

4. Nonsteady Flight over a Flat Earth

The equations governing the nonsteady flight of an aircraft over a flat earth are written as

$$\phi_1 \equiv \dot{X} - V \cos \gamma = 0, \qquad (4.1)$$

$$\phi_2 \equiv \dot{h} - V \sin \gamma = 0, \qquad (4.2)$$

$$\phi_3 \equiv \dot{V} + g \sin \gamma + \frac{D(h, V, L) - T(h, V, \alpha) \cos \epsilon}{m} = 0, \qquad (4.3)$$

$$\phi_4 \equiv \dot{\gamma} + \frac{g}{V} \cos \gamma - \frac{L + T(h, V, \alpha) \sin \epsilon}{mV} = 0, \qquad (4.4)$$

$$\phi_5 \equiv \dot{m} + \beta(h, V, \alpha) = 0, \qquad (4.5)$$

and must be completed by the inequality relative to the thrust modulus, which is equivalent to

$$\phi_6 \equiv T(h, V, \alpha) - \xi^2 = 0, \qquad (4.6)$$

$$\phi_7 \equiv T_{max}(h, V) - T(h, V, \alpha) - \eta^2 = 0. \qquad (4.7)$$

Considering the possibility of having two additional constraints of the form

$$\phi_8 \equiv A(X, h, V, \gamma, m, L, \alpha, \epsilon) = 0, \qquad (4.8)$$

$$\phi_9 \equiv B(X, h, V, \gamma, m, L, \alpha, \epsilon) = 0, \qquad (4.9)$$

we formulate the Mayer problem as follows: In the class of functions $X(t)$, $h(t)$, $V(t)$, $\gamma(t)$, $m(t)$, $L(t)$, $\alpha(t)$, $\epsilon(t)$, $\xi(t)$, $\eta(t)$ that are solutions of the system composed of equations (4.1) through (4.9), find that particular set that extremizes the difference $\Delta G = G_f - G_i$, where $G = G(X, h, V, \gamma, m, t)$.

The optimum path is described by the equations of motion in combination with the following set of Euler–Lagrange equations:

$$\dot{\lambda}_1 = \lambda_8 \frac{\partial A}{\partial X} + \lambda_9 \frac{\partial B}{\partial X}, \qquad (4.10)$$

$$\dot{\lambda}_2 = \frac{\lambda_3}{m}\left(\frac{\partial D}{\partial h} - \frac{\partial T}{\partial h}\cos \epsilon\right) - \frac{\lambda_4}{mV}\frac{\partial T}{\partial h}\sin \epsilon + \lambda_5 \frac{\partial \beta}{\partial h} + \lambda_6 \frac{\partial T}{\partial h}$$

$$+ \lambda_7 \frac{\partial}{\partial h}(T_{\max} - T) + \lambda_8 \frac{\partial A}{\partial h} + \lambda_9 \frac{\partial B}{\partial h}, \qquad (4.11)$$

$$\dot{\lambda}_3 = -\lambda_1 \cos \gamma - \lambda_2 \sin \gamma + \frac{\lambda_3}{m}\left(\frac{\partial D}{\partial V} - \frac{\partial T}{\partial V}\cos \epsilon\right)$$

$$+ \frac{\lambda_4}{V^2}\left(-g\cos \gamma + \frac{L + T\sin \epsilon}{m}\right) - \frac{\lambda_4}{mV}\frac{\partial T}{\partial V}\sin \epsilon + \lambda_5 \frac{\partial \beta}{\partial V}$$

$$+ \lambda_6 \frac{\partial T}{\partial V} + \lambda_7 \frac{\partial}{\partial V}(T_{\max} - T) + \lambda_8 \frac{\partial A}{\partial V} + \lambda_9 \frac{\partial B}{\partial V}, \qquad (4.12)$$

$$\dot{\lambda}_4 = V(\lambda_1 \sin \gamma - \lambda_2 \cos \gamma) + g\left(\lambda_3 \cos \gamma - \frac{\lambda_4}{V}\sin \gamma\right)$$

$$+ \lambda_8 \frac{\partial A}{\partial \gamma} + \lambda_9 \frac{\partial B}{\partial \gamma}, \qquad (4.13)$$

$$\dot{\lambda}_5 = \frac{\lambda_3}{m^2}(T\cos \epsilon - D) + \frac{\lambda_4}{m^2 V}(L + T\sin \epsilon) + \lambda_8 \frac{\partial A}{\partial m} + \lambda_9 \frac{\partial B}{\partial m}, \quad (4.14)$$

$$0 = \frac{1}{m}\left(\lambda_3 \frac{\partial D}{\partial L} - \frac{\lambda_4}{V}\right) + \lambda_8 \frac{\partial A}{\partial L} + \lambda_9 \frac{\partial B}{\partial L}, \qquad (4.15)$$

$$0 = -\frac{1}{m}\frac{\partial T}{\partial \alpha}\left(\lambda_3 \cos \epsilon + \frac{\lambda_4}{V}\sin \epsilon\right) + \lambda_5\frac{\partial \beta}{\partial \alpha} + (\lambda_6 - \lambda_7)\frac{\partial T}{\partial \alpha}$$

$$+ \lambda_8\frac{\partial A}{\partial \alpha} + \lambda_9\frac{\partial B}{\partial \alpha}, \tag{4.16}$$

$$0 = \frac{T}{m}\left(\lambda_3 \sin \epsilon - \frac{\lambda_4}{V}\cos \epsilon\right) + \lambda_8\frac{\partial A}{\partial \epsilon} + \lambda_9\frac{\partial B}{\partial \epsilon}, \tag{4.17}$$

$$0 = \lambda_6\xi, \tag{4.18}$$

$$0 = \lambda_7\eta. \tag{4.19}$$

These equations admit the following first integral:

$$V(\lambda_1 \cos \gamma + \lambda_2 \sin \gamma) + \lambda_3\left(\frac{T\cos\epsilon - D}{m} - g\sin\gamma\right)$$

$$+ \frac{\lambda_4}{V}\left(\frac{L + T\sin\epsilon}{m} - g\cos\gamma\right) - \lambda_5\beta = C$$

and must be solved for boundary conditions consistent with

$$[dG + \lambda_1\,dX + \lambda_2\,dh + \lambda_3\,dV + \lambda_4\,d\gamma + \lambda_5\,dm - C\,dt]'_i = 0.$$

If there are no additional constraints, that is, if $A \equiv 0$ and $B \equiv 0$, the conclusions of Section 3 concerning problems with three degrees of freedom are still valid. Thus, the optimum thrust direction is supplied by (3.20). Furthermore, the optimum thrust program is described by (3.21) and, therefore, is generally composed of coasting subarcs, maximum-thrust subarcs, and variable-thrust subarcs.

On the other hand, if additional constraints are present, the conclusions depend to a large degree on the form of the functions A and B. In this connection, several particular cases are considered below.

Vertical Ascent of a Rocket

For a rocket-powered vehicle in vertical flight with the thrust parallel to the velocity, the additional constraints are written as

$$A \equiv \gamma - \frac{\pi}{2} = 0, \tag{4.20}$$

$$B \equiv \epsilon = 0. \tag{4.21}$$

After choosing the control parameter identical to the mass flow, we may represent the engine performance by

$$T = \alpha V_E, \tag{4.22}$$

$$\beta = \alpha, \tag{4.23}$$

where V_E is the equivalent exit velocity (assumed constant). Two particular cases are now considered: minimum propellant consumption and minimum time.

Minimum propellant consumption. Consider the problem of minimizing the propellant consumption $(G \equiv -m)$ for given end-values of the velocity and altitude, the flight time being free. Employing the Euler–Lagrange equations, the equations of motion, and the transversality condition and eliminating the Lagrange multipliers, we obtain the following result: The optimum burning program includes coasting subarcs, maximum-thrust subarcs, and variable-thrust subarcs along which (see [28] and [29])

$$D\left(\frac{V}{V_E} - 1\right) + V\frac{\partial D}{\partial V} - mg = 0. \tag{4.24}$$

The way in which these subarcs are to be combined depends on the boundary conditions of the problem. For example, if both the initial and final velocities are zero (case of a sounding rocket), the initial subarc is to be flown with maximum thrust; the intermediate subarc, with variable thrust; and the final subarc, with zero thrust.

Brachistochronic burning program. The burning program minimizing the flight time $(G \equiv t)$ is now considered. Assume that the end values for the velocity, the mass, and the altitude are prescribed; then, the extremal arc is composed of coasting subarcs, maximum-thrust subarcs, and variable-thrust subarcs along which (see [30])

$$\left[D\left(\frac{V}{V_E} - 1\right) + V\frac{\partial D}{\partial V} - mg\right]\exp\left(\frac{V + gt}{V_E}\right) = \text{Const.} \tag{4.25}$$

Level Flight of a Rocket-Powered Aircraft

For a rocket-powered aircraft operating in level flight with the thrust tangent to the flight path, the additional constraints are written as

$$A \equiv \gamma = 0, \tag{4.26}$$

$$B \equiv \epsilon = 0. \tag{4.27}$$

Two particular problems are now considered: maximum range and maximum endurance.

Maximum range. If the range is to be maximized ($G \equiv -X$) for a given propellant mass and given end-velocities, the flight time being free, then the optimum burning program includes coasting subarcs, maximum-thrust subarcs, and variable-thrust subarcs along which (see [31])

$$V\left(D + V_E \frac{\partial D}{\partial V} - mg \frac{\partial D}{\partial L}\right) - V_E D = 0. \qquad (4.28)$$

For a parabolic polar with constant coefficients, (4.28) yields, as a particular case, the results derived in [32].

Maximum endurance. A modification of the previous problem consists of maximizing the flight time ($G \equiv -t$) for a given propellant mass and given end-velocities, the range being free. The optimum burning program includes coasting subarcs, maximum-thrust subarcs, and variable-thrust subarcs along which (see [33])

$$D + V_E \frac{\partial D}{\partial V} - mg \frac{\partial D}{\partial L} = 0. \qquad (4.29)$$

Nonlifting Rocket Trajectories

For the class of nonlifting paths flown with the thrust tangent to the flight path, the additional constraints are written as

$$A \equiv L = 0, \qquad (4.30)$$
$$B \equiv \epsilon = 0. \qquad (4.31)$$

The optimum burning program associated with these paths was determined for problems with no time condition imposed in [34] and for problems in which a condition is imposed on the flight time in [17].

Simplified Analysis of the Climbing Flight of Turbojet Aircraft

A simplified approach to the problem of the optimum climbing technique for a turbojet-powered aircraft is now presented. It is assumed that the power setting is specified and that the thrust is tangent to the flight path, so that the additional constraints take the form

$$A \equiv \alpha - \text{Const} = 0, \qquad (4.32)$$
$$B \equiv \epsilon = 0. \qquad (4.33)$$

It is also stipulated that the variations in the weight of the aircraft due to the fuel consumption are negligible and that the induced drag is calculated by approximating the lift with the weight, that is,

$$D_i(h, V, L) \cong D_i(h, V, mg). \tag{4.34}$$

Two particular problems are now considered: minimum time and minimum fuel consumption.

Brachistochronic climb. Consider the problem of minimizing the flight time $(G \equiv t)$ for given end-values for the velocity and altitude, the horizontal distance being free. The extremal arc is composed of subarcs of three kinds: vertical dives, vertical climbs, and subarcs flown with variable path inclination along which (see [8] and [9])

$$\frac{\partial}{\partial V}(TV - DV) - \frac{V}{g}\frac{\partial}{\partial h}(TV - DV) = 0. \tag{4.35}$$

After defining the energy-height as

$$H = h + \frac{V^2}{2g},$$

and transforming (4.35) from the Vh-domain into the VH-domain, one obtains the well-known result ([8], [35]) that

$$\left[\frac{\partial}{\partial V}(TV - DV)\right]_{H=\text{Const}} = 0,$$

which is the basis of the energy-height method commonly used by aircraft manufacturers. This method consists of plotting the net power as a function of the velocity for constant values of the energy-height and of finding the point at which the net power is a maximum.

Most economic climb. A modification of the previous problem consists of minimizing the fuel consumed $(G \equiv -m)$ for the case in which the time and the horizontal distance are free. The optimum climbing program includes subarcs of three kinds: vertical dives, vertical climbs, and variable path-inclination subarcs along which (see [9])

$$\frac{\partial}{\partial V}\left(\frac{TV - DV}{cT}\right) - \frac{V}{g}\frac{\partial}{\partial h}\left(\frac{TV - DV}{cT}\right) = 0. \tag{4.36}$$

If the problem is transformed into the velocity-energy height domain, this equation can be rewritten as

$$\left[\frac{\partial}{\partial V}\left(\frac{TV - DV}{cT}\right)\right]_{H=\text{Const}} = 0.$$

More General Investigations of Climbing Flight

The preceding investigations were carried out under particular hypotheses, of which the essential analytical objective was to simplify the calculation of the part of the drag that depends on the lift, that is, the induced drag. When we remove the above restrictions, the problem of the optimum climbing program no longer yields analytical solutions (see [36]–[40]).

As an example, consider the problem of extremizing the flight time ($G \equiv t$) for given end-values for the velocity, the altitude, and the path inclination, the horizontal distance being free. If we retain hypotheses (4.32) and (4.33) and neglect the variation in the weight of the aircraft, the optimum path is described by the following equations of motion:

$$\dot{h} = V \sin \gamma, \tag{4.37}$$

$$\dot{V} = \frac{T - D}{m} - g \sin \gamma, \tag{4.38}$$

$$\dot{\gamma} = \frac{1}{V}\left(\frac{L}{m} - g \cos \gamma\right), \tag{4.39}$$

and by the optimum conditions:

$$\lambda_2 = \frac{\lambda_3}{m}\frac{\partial}{\partial h}(D - T), \tag{4.40}$$

$$\lambda_3 = -\lambda_2 \sin \gamma + \lambda_3 \left[\frac{1}{m}\frac{\partial}{\partial V}(D - T)\right.$$

$$\left. + \frac{1}{V}\frac{\partial D}{\partial L}\left(\frac{L}{m} - g \cos \gamma\right)\right], \tag{4.41}$$

$$\lambda_2 V \sin \gamma + \lambda_3\left[\frac{T - D}{m} - g \sin \gamma + \frac{\partial D}{\partial L}\left(\frac{L}{m} - g \cos \gamma\right)\right] = 1. \tag{4.42}$$

The system composed of equations (4.37) through (4.42) involves the six unknown functions $h(t)$, $V(t)$, $\gamma(t)$, $L(t)$, $\lambda_2(t)$, $\lambda_3(t)$ and must be integrated with the help of digital computing equipment. An important complication arises from the fact that this is a boundary-value problem, that is, a problem with conditions prescribed in part at the initial point and in part at the final point. Thus, the use of trial-and-error techniques is an unavoidable necessity. More specifically, the integration of equations (4.37) through (4.42) requires that the following initial values be specified:

$$h(0) = h_i, \qquad V(0) = V_i,$$
$$\gamma(0) = \gamma_i, \qquad L(0) = L_i,$$
$$\lambda_2(0) = \lambda_{2i}, \qquad \lambda_3(0) = \lambda_{3i}.$$

Of these, three (h_i, V_i, γ_i) are known from the initial conditions of the problem, two $(\lambda_{2i}, \lambda_{3i})$ must be guessed, and one (L_i) is to be determined by solving (4.42). As a conclusion, if the multipliers λ_2 and λ_3 are varied at the initial point, a two-parameter family of extremal solutions can be generated. The boundary-value problem consists of determining the particular member of this family that satisfies *all* the conditions prescribed at the final point.

Flight in a Vacuum

The case of a rocket-powered vehicle operating in a vacuum is now considered. Because of the absence of aerodynamic forces, steering can be accomplished only by varying the direction of thrust. The equations of motion and the Euler–Lagrange equations relevant to this problem are obtained from equations (4.1) through (4.19) by setting $A \equiv L = 0$ and $D = 0$.

Problems with two degrees of freedom. If the second additional constraint does not exist $(B \equiv 0)$, several general conclusions can be derived. The optimum thrust direction is supplied by

$$\tan \psi = \frac{C_1 + C_2 t}{C_3 + C_4 t}, \tag{4.43}$$

where $\psi = \gamma + \epsilon$ is the inclination of the thrust with respect to the horizon, and C_1 through C_4 are integration constants. Thus, the inclination of the thrust with respect to the horizon is a bilinear function of time [41].

Concerning the thrust modulus, the optimum flight program includes subarcs of only two kinds: coasting subarcs and maximum-thrust subarcs. No variable-thrust subarc may appear in the composition of the extremal arc. While this result was independently surmised for particular cases ([42]–[44]), the conclusive proof can be found in [15]. In [15], it was also concluded that the extremal path may be composed of no more than three subarcs. The way in which these subarcs must be combined depends on the nature of the function G and the boundary conditions of the problem.

Maximum range. Consider the problem of maximizing the range $(G \equiv -X)$, assuming that the propellant mass is given, that the initial velocity is zero, and that the final altitude is equal to the initial altitude. Assume, also, that the velocity modulus at the final point, the path inclination at the final point, and the time are free. Under these conditions the extremal arc includes only two subarcs, that is, an initial subarc flown with maximum thrust and a final subarc flown by coasting. During the powered part of the flight trajectory, the thrust is inclined at a *constant angle* with respect to the horizon and is perpendicular to the velocity at the final point [45], [46].

Maximum burnout velocity. Consider the problem of maximizing the burnout velocity $(G \equiv -V)$, assuming that the propellant mass is given, the initial velocity is zero, the final altitude is given, the inclination of the final velocity is zero, and the range is free. Concerning the thrust modulus, the trajectory is composed of a maximum-thrust subarc followed by a coasting subarc. Along the maximum-thrust subarc the thrust direction is to be programmed as follows (see [47]):

$$\tan \psi = C_5 + C_6 t. \tag{4.44}$$

Vertical flight. If the flight path is vertical, the additional constraint

$$B \equiv \gamma - \frac{\pi}{2} = 0$$

is to be considered. Consequently, one degree of freedom remains—that associated with the optimization of the thrust modulus.

Consider, now, the problem of maximizing the increase in altitude $(G \equiv -h)$ for given end-velocities and a given propellant mass, the flight time being free. The extremal arc for this problem is composed of only two subarcs, that is, an initial subarc flown with maximum engine output followed by a final subarc flown by coasting.

An interesting case occurs when the flight time is to be extremized $(G \equiv t)$ for given end-velocities and a given propellant mass, the increase in altitude being free. This is a degenerate case, insofar as any arbitrary thrust program $\beta(t)$ is a solution of the Euler–Lagrange equations. Consequently, the flight time is independent of the mode of propellant expenditure.

5. Nonsteady Flight over a Spherical Earth

The equations governing the nonsteady flight of an aircraft in a great circle plane are given by

$$\phi_1 \equiv \dot{X} - \frac{R}{R+h} V \cos \gamma = 0, \tag{5.1}$$

$$\phi_2 \equiv \dot{h} - V \sin \gamma = 0, \tag{5.2}$$

$$\phi_3 \equiv \dot{V} + g\left(\frac{R}{R+h}\right)^2 \sin \gamma + \frac{D(h, V, L) - T(h, V, \alpha) \cos \epsilon}{m} = 0, \tag{5.3}$$

$$\phi_4 \equiv \dot{\gamma} - \frac{V \cos \gamma}{R+h} + \frac{g}{V}\left(\frac{R}{R+h}\right)^2 \cos \gamma - \frac{L + T(h, V, \alpha) \sin \epsilon}{mV}$$

$$+ 2\omega \cos \varphi = 0, \tag{5.4}$$

$$\phi_5 \equiv \dot{m} + \beta(h, V, \alpha) = 0, \tag{5.5}$$

$$\phi_6 \equiv T(h, V, \alpha) - \xi^2 = 0, \tag{5.6}$$

$$\phi_7 \equiv T_{\max}(h, V) - T(h, V, \alpha) - \eta^2 = 0 \tag{5.7}$$

$$\phi_8 \equiv A(X, h, V, \gamma, m, L, \alpha, \epsilon) = 0, \tag{5.8}$$

$$\phi_9 \equiv B(X, h, V, \gamma, m, L, \alpha, \epsilon) = 0, \tag{5.9}$$

where g denotes the acceleration of gravity at sea level, R the radius of the earth, X a curvilinear coordinate measured on the surface of the earth, h the altitude above sea level, ω the angular velocity of the earth, and φ the smaller of the two angles that the polar axis forms with the perpendicular to the plane of the motion.

For the problem of extremizing the difference ΔG between the end-values of an arbitrarily specified function $G(X, h, V, \gamma, m, t)$, the Euler–Lagrange equations are written as follows:

$$\lambda_1 = \lambda_8 \frac{\partial A}{\partial X} + \lambda_9 \frac{\partial B}{\partial X}, \tag{5.10}$$

$$\lambda_2 = \lambda_1 \frac{R}{(R+h)^2} V \cos \gamma$$

$$+ \lambda_3\left[\frac{1}{m}\left(\frac{\partial D}{\partial h} - \frac{\partial T}{\partial h} \cos \epsilon\right) - 2g \frac{R^2}{(R+h)^3} \sin \gamma\right]$$

$$+ \lambda_4\left[-\frac{\sin \epsilon}{mV} \frac{\partial T}{\partial h} + \frac{\cos \gamma}{(R+h)^2}\left(V - \frac{2g}{V}\frac{R^2}{R+h}\right)\right]$$

$$+ \lambda_5 \frac{\partial \beta}{\partial h} + \lambda_6 \frac{\partial T}{\partial h} + \lambda_7 \frac{\partial(T_{\max} - T)}{\partial h} + \lambda_8 \frac{\partial A}{\partial h} + \lambda_9 \frac{\partial B}{\partial h}, \tag{5.11}$$

$$\lambda_3 = -\lambda_1 \frac{R}{R+h} \cos\gamma - \lambda_2 \sin\gamma + \frac{\lambda_3}{m}\left(\frac{\partial D}{\partial V} - \frac{\partial T}{\partial V} \cos\epsilon\right)$$

$$+ \lambda_4\left[\frac{L + T\sin\epsilon}{mV^2} - \frac{\sin\epsilon}{mV}\frac{\partial T}{\partial V} - \frac{\cos\gamma}{R+h}\left(1 + \frac{g}{V^2}\frac{R^2}{R+h}\right)\right]$$

$$+ \lambda_5 \frac{\partial\beta}{\partial V} + \lambda_6 \frac{\partial T}{\partial V} + \lambda_7 \frac{\partial}{\partial V}(T_{\max} - T) + \lambda_8 \frac{\partial A}{\partial V} + \lambda_9 \frac{\partial B}{\partial V}, \quad (5.12)$$

$$\lambda_4 = \lambda_1 \frac{R}{R+h} V\sin\gamma - \lambda_2 V\cos\gamma + \lambda_3 g\left(\frac{R}{R+h}\right)^2 \cos\gamma$$

$$+ \lambda_4 \frac{\sin\gamma}{R+h}\left(V - \frac{g}{V}\frac{R^2}{R+h}\right) + \lambda_8 \frac{\partial A}{\partial\gamma} + \lambda_9 \frac{\partial B}{\partial\gamma}, \quad (5.13)$$

$$\dot\lambda_5 = \frac{\lambda_3}{m^2}\left(T\cos\epsilon - D\right) + \frac{\lambda_4}{m^2 V}(L + T\sin\epsilon) + \lambda_8 \frac{\partial A}{\partial m}$$

$$+ \lambda_9 \frac{\partial B}{\partial m}, \quad (5.14)$$

$$0 = \frac{1}{m}\left(\lambda_3 \frac{\partial D}{\partial L} - \frac{\lambda_4}{V}\right) + \lambda_8 \frac{\partial A}{\partial L} + \lambda_9 \frac{\partial B}{\partial L}, \quad (5.15)$$

$$0 = -\frac{1}{m}\frac{\partial T}{\partial\alpha}\left(\lambda_3 \cos\epsilon + \frac{\lambda_4}{V}\sin\epsilon\right) + \lambda_5 \frac{\partial\beta}{\partial\alpha}$$

$$+ (\lambda_6 - \lambda_7)\frac{\partial T}{\partial\alpha} + \lambda_8 \frac{\partial A}{\partial\alpha} + \lambda_9 \frac{\partial B}{\partial\alpha}, \quad (5.16)$$

$$0 = \frac{T}{m}\left(\lambda_3 \sin\epsilon - \frac{\lambda_4}{V}\cos\epsilon\right) + \lambda_8 \frac{\partial A}{\partial\epsilon} + \lambda_9 \frac{\partial B}{\partial\epsilon}, \quad (5.17)$$

$$0 = \lambda_6\xi, \quad (5.18)$$

$$0 = \lambda_7\eta. \quad (5.19)$$

These equations admit the following first integral:

$$V\left(\lambda_1 \frac{R}{R+h}\cos\gamma + \lambda_2 \sin\gamma\right) + \lambda_3\left[\frac{T\cos\epsilon - D}{m} - g\left(\frac{R}{R+h}\right)^2 \sin\gamma\right]$$

$$+ \lambda_4\left[\frac{L + T\sin\epsilon}{mV} - \frac{g}{V}\left(\frac{R}{R+h}\right)^2 \cos\gamma + \frac{V\cos\gamma}{R+h} - 2\omega\cos\varphi\right]$$

$$- \lambda_5\beta = C.$$

They must be solved for boundary conditions consistent with

$$[dG + \lambda_1 \, dX + \lambda_2 \, dh + \lambda_3 \, dV + \lambda_4 \, d\gamma + \lambda_5 \, dm - C \, dt]'_i = 0.$$

For problems with three degrees of freedom ($A \equiv 0$, $B \equiv 0$), the general conclusions relative to the optimum thrust direction and the optimum thrust program are identical to those obtained in the flat-earth case.

Generally speaking, the equations of motion and the Euler–Lagrange equations must be integrated with the help of digital computing equipment. In a few particular cases, the optimizing condition can be expressed in an explicit form, that is, in a form not involving multipliers. Some of these particular cases are discussed below.

Optimum Thrust Program for Vertical Flight

The additional constraints for a vertically ascending rocket with thrust tangent to the flight path are expressed by (4.20) and (4.21). For the problem of minimizing the propellant consumption ($G \equiv -m$) for given end-values of the velocity and altitude, the burning program is composed of coasting subarcs, maximum-thrust subarcs, and variable-thrust subarcs. If the effects due to the earth's rotation are neglected, the optimum condition for the variable-thrust subarcs is written as (see [48]):

$$D\left(\frac{V}{V_E} - 1\right) + V \frac{\partial D}{\partial V} - mg \left(\frac{R}{R+h}\right)^2 = 0. \qquad (5.20)$$

Optimum Thrust Program for Level Flight

For a rocket-powered aircraft in level flight with the thrust tangent to the flight path, the additional constraints are expressed by (4.26) and (4.27). Consider the problem of maximizing the range ($G \equiv -X$) for a given propellant mass and given end-velocities (free flight time); the burning program includes $T = 0$ subarc, $T = T_{\max}$ subarcs, and variable-thrust subarcs along which (see [46])

$$\left(D - L \frac{\partial D}{\partial L}\right)\left(\frac{V}{V_E} - 1\right) + V \frac{\partial D}{\partial V}$$

$$- mg \frac{\partial D}{\partial L}\left[\frac{V^2}{g(R+h)} + \left(\frac{R}{R+h}\right)^2\right] = 0. \quad (5.21)$$

Optimum Flight Program for Equilibrium Paths

A modification of the previous problem consists of eliminating the altitude constraint and of simultaneously optimizing the burning program and the angle-of-attack program. If the thrust is assumed tangent to the flight path, the first additional constraint is written as

$$A \equiv \epsilon = 0, \tag{5.22}$$

while the second additional constraint is $B \equiv 0$. For a hypervelocity glider boosted by rockets, the following simplifications are permissible:

a. The altitude of the vehicle above sea level is such that

$$\frac{h}{R} \ll 1.$$

b. The slope of the flight path with respect to the local horizon is such that

$$\cos \gamma \cong 1,$$
$$\sin \gamma \cong \gamma,$$
$$mg \sin \gamma \ll D.$$

c. Both the Coriolis acceleration and the part of the centripetal acceleration that is due to the time rate of change of the inclination of the velocity with respect to the horizon are neglected in the equations of motion on the normal to the flight path.†

For the problem of maximizing the range $(G \equiv -X)$ with a given propellant mass and given end-velocity (free time), the angle-of-attack program is such that [49]

$$\frac{\partial D}{\partial h} = 0. \tag{5.23}$$

Therefore, for each instantaneous velocity, the flight altitude is to be adjusted in such a way that the over-all drag is a minimum. Concerning the burning program, the optimum path includes only two subarcs, that is, a maximum-thrust subarc followed by a coasting subarc. No variable-thrust subarc appears in the composition of the extremal arc.

---

† The resulting trajectory is called an equilibrium trajectory, since the weight is balanced by the aerodynamic lift plus the portion of the centrifugal force that is due to the curvature of the earth.

6. Conclusions

It is clear from the present survey that much has been achieved in recent years in the field of terrestrial flight mechanics. Many problems have been conquered. Nevertheless, an even larger domain is still unexplored both from a theoretical standpoint and with regard to practical engineering applications.

There is an immediate need for improved methods for integrating the system of Euler equations and constraining equations and for solving the associated boundary-value problems. An extension of the available closed-form solutions would be of great value for engineering applications. In view of the rather weak character of the maxima and minima of the mechanics of flight, the finding of short cuts and simplifications applicable to particular problems would also be valuable.

At the present time, the work in the area of sufficient conditions for an extremum lags far behind the work accomplished in obtaining necessary conditions. These sufficiency conditions have given rise to questions, some with answers still incomplete or unknown, especially in connection with discontinuous extremal solutions.

In the era of supersonic interceptors, intercontinental missiles, satellites, and interplanetary vehicles, variational methods constitute a much-needed and important step forward in advance performance calculations. It is the opinion of the writer that, as the industry progresses toward faster and faster vehicles, the calculus of variations will become the standard, rather than the specialized, tool for optimum performance analysis of aircraft and missiles.

References

1. Bliss, G. A., *Lectures on the Calculus of Variations*, University of Chicago Press, Chicago, Ill., 1946.
2. Miele, A., "Interrelationship of Calculus of Variations and Ordinary Theory of Maxima and Minima for Flight Mechanics Applications," *ARS Journal*, Vol. 29, No. 1, 1959, pp. 75–76.
3. Miele, A., *Minimal Maneuvers of High-performance Aircraft in a Vertical Plane*, NASA T.N. D-155, 1959.
4. Miele, A., "Lagrange Multipliers and Quasi-steady Flight Mechanics," *J. Aero/Space Sci.*, Vol. 26, No. 9, 1959, pp. 592–598.
5. Goddard, R. H., "A Method for Reaching Extreme Altitudes," *Smithsonian Misc. Collections*, Vol. 71, No. 2, 1921.
6. Hamel, G., "Uber eine mit dem Problem der Rakete zusammenhängende Aufgabe der Variationsrechnung," *Z. Angew. Math. Mech.*, Vol. 7, No. 6, 1927.
7. Lippisch, A., *Performance Theory of Airplanes with Jet Propulsion*,

Headquarters Air Materiel Command, Translation Report F-TS-685-RE, 1946.

8. Lush, K. J., *A Review of the Problem of Choosing a Climb Technique, with Proposals for a New Climb Technique for High Performance Aircraft*, Aeronautical Research Council, R.M. 2557, 1951.

9. Miele, A., "Problemi di Minimo Tempo nel Volo Non-Stazionario degli Aeroplani," *Atti Accad. Sci. Torino*, Vol. 85, 1950–1951.

10. Hestenes, M. R., *A General Problem in the Calculus of Variations with Applications to Paths of Least Time*, The RAND Corporation, RM-100, 1950.

11. Garfinkel, B., "Minimal Problems in Airplane Performance," *Quart. Appl. Math.*, Vol. 9, No. 2, 1951, pp. 149–162.

12. Cicala, P., "Le Evoluzioni Ottime di un Aereo," *Atti Accad. Sci. Torino*, Vol. 89, 1954–1955. Also published as Torino Polytech. Inst., Aero Lab, Monograph 361, 1956.

13. Cicala, P., "Soluzioni Discontinue nei Problemi di Volo Ottimo," *Atti Accad. Sci. Torino*, Vol. 90, 1955–1956, pp. 533–551.

14. Breakwell, J. V., "The Optimization of Trajectories," *J. Soc. Indust. Appl. Math.*, Vol. 7, No. 2, 1959, pp. 215–247.

15. Leitmann, G., "On a Class of Variational Problems in Rocket Flight," *J. Aero/Space Sci.*, Vol. 26, No. 9, 1959, pp. 586–591.

16. Fried, B. D., "General Formulation of Powered Flight Trajectory Optimization Problems," *J. Appl. Phys.*, Vol. 29, No. 8, 1958, pp. 1203–1209.

17. Miele, A., "General Variational Theory of the Flight Paths of Rocket-powered Aircraft, Missiles and Satellite Carriers," *Astronaut. Acta*, Vol. 4, No. 4, 1958, pp. 264–288.

18. Bellman, R., *Dynamic Programming of Continuous Processes*, The RAND Corporation, R-271, 1954.

19. Cartaino, T. F., and S. E. Dreyfus, *Application of Dynamic Programming to the Airplane Minimum Time-to-climb Problem*, The RAND Corporation, P-834, 1957.

20. Bellman, R., and S. Dreyfus, "An Application of Dynamic Programming to the Determination of Optimal Satellite Trajectories," *J. Brit. Interplanet. Soc.*, Vol. 17, Nos. 3–4, 1959, pp. 78–83.

21. Miele, A., *General Solutions of Optimum Problems in Nonstationary Flight*, NACA T.M. 1388, 1955.

22. Miele, A., "Flight Mechanics and Variational Problems of a Linear Type," *J. Aero/Space Sci.*, Vol. 25, No. 9, 1958, pp. 581–590.

23. Kelley, H. J., "Gradient Theory of Optimal Flight Paths," *ARS Journal*, Vol. 30, No. 10, 1960, pp. 947–954.

24. Santangelo, G., "Sulle Caratteristiche di Volo degli Aeroplani con Turboreattore," *L'Aerotecnica*, Vol. 27, No. 5, 1947.

25. Ashkenas, I. L., "Range Performance of Turbo-jet Airplanes," *J. Aero. Sci.*, Vol. 15, No. 2, 1948, pp. 97–101.

26. Miele, A., "Variational Approach to the Stratospheric Cruise of a Turbo-jet Powered Aircraft," *Z. Flugwiss.*, Vol. 6, No. 9, 1958, pp. 253–257.

27. Page, R. K., "Performance Calculation for Jet-propelled Aircraft," *J. Roy. Aero. Soc.*, Vol. 51, No. 437, 1947, pp. 440–450.

28. Tsien, H. S., and R. L. Evans, "Optimum Thrust Programming for a Sounding Rocket," *ARS Journal*, Vol. 21, No. 5, 1951, pp. 99–107.

29. Leitmann, G., "Stationary Trajectories for a High-Altitude Rocket with Drop-away Booster," *Astronaut. Acta*, Vol. 2, No. 3, 1956, pp. 119–124.

30. Miele, A., "Generalized Variational Approach to the Optimum Thrust Programming for the Vertical Flight of a Rocket, Part I, Necessary Conditions for the Extremum," *Z. Flugwiss.*, Vol. 6, No. 3, 1958, pp. 69–77.

31. Cicala, P., and A. Miele, "Generalized Theory of the Optimum Thrust Programming for the Level Flight of a Rocket-powered Aircraft," *Jet Propulsion*, Vol. 26, No. 6, 1956, pp. 443–455.

32. Hibbs, A. R., "Optimum Burning Program for Horizontal Flight," *ARS Journal*, Vol. 22, No. 4, 1952, pp. 204–212.

33. Miele, A., "An Extension of the Theory of the Optimum Burning Program for the Level Flight of a Rocket-powered Aircraft," *J. Aero. Sci.*, Vol. 24, No. 12, 1957, pp. 874–884.

34. Bryson, A. E., and S. E. Ross, "Optimum Rocket Trajectories with Aerodynamic Drag," *Jet Propulsion*, Vol. 28, No. 7, 1958, pp. 465–469.

35. Rutowski, E. S., "Energy Approach to the General Aircraft Performance Problem," *J. Aero. Sci.*, Vol. 21, No. 3, 1954, pp. 187–195.

36. Cicala, P., and A. Miele, "Brachistocronic Maneuvers of a Variable Mass Aircraft in a Vertical Plane," *J. Aero. Sci.*, Vol. 22, No. 8, 1955, pp. 577–578.

37. Behrbohm, H., "Optimal Trajectories in the Vertical Plane," SAAB T.N. 34, 1955.

38. Carstoiu, J., "On a Minimum-time Flight Path of a Jet Aircraft," *J. Aero. Sci.*, Vol. 24, No. 9, 1957, pp. 704–706.

39. Kelley, H. J., "An Investigation of Optimal Zoom Climb Techniques," *J. Aero/Space Sci.*, Vol. 26, No. 12, 1959, pp. 794–802.

40. Theodorsen, T., "Optimum Path of an Airplane—Minimum Time to Climb," *J. Aero/Space Sci.*, Vol. 26, No. 10, 1959, pp. 637–642.

41. Lawden, D. F., "Optimal Rocket Trajectories," *Jet Propulsion*, Vol. 27, No. 12, 1957, p. 1263.

42. Okhotsimskii, D. E., and T. M. Eneev, "Some Variation Problems Connected with the Launching of Artificial Satellites of the Earth," *J. Brit. Interplanet. Soc.*, Vol. 16, No. 5, 1958, pp. 261–262.

43. Miele, A., and J. O. Cappellari, "Topics in Dynamic Programming for Rockets," *Z. Flugwiss.*, Vol. 7, No. 1, 1959, pp. 14–21.

44. Lawden, D. F., "Dynamic Problems of Interplanetary Flight," *Aero. Quart.*, Vol. 6, Nos. 1–4, 1955, pp. 165–180.

45. Fried, B. D., and J. M. Richardson, "Optimum Rocket Trajectories," *J. Appl. Phys.*, Vol. 27, No. 8, 1956, pp. 955–961.

46. Newton, R. R., "On the Optimum Trajectory of a Rocket," *J. Franklin Inst.*, Vol. 266, No. 3, 1958, pp. 155–187.

47. Fried, B. D., "On the Powered Flight Trajectory of an Earth Satellite," *Jet Propulsion*, Vol. 27, No. 6, 1957, pp. 641–643.

48. Miele, A., and J. O. Cappellari, *Some Variational Solutions to Rocket*

Trajectories over a Spherical Earth, Purdue University, School of
Aeronautical Engineering, Report No. A-58-9, 1958.
49. Miele, A., "On the Flight Path of a Hypervelocity Glider Boosted by
Rockets," *Astronaut. Acta*, Vol. 5, No. 6, 1959, pp. 367–379.

For some additional reading, see:

50. Lawden, D. F., "Maximum Ranges of Intercontinental Missiles,"
Aero. Quart., Vol. 8, Nos. 1–4, 1957, pp. 269–278.
51. Lawden, D. F., "Optimum Launching of a Rocket into an Orbit about
the Earth," *Astronaut. Acta*, Vol. 1, No. 4, 1955, pp. 185–190.
52. Fraeijs de Veubeke, B., "Le Problème du Maximum de Rayon d'Action
dans un Champ de Gravitation Uniforme," *Astronaut. Acta*, Vol. 4,
No. 1, 1958, pp. 1–14.
53. Faulkner, F. D., "The Problem of Goddard and Optimum Thrust
Programming," *Proc. Amer. Astronaut. Soc.*, 1956.
54. Ross, S., "Minimality for Problems in Vertical and Horizontal Rocket
Flight," *Jet Propulsion*, Vol. 28, No. 1, 1958, pp. 55–56.
55. Behrbohm, H., "Optimal Trajectories in the Horizontal Plane,"
SAAB T.N. 33, 1955.
56. Leitmann, G., "Trajectory Programming for Maximum Range," *J.
Franklin Inst.*, Vol. 264, No. 6, 1957, pp. 443–452.
57. Baker, G. A., K. W. Ford, and C. E. Porter, "Optimal Accuracy
Rocket Trajectories," *J. Appl. Phys.*, Vol. 30, No. 12, 1959, pp.
1925–1932.
58. Leitmann, G., "The Optimization of Rocket Trajectories: A Survey"
(to be published).
59. Edelbaum, T., *Maximum Range Flight Paths*, United Aircraft Corporation, Report No. R-22465-24, 1955.
60. Irving, J. H., and E. K. Blum, "Comparative Performance of Ballistic
and Low-Thrust Vehicles for Flight to Mars," in M. Alperin and
H. F. Gregory (eds.), *Vistas in Astronautics*, Vol. II, Pergamon Press,
New York, 1959.
61. Miele, A., *Flight Mechanics, Vol. 1: Theory of Flight Paths*, Addison-
Wesley Publishing Company, Inc., Reading, Massachusetts, 1962.
62. Miele, A., *Flight Mechanics, Vol. 2: Theory of Optimum Flight Paths*,
Addison-Wesley Publishing Company, Inc., Reading, Massachusetts
(in preparation).
63. Miele, A., *Extreme Problems in Aerodynamics*, Academic Press, New
York, 1963.

Chapter 2

Estimating Performance Capabilities
of Boost Rockets

P. DERGARABEDIAN
AND R. P. TEN DYKE

1. Introduction

Before an optimization problem can be solved, it is necessary to define
an objective function, a cost function, and a set of constraints. This
chapter reports results of a parametric study of boost rockets. The term
boost rocket includes rockets launched from the surface of the earth for
the purpose of achieving near-orbital or greater velocities. The significant benefit of this study is the derivation of objective functions for use
in problems of ballistic missile preliminary design.

The parameters studied can be divided into two categories: vehicle-
design parameters and trajectory parameters. Vehicle-design parameters describe the physical rocket and include such quantities as weights,
thrusts, propellant flow rates, drag coefficients, and the like. A set of
these parameters would serve as a basic set of specifications with which
to design a vehicle. Trajectory parameters include such quantities as
impact range, apogee altitude, and burnout velocity. Trajectory
parameters can also serve, though not uniquely, as specifications for a
missile system. A particular vehicle system can perform many missions,
and any one mission can be performed by many vehicles. We usually
think of missions in terms of trajectory parameters and vehicles in
terms of design parameters, and the problem becomes one of relating
the two.

The simplest relation is found in the well-known equation:

$$V_i = I_i g \ln r_i, \tag{1.1}$$

where I_i = stage i specific impulse (thrust divided by flow rate of fuel),

g = gravitational constant = 32.2 ft/sec²,

r_i = stage i burnout mass ratio (initial mass divided by burnout mass),

V_i = velocity added during stage i.

[A list of the symbols used in this chapter is provided in Appendix C.]

If several stages are used, the total velocity is the sum of the velocities added during each stage. Certain assumptions used in the derivation of the rocket equation limit its usefulness for boost rockets. They are (a) no gravitational acceleration, (b) no drag, and (c) constant specific impulse. When it becomes necessary to include these effects, the most frequent technique is to solve the differential equations of motion by use of a computing machine. Since some of the inputs to the problem are not analytic, such as drag coefficient as a function of Mach number, the machine uses an integration technique that virtually "flies" the missile on the computer. In this manner, such variables as impact range, apogee altitude, burnout velocity, and burnout altitude can be determined as functions of vehicle-design parameters.

The same vehicle can be flown on many paths, so it is necessary to provide the machine with some sort of steering program. The most frequently used program for the atmospheric part of flight is the "zero-lift" turn. On the assumption that the rocket-thrust vector is aligned with the vehicle longitudinal axis, the vehicle attitude is programmed to coincide with the rocket-velocity vector. For this reason, the zero-lift trajectory is frequently referred to as the "gravity turn." If a rotating earth is used, the thrust is aligned with the velocity vector as computed in a rotating coordinate system. Since the missile is launched with zero initial velocity, a singularity exists for the velocity angle at the instant of launch. All gravity-turn trajectories, regardless of burnout angle, must initiate vertically. For that reason, a mathematical artifice (an initial "kick" angle) is applied to the velocity vector a few seconds after launching to start the turn.

Most problems can be solved very quickly by the computer, and the accuracy of the results is almost beyond question; but there are also disadvantages. First, the actual computer time consumed may be small, but the time required to prepare the input data and arrange for computer time can be quite long in comparison. Second, the degree of accuracy required of results for preliminary design purposes is quite different from that required for, let us say, targeting purposes; the high accuracy offered by the digital machine frequently goes to waste. Finally, while it may be possible to feed the computer one set of data and receive a set of answers, it may at times be preferable to be able to view an analytic relation or graph and get a "feel" for the system as a whole. For these reasons, simplified—even if approximate—solutions to the problem of determining trajectory parameters for boost vehicles are quite useful.

Two techniques may be employed to determine approximate solutions to the differential equations of motion. One technique uses approximation *before* the equations are solved. The original model is transformed into a simpler one for which the solutions are known. In this case one must make a priori guesses as to the accuracy lost in simplification. The digital computer, however, has provided the tool for making approximations *after* solution. The model to be simplified is the solution of the set of differential equations, not the set itself, and the accuracy of the approximations can be readily observed. The latter technique has been employed in this study.

The differential equations are helpful in showing which are the important variables to consider. A short theoretical analysis (Appendix B) shows that the following missile-design parameters, together with a burnout velocity angle, determine a trajectory:

I = vacuum specific impulse, i.e., vacuum thrust divided by flow rate of fuel,

r = burnout mass ratio,

N_0 = ratio of initial (launch) thrust to liftoff weight,

$\dfrac{C_{DM}A}{W_0}$ = drag parameter [C_{DM} is the maximum value for drag coefficient (function of Mach number), A is the reference area, and W_0 is the liftoff weight of missile],

$\dfrac{I_s}{I}$ = ratio of initial (sea-level) specific impulse to vacuum specific impulse,

t_b = burning time = $I_s/N_0(1 - 1/r)$ for constant-weight flow rate.

The trajectory parameters studied are the following:

V_b = burnout velocity,

β_b = velocity burnout angle (with respect to local vertical),

h_b = burnout altitude from the earth's surface,

x_b = surface range at burnout,

R = impact range.

It is clear from the number of parameters studied that it would be impossible simply to plot the results. Therefore, simplification and codification of the results have been a significant part of the study. Results are presented in two forms: (a) a set of general equations for determining V_b, h_b, and x_b as functions of β_b for selected ranges of missile-design parameters, with necessary "constants" used in the equations presented in graphical form; and (b) a simple equation for maximum impact range as a function of missile parameters, together with many of its derivatives.

In addition, a table of equations of several free-flight trajectory parameters based on the Kepler ellipse is included in Appendix A. These equations are well known but are included for convenience. The formulas, together with burnout conditions determined from the computer study, will aid in the solution of a large variety of the problems frequently encountered in preliminary design.

The free-flight trajectory for a vehicle is defined by the velocity and position vectors at burnout. The velocity vector† is defined in terms of its magnitude, V_b, and of the angle, β_b, between it and the local vertical. The position vector is defined by an altitude h and surface range x_b. The quantities V_b, h_b, and x_b are determined as functions of β_b and the vehicle-design parameters.

2. Velocity versus Burnout Angle

The "theoretical" burnout velocity for a vehicle may be determined by equation (2.1). We define the quantity V_L as being the loss in velocity caused by gravitation and atmosphere. Then

$$V_b = V^* - V_L, \tag{2.1}$$

where

$$V^* = \sum V_i = \sum I_i g \ln r_i. \tag{2.2}$$

The following empirical equation for V_L in terms of vehicle-design parameters has been derived by comparing results of several hundred machine trajectory calculations, assuming single-stage vehicles, a gravity turn, and a spherical, nonrotating earth:

$$V_L = (g t_b - K_{g0})\left[1 - K_g\left(1 - \frac{1}{r}\right)\left(\frac{\beta_b}{90°}\right)^2\right]$$

$$+ K_D \frac{C_{DM} A}{W_0} + K_a. \tag{2.3}$$

---

† The term *velocity* will refer to the magnitude of the velocity vector. If the vector is meant, *velocity vector* will be used.

It will be convenient to discuss this equation term by term, so we shall designate the three components as follows:

V_g = gravitational loss

$$= (gt_b - K_{gg})\left[1 - K_g\left(1 - \frac{1}{r}\right)\left(\frac{\beta_b}{90°}\right)^2\right], \tag{2.4}$$

$$V_D = \text{drag loss} = K_D\frac{C_{DM}A}{W_0}, \tag{2.5}$$

$$V_a = \text{nozzle-pressure loss} = K_a \tag{2.6}$$

Gravitational Loss

The gravitational loss was determined by setting the drag equal to zero and flying the vehicle to several burnout angles. The term gt_b is the gravitational loss to be expected from a vertical flight in a constant gravitational field. A realistic gravitational field varies as the inverse square of the distance from the earth's center, so the term actually overestimates this loss. For ranges of vehicles using currently available propellants, the differences between the amount gt_b and the correct gravity loss will be small; and for this equation the difference has been included as the constant K_{gg}. The term

$$\left[1 - K_g\left(1 - \frac{1}{r}\right)\left(\frac{\beta_b}{90°}\right)^2\right]$$

fits a curve as a function of β_b. The constant K_g was determined by a least-squares curve-fitting technique and usually resulted in a curve fit that was within 30 ft/sec of the machine results. The above form was found to fit actual results better than a more obvious choice, $K \cos \beta_b$, which resulted in maximum differences of 300 ft/sec. Curves for K_g as a function of I and N_0 are found in Figure 1, and a curve for K_{gg} as a function of I is found in Figure 2.

Drag Velocity Loss

The velocity lost to drag is proportional to the quantity $C_{DM}A/W_0$, in which C_{DM} has been chosen as a single parameter to define all drag curves. The reasons for this choice are that (a) most realistic drag curves have approximately the same form, except for the absolute magnitudes of the values, and (b) the greater portion of the drag loss occurs early in powered flight, where C_D attains a maximum. The actual drag curve used in the machine trajectory calculation is shown in Figure 3. The empirical constant K_D was obtained by computing the

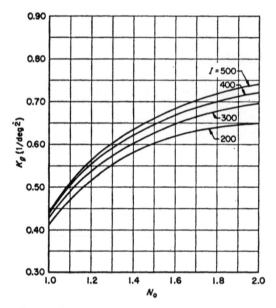

Fig. 1. K_g as a function of vacuum specific impulse, I, and initial thrust-to-weight ratio, N_0.

difference between burnout velocities for similar vehicles with and without drag. All comparisons were made for identical burnout angles. The constant was found to be a function of I_s/N_0, β_b, and N_0. The function K_D was so weakly dependent upon N_0, however, that this effect was disregarded for simplicity in presenting the results; K_D is shown in Figure 4 as a function of I_s/N_0 and β_b.

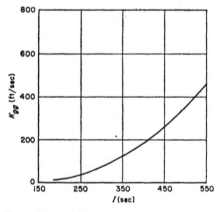

Fig. 2. K_{gg} as a function of vacuum specific impulse, I.

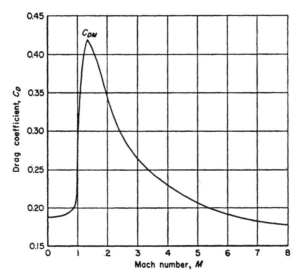

Fig. 3. Drag coefficient, C_D, as a function of Mach number.

Nozzle-Pressure Loss

For the same propellant flow rate, the effective thrust at sea-level ambient pressure is less than in a vacuum. This may be thought of as

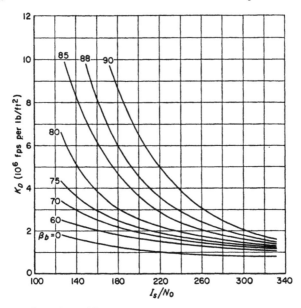

Fig. 4. K_D as a function of burnout velocity angle, β_b, and ratio of sea-level specific impulse to initial thrust-to-weight ratio, I_s/N_0.

a change in specific impulse. The ratio of sea-level specific impulse to vacuum specific impulse is dependent upon the chamber pressure, nozzle area-expansion ratio, and ratio of specific heats for the combustion products. Thrust coefficient tables are readily available to provide this information. It was again assumed that the greater portion of the losses would occur early in flight, and all losses were computed for vertical trajectories. The results are given in Figure 5, where K_a is plotted as a function of I_s/I.

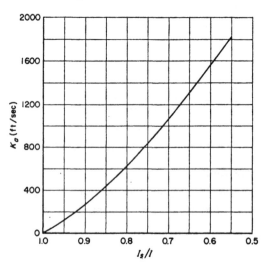

Fig. 5. K_a as a function of ratio of sea-level to vacuum specific impulse, I_s/I.

Accuracy of Results

Accuracies to within 150 ft/sec should be expected with the above results. Occasionally, cases may occur exceeding these limits. First, drag curves may not actually be similar to the one selected for this study. Second, simplification of the results to a form facilitating rapid computation has necessitated several approximations. It is believed that the results as presented will be more useful in preliminary design than extremely accurate results would be. On the assumption that the typical first stage is designed to achieve about 10,000 ft/sec, the accuracy of 150 ft/sec amounts to 1.5 per cent.

Application to More Than One Stage

All computations were performed for single-stage vehicles, but the results may be applied to multistage vehicles.

If the first stage can be assumed to burn out at greater than 200,000 ft at a velocity angle less than 75°, the drag losses may be assumed to have occurred during first stage. It is important to note that the constant K_D is determined on the basis of the velocity burnout angle for the first stage. For multistage vehicles, this angle may be 5° to 15° less than the angle at final-stage burnout; but for β_b less than 75°, the drag losses are relatively insensitive to β_b, and any reasonable estimate will probably be satisfactory.

Under almost any circumstances, the nozzle-pressure loss can be considered to occur during the first stage. Constants applicable to the first stage should be used.

The most significant velocity loss from succeeding stages is gravitational loss. Since the velocity angle is more constant during succeeding stages, it is usually satisfactory to assume a constant value between the assumed burnout of the first stage and the desired final burnout angle. Then the velocity loss for succeeding stages may be computed as

$$V_{L2} = g \left(\frac{R_e}{R_e + \bar{h}} \right)^2 t_{b2} \cos \bar{\beta}, \qquad (2.7)$$

where $\bar{\beta}$ is an intermediate velocity angle, $\bar{h}$ is an "average" altitude for second-stage powered flight, and the subscript 2 refers to succeeding stages. The difference between burnout angles of the first stage and that for the final burnout will depend on the thrust pitch program selected for succeeding stages. Several authors have discussed the optimum pitch program for a variety of missions assuming powered flight in a vacuum (see [1]–[4]). For a ballistic missile, with impact range the desired result, holding the thrust vector constant with respect to a stationary inertial coordinate system has been found to yield greater ranges than the gravity turn. For this case, the change in β from first-stage burnout to final burnout will be comparatively small. In contrast, many satellite missions require that burnout angles approach or equal 90°. Under these circumstances, a gravity turn or one in which the vehicle is pitched downward is a more likely trajectory. The resulting difference in burnout angles between first and final stages will be quite large.

In any trajectory in which thrust is not aligned with velocity, some energy is expended in "turning" the velocity vector. The proportion of the thrust that goes to increasing the velocity varies as the cosine of the angle of the attack; hence for small angles of attack the loss is small.

Effect of the Earth's Rotation

The significant parameter in determining performance is the inertial velocity. Thus, the velocity of the launch point must be considered in any realistic calculation. A simple, albeit approximate, correction may be made by taking the vector sum of the inertial velocity vector of the launch point and the vehicle velocity vector at burnout. In several comparisons between this approximate technique and that of a machine-determined trajectory for an eastward launch on a rotating earth, this approximation underestimated the actual burnout velocity. It has not been determined whether this is generally true; but the few comparisons indicate that the approximation tends toward conservative results.

3. Burnout Altitude versus Burnout Angle

The burnout altitude is a particularly important parameter in determining payload capabilities for low-altitude satellites with circular orbits. As with the rocket equation, a closed-form expression may be derived for the distance traversed by an ideal rocket in vertical flight (constant g, no drag, constant specific impulse), namely

$$h^* = gIt_b\left(1 - \frac{\ln r}{r - 1}\right) - \frac{gt_b^2}{2}. \tag{3.1}$$

It was found that the above form could be modified to account for drag, nozzle pressure, and burnout angle as follows:

$$h_b = \left[h^* - \frac{(V_D + V_a)t_b}{2}\right]\left[1 - \left(\frac{\beta}{K_h}\right)^2\right], \tag{3.2}$$

where

$$K_h = 93 + \frac{28}{r}[1 + 5(2 - N_0)^2], \qquad 1 \le N_0 \le 2. \tag{3.3}$$

Equation (3.2) assumes that the drag and nozzle-pressure losses are averaged over the duration of flight. This is not exactly true, but the approximation has proven to be satisfactory because the correction is small. The constant K_h has been determined empirically. Accuracies for (3.2) have been found to agree with machine calculations to about 20,000 ft.

In calculating values for multistage vehicles, (3.2) yields the altitude of burnout for the first stage. The additional altitude achieved during succeeding stages may be calculated by using the first-stage burnout velocity as computed from (2.3) and the following relation, derived by integrating $Ig \ln r - gt \cos \beta$ at a constant, average flight path angle, $\bar{\beta}$:

$$h_{b2} = h_{b1} + V_{b1}t_{b2} \cos \bar{\beta}$$

$$+ \left\{gI_2t_{b2}\left(1 - \frac{\ln r_2}{r_2 - 1}\right) - \frac{gt_{b2}^2 \cos \bar{\beta}}{2}\right\} \cos \bar{\beta}, \tag{3.4}$$

where the subscripts 1 and 2 refer to the first and second stage, respectively. The above form may be extended to cover additional stages. Again, an intermediate value for the flight path angle β may be selected between the estimated first-stage burnout flight path angle and the desired final burnout angle.

No correction is suggested for use with a rotating earth. In several comparisons with machine trajectories assuming an eastward launch on a rotating earth, the altitude value for the nonrotating earth was approximately equal to that for the rotating earth.

4. Burnout Surface Range

The surface range at burnout may be determined by the following empirical expression:

$$x_b = 1.1h^* \left(\frac{\beta_b}{90°} \right). \tag{4.1}$$

The surface range is the least important of the trajectory parameters in determining gross vehicle performance. It is important, however, in that it adds to the impact range of a surface-to-surface ballistic missile. Again, no correction is offered for the rotating earth because, for reasonably short flight duration, the increased inertial velocity of the vehicle and the velocity of the launch point may be assumed to cancel. Equation (4.1) has been found to yield surface range at burnout within an accuracy of about 10 per cent.

For multistage vehicles, the same technique used in determining altitude may be applied, thus:

$$x_{b2} = x_{b1} + (h_{b2} - h_{b1}) \tan \bar{\beta}. \tag{4.2}$$

5. Free-Flight Trajectory

The calculation of the burnout conditions of a vehicle is only an intermediate step in determining its performance. Performance is usually measured in terms of impact range, apogee altitude, or some other end condition. Since all vehicles in free-flight follow a Kepler ellipse, values for range, apogee altitude, and the like may be determined from the burnout conditions by using equations yielding these values in closed form. A number of these equations are listed in the first part of Appendix A.

6. Range Equation

Experience in the optimization of performance of medium- and long-range missiles at Space Technology Laboratories has shown that the trajectory consisting of a short period of vertical flight, followed by a gravity turn to staging and a constant attitude (thrust angle with respect to launch coordinate system) throughout subsequent stages of flight, yields a near-optimum range trajectory.

In the case of a single-stage missile, the constant-attitude part of the trajectory is initiated at an altitude of approximately 150,000 ft. The velocity angle of the missile at burnout is optimized for maximum range. An examination of the trajectory equations shows that the range of a missile is determined by specifying the same vehicle-design parameters that were investigated in the previous section. (In determining the empirical equation, however, only one value of the ratio I_e/I was used, based on a chamber pressure of 500 psi, an expansion ratio of 8, and a γ of 1.24.) This study was performed at a different time from that in the preceding section, and a slightly different drag curve was assumed, but it is not expected that the results will be significantly different for this reason.

Machine calculations were performed to determine maximum range of vehicles launched from a spherical, nonrotating earth. Here, impact range is measured from the launch point rather than from the burnout point. Computer data have been used to plot a curve showing the quantity V^* as a function of missile range. Even with a large variation in vacuum specific impulse, varying from 200 to 1,000 sec, all of the data points fall essentially on a single curve for a given N_0 and $C_{DM}A/W_0$. For any other values of N_0 and $C_{DM}A/W_0$ similar results are obtained. Figure 6 shows the mean curve obtained for $N_0 = 1.5$ and $C_{DM}A/W_0 = 0.000265$.

The results of Figure 6 have been replotted in Figure 7 on semilog paper, together with a curve given by

$$R = D(e^{V^*/Bg} - 1). \tag{6.1}$$

For ranges varying from 400 to 6,000 n mi, it can be seen that (6.1) quite accurately represents the curve obtained from the machine calculations. We have found that the parameter B is very insensitive to changes in N_0 and $C_{DM}A/W_0$, while the parameter D is fairly sensitive to such changes. The values of the parameters in Figure 6 are $D=80$ and $B=208$.

The parameter B determines the slope of the fitted curve, and the parameter D determines the displacement. The two constants, how-

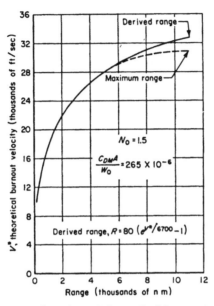

Fig. 6. Range as a function of theoretical burnout velocity, V^*.

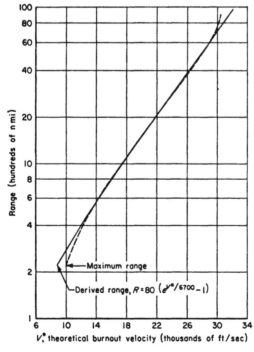

Fig. 7. Range as a function of theoretical burnout velocity, V^*.

ever, must be treated as a pair. Many curves might be fitted to the empirical data, giving better accuracies in some ranges and poorer accuracies in others. We have arbitrarily selected the value of 208 sec for B, and all values of D have been determined on this basis. If another value for B is selected, new values for D must be derived. Figure 8 shows D as a function of N_0 for various values of $C_{DM}A/W_0$.

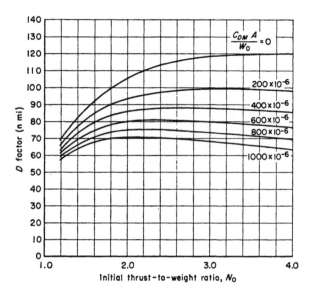

Fig. 8. D factor as a function of initial thrust-to-weight ratio for various $C_{DM}A/W_0$.

The results of (6.1) can be extended for use from 400 to 10,800 n mi (halfway around the earth) by the following argument. Burnout angles were selected to maximize range. For ranges beyond 6,000 n mi, the use of maximum-range trajectories results in very large range misses for errors in burnout speed. This can be seen by the slope of the curve in Figure 7. Lofting the trajectories so that the burnout velocity increases as determined by equation (6.1) results in an increase of about 5 per cent above the maximum-range burnout velocity for the 10,800 n mi range. At the same time, the lofting decreases the miss from about 10 n mi to less than 2 n mi for an error in the burnout speed of 1 ft/sec. For design purposes, deviation from the maximum-range trajectory for ranges beyond 6,000 n mi is reasonable and, in fact, desirable.

In the case of two-stage missiles we note that

$$V^* = I_1 g \ln r_1 + I_2 g \ln r_2.$$

Thus, for two-stage missiles, (6.1) becomes

$$R = D(r_1^{I_1/B} r_2^{I_2/B} - 1). \tag{6.2}$$

By differentiating (6.2) one may obtain a number of exchange ratios, some of which have been derived and are presented in Appendix A.

Equation (6.2) has been checked many times against results of machine computation. To date, the equation has been accurate to about 5 per cent of the range. It has been found that the equation is useful in two ways. First, if the missile under study has no close counterpart and no machine data are available, a value for D as found in Figure 8 is used. Frequently, however, a vehicle is studied for which a small amount of machine data is or can be made available. In this case, the value of D is derived by solving (6.2) "backward." Once a value of D has been determined for the particular missile system, the calculation of perturbations of this missile system may be made by using (6.2) and the D value thus derived.

Appendix A: Formulas

Miscellaneous Formulas for Kepler Ellipse

$$\sigma = \frac{R_e + h_b}{R_e} \qquad \epsilon = \sqrt{1 - 2\lambda \sin^2 \beta + \lambda^2 \sin^2 \beta}$$

$$\lambda = \frac{V_b^2 \sigma_b}{gR_e} \qquad R_e = \text{earth radius} = 20.9 \times 10^6 \text{ ft}$$

Conservation of Energy

$$V^2[t] - \frac{2gR_e^2}{z} = \text{const}$$

Conservation of Angular Momentum
$$Vz \sin \beta = \text{const}$$

Impact Range Angle from Burnout

$$\psi = \pi - \sin^{-1}\left(\frac{1 - \lambda\sigma \sin^2 \beta_b}{\epsilon}\right) - \sin^{-1}\left(\frac{1 - \lambda \sin^2 \beta_b}{\epsilon}\right)$$

Velocity Required To Obtain Impact Range

$$V_b = \left[\frac{gR_e}{\sigma} \frac{1 - \cos \psi}{\sigma \sin^2 \beta_b + \sin \beta_b \sin (\psi - \beta_b)}\right]^{1/2}$$

Apogee Altitude

$$h_a = \frac{\sigma R_e \lambda \sin^2 \beta_b}{1 - \epsilon} - R_e$$

Velocity Required To Obtain Apogee Altitude

$$V_b = \left[\frac{2gR_e}{\sigma} \frac{1 - \dfrac{\sigma}{\sigma_a}}{1 - \left(\dfrac{\sigma}{\sigma_a} \sin \beta_b\right)^2} \right]^{1/2} ; \qquad \sigma_a = \frac{R_e + h_a}{R_e}$$

Period for Complete Elliptic Orbit

$$T = 2\pi \frac{r_b^{3/2}}{(2 - \lambda)^{3/2}(gR_e^2)^{1/2}}$$

Time to Apogee from Burnout

$$t_a = \frac{T}{2\pi}\left[\sqrt{1 - \epsilon^2}\, \cot \beta_b + \cos^{-1}\left(\frac{1 - \lambda}{\epsilon}\right) \right]$$

Exchange Ratios for Single-Stage Vehicles

$$R(\text{n mi}) = D(r^{I/B} - 1)$$

$$\frac{\partial R}{\partial V_b} = \frac{R + D}{Bg}$$

$$\left.\frac{\partial R}{\partial W_b}\right|_{W_0} = -\frac{I(R + D)}{BW_b}$$

$$\frac{\partial R}{\partial W_0} = \frac{I(R + D)}{BW_0}$$

$$\frac{\partial R}{\partial I} = \frac{(R + D)\ln r}{B}$$

$$\left.\frac{\partial W_0}{\partial W_b}\right|_{C_{DM}A/W_0} = r\left[1 - \frac{B}{I}\frac{RN_0}{D(R + D)}\frac{\partial D}{\partial N_0} \right]^{-1}$$

$$\left.\frac{\partial W_0}{\partial W_b}\right|_{N_0, C_{DM}A/W_0} = r$$

$$\frac{\partial W_0}{\partial W_L} = \frac{W_0}{W_L}$$

Exchange Ratios for Two-Stage Vehicles†

$$R(\text{n mi}) = D(r_1^{I_1/B} r_2^{I_2/B} - 1)$$

$$\frac{\partial R}{\partial V_b} = \frac{(R + D)}{Bg}$$

$$\frac{\partial R}{\partial W_j} = - \frac{I_1(R + D)}{BW_{b1}}$$

$$\frac{\partial R}{\partial W_{b2}} = - \frac{I_2(R + D)}{BW_{b2}} + \frac{\partial R}{\partial W_j}$$

$$\left.\frac{\partial R}{\partial W_{01}}\right|_{r_1/r_2} = \frac{I_2(R + D)}{BW_{01}}$$

$$\left.\frac{\partial R}{\partial W_{01}}\right|_{W_{02}} = \frac{I_1(R + D)}{BW_{01}}$$

$$\left.\frac{\partial R}{\partial W_{01}}\right|_{W_{01}-W_{02}} = \frac{I_1(R + D)}{BW_{01}}\left(1 - r_1 + \frac{I_2 r_1}{I_1}\right)$$

$$\frac{\partial R}{\partial I_1} = \frac{(R + D)\ln r_1}{B}$$

$$\frac{\partial R}{\partial I_2} = \frac{(R + D)\ln r_2}{B}$$

$$\left.\frac{\partial W_{01}}{\partial W_j}\right|_{C_{DM}A/W_0} = r_1\left[1 - \frac{B}{I_1}\frac{RN_0}{D(R + D)}\frac{\partial D}{\partial N_0}\right]^{-1}$$

$$\left.\frac{\partial W_{01}}{\partial W_{b2}}\right|_{C_{DM}A/W_0} = r_1 r_2\left[1 - \frac{B}{I_2}\frac{RN_0}{D(R + D)}\frac{\partial D}{\partial N_0}\right]^{-1}$$

$$\left.\frac{\partial W_{01}}{\partial W_j}\right|_{N_0,C_{DM}A/W_0} = r_1$$

$$\left.\frac{\partial W_{01}}{\partial W_{b2}}\right|_{N_0,C_{DM}A/W_0} = r_1 r_2$$

$$\frac{\partial W_{01}}{\partial W_L} = \frac{W_{01}}{W_L}$$

---

† Numerical subscripts refer to stages and are in order of burning period.

Appendix B: Analysis

This appendix describes the theoretical analysis that determined the selection of missile-design parameters for this study. This analysis also suggested the use of the V_L concept in reducing the computer output data to a manageable form.

Equations of Motion

In determining the performance of a rocket, one is confronted with complicated differential equations of motion. Accurate solutions are obtained only by using a digital computer. However, without a computer one can obtain a large amount of information about such factors as gravitational and atmospheric effects on the performance of boost rockets by examining the individual terms in the equations. The basic equation of motion is:

$$\ddot{z} = n[z,\, t]\kappa[t] + \alpha[z,\, \ddot{z},\, m], \qquad (\text{B.1})$$

where $\quad z =$ radius vector from earth center to missile,

$$n = \text{thrust-to-mass ratio} = \frac{F[z]}{m[t]},$$

$t =$ time,
$m[t] =$ mass of missile,
$\kappa =$ unit vector in the direction of thrust,
$\alpha = \alpha[\text{gravitation}] + \alpha[\text{drag}].$
For α, we use

$$\alpha[\text{gravitation}] = -\frac{gR_e^2}{z^2}\frac{z}{z}, \qquad (\text{B.2})$$

$$\alpha[\text{drag}] = -\frac{1}{2}\left(\frac{\rho V^2 C_D A}{m}\right)\left(\frac{V}{V}\right), \qquad (\text{B.3})$$

where $C_D =$ drag coefficient, a function of Mach number,
$R_e =$ radius of earth,
$V =$ vehicle velocity vector relative to the atmosphere,
$A =$ reference area,
$\rho =$ air density.

Replacing α with the terms for $\alpha[\text{gravitation}]$ and $\alpha[\text{drag}]$ and dividing by g, we obtain

$$\frac{\ddot{z}}{g} = \frac{F\,[z]}{W\,[t]}\,\kappa[t] - \left(\frac{R_e^2}{Z^2}\right)\left(\frac{Z}{Z}\right) - \frac{1}{2}\rho V^2\left(\frac{C_D A}{W[t]}\right)\left(\frac{V}{V}\right). \qquad (\text{B}.4)$$

We assume that thrust in a vacuum is proportional to the weight flow rate. Thrust as a function of altitude is taken as the vacuum thrust corrected for ambient pressure, as follows:

$$F[z = \infty] = F_\infty = I\dot{W}, \qquad (\text{B}.5)$$

$$F(z) = F_\infty\left[1 - \frac{p[z]}{p_s}\left(1 - \frac{I_s}{I}\right)\right], \qquad (\text{B}.6)$$

where $p[z]$ = ambient pressure,
       p_s = ambient pressure at sea level,
   I_s/I = ratio of sea-level thrust to vacuum thrust for identical flow rates.

Values for I_s/I may be calculated from tables showing thrust coefficient versus expansion-area ratio, ratio of specific heats for exhaust products, and chamber pressure. Defining N_0 as ratio of initial thrust to initial weight and assuming constant $\dot{W}$, we can write the equation of motion in terms of missile-design parameters as follows:

$$\frac{\ddot{z}}{g} = N_0\,\frac{I}{I_s}\,\frac{1}{1 - \dfrac{N_0}{I_s}t}\,\kappa - N_0\,\frac{\left(\dfrac{I}{I_s} - 1\right)\dfrac{p}{p_s}}{1 - \dfrac{N_0}{I_s}t}\,\kappa$$

$$- \frac{R_e^2}{z^2}\frac{z}{z} - \frac{1}{2}\rho V^2\,\frac{C_D A}{W_0\left(1 - \dfrac{N_0}{I_s}t\right)}\,\frac{V}{V}. \qquad (\text{B}.7)$$

In some cases, flow rate will not be constant, but we assume it to be so during the first several seconds of flight. Forming the dot product of V/V with $\ddot{z}$, integrating for a gravity turn (thrust aligned with velocity), and assuming a spherical, nonrotating earth, we obtain

$$\frac{V_b}{g} = I \ln r - \int_0^{tb} \frac{N_0\left(\dfrac{I}{I_s} - 1\right)\dfrac{p}{p_s}}{1 - \dfrac{N_0}{I_s}t}\,dt - Q, \qquad (\text{B}.8)$$

where

$$Q = \int_0^{tb} \frac{R_e^2}{z^2} \cos \beta \, dt + \int_0^{tb} \frac{\frac{1}{2}\rho V^2 C_D A}{W_0\left(1 - \frac{N_0}{I_s}t\right)} \, dt.$$

We define the velocity lost to gravitation, drag, and atmosphere as:

$$V_g = g \int_0^{tb} \frac{R_e^2}{z^2} \cos \beta \, dt, \tag{B.9}$$

$$V_D = g \int_0^{tb} \frac{\frac{1}{2}\rho V^2 C_D A}{W_0\left(1 - \frac{N_0}{I_s}t\right)} \, dt, \tag{B.10}$$

$$V_a = g \int_0^{tb} \frac{N_0\left(\frac{I}{I_s} - 1\right)\frac{p}{p_s}}{1 - \frac{N_0}{I_s}t} \, dt. \tag{B.11}$$

The design velocity is given by

$$V^* = Ig \ln r.$$

Hence the burnout velocity becomes

$$V_b = V^* - V_g - V_D - V_a. \tag{B.12}$$

It is apparent that the velocity lost is intimately tied in with the trajectory itself. Forming the dot product of $\ddot{z}/g$ with a unit vector normal to the velocity, and again assuming a gravity turn and nonrotating earth, we obtain

$$\dot{\beta} = \frac{g}{V}\frac{R_e^2}{z^2}\sin \beta - \frac{V \sin \beta}{z}. \tag{B.13}$$

For low velocity, the turning rate is large, and the greater portion of turning is to be expected early in the trajectory. The amount of turning to be achieved is limited, however, by the desired burnout angle. Therefore, it is frequently necessary to keep $\sin \beta$ (therefore β) quite small during the early part of the trajectory to prevent too much turning. The trajectory can be thought of as consisting of three segments: (a) a segment during which the vehicle flies steeply, (b) a period of turning, and (c) a segment in which the velocity angle remains relatively constant.

For the early segment of flight, the velocity can be approximated by

$$V \doteq (N_0 - 1)gt. \tag{B.14}$$

For a given burnout angle, the start of the period of turning depends primarily on the initial thrust-to-weight ratio, N_0. Thus, for low values of N_0 (near 1.0), the initial segment of flight is at lower velocity and the turning rate is increased. To achieve the same burnout angle as that for a higher value of N_0, the initial segment of the trajectory must be steeper (smaller β).

The turning rate for a gravity turn is zero when the vehicle velocity equals that required for a circular satellite orbit at the same altitude.

Gravity Loss

We can use the foregoing to gain insight into the behavior of the velocity lost to gravitation and atmosphere. In vertical flight the gravity loss should be proportional to t_b. For a missile burning out in its trajectory at angle β_b, the gravity loss is some fraction of that lost in purely vertical flight; and we would expect that fraction to depend on N_0 and the proportion of total mass consumed as propellant $(1 - 1/r)$.

It is sometimes proposed that the velocity lost to gravitation is not really lost at all but converted into potential energy. It may be observed, however, that a vehicle in powered flight is not a conversative system. A ballistic missile does not burn impulsively (i.e., all the propellant is not burned on the ground). Some of the fuel is used to lift the unburned fuel, so that the vehicle always ends up at some altitude. Energy is imparted to the expended propellant by raising the unburned propellant to some finite altitude.

One way to see what happens is to consider the following comparison of two single-stage vehicles that are identical in all respects except thrust. Figure 9 compares vertical trajectories for the two vehicles. With vehicle (1) we assume an infinite thrust (impulsive burning) and with vehicle (2) a finite thrust. Vehicle (1) burns out all its propellants at the surface of the earth, achieves a theoretical velocity V^*, rises, returns to earth, and impacts at the same velocity. For vehicle (2), at height h_b, we have

$$V_b = V^* - gt_b, \tag{B.15}$$

$$V_{\text{impact}} = V^* - gt_b + gt_j \tag{B.16}$$

$$V_{\text{impact}} = V^* - gt_b\left(1 - \frac{t_j}{t_b}\right), \tag{B.17}$$

where t_b = burning time for vehicle (2),

$\quad t_j$ = time from burnout altitude to impact on reëntry.

The time t_j is less than t_b, for it takes a time t_b to get from a velocity of 0 to V_b, whereas it takes a time t_j to get from a velocity of V_b to V_{impact}, where $V_{\text{impact}} > V_b$. The kinetic energy of vehicle (1) at impact is

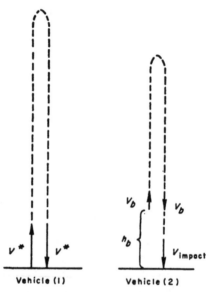

Fig. 9. Comparison of impulsive and finite thrust for vertical trajectory (constant gravitational field and no atmosphere).

essentially proportional to the square of its impact velocity, V^*. The kinetic energy of vehicle (2) is essentially proportional to the square of its impact velocity, and V_{impact} is smaller than the theoretical velocity of vehicle (1). As the thrust-to-weight ratio of vehicle (2) increases, V_{impact} gets closer to V^*, and hence gravity losses decrease. In actual missiles the thrust-to-weight ratio is closer to 1 than to infinity because the weights of engines and structural components increase with increased thrust. We reach a point where the advantage of higher thrust in terms of velocity losses is offset by increase in burnout weight.

We see that not all of the velocity loss goes into gaining altitude; some is lost to the expended propellants. By substituting the appropriate numerical values for an existing vehicle in these equations, it was determined that approximately 25 per cent of the velocity loss went into gaining altitude; 75 per cent was lost as equivalent energy to the expended propellants. This calculation presents a good argument for holding the burnout altitude as low as possible. It is true that low burnout altitudes mean larger drag effects, but these are relatively small when compared with gravity losses. Aerodynamic effects, of course, lead to heating, and heating often means an increase in structural weight, but gravitational losses are still a prime concern.

Drag Loss

The drag loss (B.10) is dependent on the ratio $C_D A/W_0$, ρ, and V^2. The air density ρ is dependent on altitude and, for qualitative purposes, can be considered to decay exponentially with altitude according to the following equation:

$$\rho(z) = \rho_s e^{-k(z-R_e)}. \tag{B.18}$$

The dependency of the drag losses on V^2 is significant during late segments of flight if the trajectory is flat (low) and if high velocities are achieved below, say, 150,000 ft. The effect of V^2 for most "normal" trajectories is not important, because these values occur when the vehicle is beyond the atmosphere. The greatest erosion of velocity occurs when C_D is near its peak and early in flight when ρ is of the same magnitude as ρ_s. In a typical trajectory with an initial thrust-to-weight ratio of 1.2, the vehicle achieves Mach 1 in 80 sec at about 30,000 ft, where the density is approximately 0.37 times that at the earth's surface.

It would be expected that for equivalent trajectories the velocity loss due to drag would be sensitive to N_0. There are two effects, however: for high N_0, higher velocities are achieved at lower altitudes, and hence the density for Mach 1 velocity is large; but for high N_0, the duration of time through which the drag forces are acting is reduced, and the effects tend to cancel. Thus, drag losses are very insensitive to N_0.

For the most part, V_D depends on $C_{DM}A/W_0$, N_0/I_s, and β at burnout. Because the trajectory changes little with variation in $C_{DM}A/W_0$, the losses can be expected to be proportional to this quantity. The C_{DM} (the maximum C_D) is the single parameter selected to be characteristic of all drag curves, for reasons stated elsewhere in this chapter. The term I_s/N_0 is equivalent to $W_0/\dot{W}$, which determines the change in $C_{DM}A/W[t]$ with time. For the same initial weight, the missile with lower I_s/N_0 has less weight at the time when the drag forces become most important. The burnout angle β is a measure of the proportion of the total trajectory contained in the atmosphere. As β is increased, the density associated with each velocity is increased, and the resulting velocity loss is greater.

As the trajectory becomes very flat and high velocities are achieved at low altitudes, the effect of V^2 and the long duration of the drag force combine to increase the drag loss to very high values. It is not expected that such trajectories are realistic, as aerodynamic heating may preclude extremely flat burnout angles. Flat burnout angles may be achieved if the thrust is reduced to increase the total time of powered flight. Usually, thrust levels that are sufficient to boost the vehicle at launch yield comparatively short over-all burning times. Thrust may be reduced by throttling a single-stage vehicle or, more profitably, by staging. If either of these techniques is not sufficient, and if flat burnout velocities are required, a coasting period may be inserted between burning periods. If restart capabilities are not available or not desirable, the remaining alternative is to fly the vehicle steeply during an early segment of flight and pitch down after sufficient altitude has been achieved, yielding a negative angle of attack. In this type of trajectory, considerable velocity (and payload) is lost in turning the velocity vector when the magnitude of the velocity is high. To date, no approximation has been found to determine these "turning losses"; the only realistic approach has been to use a computing machine.

Nozzle-Pressure Loss

The term V_a results from the fact that thrust is lost when the nozzle pressure in the exit plane is less than the ambient pressure. This loss is frequently thought of in terms of a reduction in specific impulse. The amount of the thrust loss as a function of trajectory parameters is dependent only on the ambient pressure; hence the total velocity loss occurs early in powered flight.

The integral in equation (B.11) shows that the nozzle-pressure loss should also be proportional to N_0. An increase in N_0 increases the rate at which altitude is achieved, however, and reduces the duration of flight time at high ambient pressure by an amount also dependent on N_0; the two effects tend to cancel. The effect of I_s/N_0, or the change in vehicle weight with time, is less significant with the nozzle-pressure loss than with the drag loss because the largest percentage of nozzle-pressure loss occurs early in powered flight.

Appendix C: Symbols

A vehicle reference area for drag calculations
B empirical parameter in simplified range equation

C_D drag coefficient, function of Mach number
C_{DM} maximum drag coefficient
D empirical parameter in simplified range equation
F thrust (lb)
g gravitational constant, 32.2 ft/sec^2
h_b burnout altitude measured from earth's surface
$\bar{h}$ intermediate altitude between first-stage burnout and final
             burnout, used in computing velocity loss in succeeding stages
h^* burnout altitude for vertical trajectory, neglecting atmos-
             pheric effects
i index denoting stage measured from launch
I vacuum specific impulse, i.e., vacuum thrust divided by flow
             rate of fuel
I_s sea-level specific impulse, i.e., sea-level thrust divided by flow
             rate of fuel
K_a empirical constant used to determine V_a
K_D empirical constant used to determine V_D
K_g empirical constant used to determine V_g
K_{gg} empirical constant used to determine V_g
$\ln$ natural logarithm
m mass of vehicle
N_0 initial thrust-to-weight ratio, i.e., launch thrust divided by
             launch weight
n thrust-to-mass ratio, a function of time
p atmospheric pressure, a function of altitude
p_s atmospheric pressure at sea level
R impact range
R_e radius of earth = 20.9×10^6 ft
r burnout mass ratio = stage initial weight (mass) divided by
             stage burnout weight (mass)
T total period of elliptic orbit
t time
t_a time from selected trajectory conditions to apogee
t_b burning time
t_f time from reaching burnout altitude to impact on reëntry
$\mathbf{V}$ vehicle velocity vector
$V[t]$ magnitude of velocity as function of time
V_a velocity lost to nozzle pressure
V_b magnitude of burnout velocity
V_D velocity lost to drag
V_{impact} velocity at impact
V_g velocity lost to gravitation
V_L total velocity lost = $V_a + V_D + V_g$

V^*	theoretical velocity as determined by rocket equation
$W[t]$	weight of vehicle, a function of time
W_0	vehicle initial weight
W_b	vehicle final (burnout) weight
W_j	weight jettisoned between stages of two-stage vehicle
W_p	weight of payload (includes guidance and other weights that do not vary with last-stage size)
x_b	surface range at burnout
$\mathbf{z}$	radius vector from earth center to vehicle
z	magnitude of $\mathbf{z}$
β_b	angle between vehicle velocity vector at burnout and local vertical
$\bar{\beta}$	selected β between those for first-stage and final-stage burnout to be used in determining velocity losses and altitude gains
ϵ	eccentricity of free-flight ellipse
$\boldsymbol{\kappa}$	unit vector aligned with thrust
λ	nondimensional parameter $= V_b{}^2\sigma/gR_e$
γ	ratio of specific heats of combustion products
ρ	atmospheric density, function of altitude
$[\ \ ]$	brackets indicate functional notation
σ	nondimensional parameter $= R_e + h_b/R_e$
ψ	impact range angle
$\dfrac{\partial u}{\partial v_w}$	partial derivative of u with respect to v with w held constant

References

1. Fried, B. D., "On the Powered Flight Trajectory of an Earth Satellite," *Jet Propulsion*, Vol. 27, No. 6, 1957, pp. 641–643.
2. Fried, B. D., and J. M. Richardson, "Optimum Rocket Trajectories," *J. Appl. Phys.*, Vol. 27, No. 8, 1956, pp. 955–961.
3. Fried, B. D., "Corrections to Comments on the Powered Flight Trajectory of a Satellite," *Jet Propulsion*, Vol. 28, No. 5, 1958, pp. 342–343,
4. Lawden, D. F., "Optimal Rocket Trajectories," *Jet Propulsion*, Vol. 27. No. 12, 1957, p. 1263.

Chapter 3

The Optimum Spacing of Corrective Thrusts in Interplanetary Navigation[†]

J. V. BREAKWELL

1. The Problem of Corrective Thrusts

Suppose that a spaceship is in free (i.e., unpowered) flight on its way from Earth to Mars. Except in the immediate vicinity of Earth and Mars, its trajectory is essentially a heliocentric ellipse. Let us pretend that the orbits of Earth and Mars and the "transfer ellipse" are co-planar. Now, the actual transfer trajectory, if uncorrected beyond some point P_n, will miss the destination planet Mars by a distance D_{n-1}, to which we may attach a sign (e.g., $\pm$) according to whether the space-ship passes to the left or to the right of Mars (see Fig. 1). Suppose that a corrective velocity impulse $\mathbf{v}_n^{(c)}$ is to be applied at the point P_n on the transfer trajectory. This, of course, effectively includes the appli-cation of a finite thrust over a duration very much smaller than the flight duration. The amount of velocity correction $\mathbf{v}_n^{(c)}$ is computed as follows: (a) Make an estimate $\hat{D}_{n-1}$ of D_{n-1} based on measurements (probably angular) determining present and past positions; (b) Com-pute $\mathbf{v}_n^{(c)}$ so that, in a linearized error theory,

$$\frac{\partial D}{\partial \mathbf{v}}(t_n) \cdot \mathbf{v}_n^{(c)} = - \hat{D}_{n-1}. \tag{1.1}$$

It should be mentioned here that errors in the estimate $\hat{D}_{n-1}$ may include biases in the subsequent trajectory calculation due, for example, to oversimplifying the computation. Our main concern, however, will

---

[†] A version of this paper appears as Chapter 12 in George Leitmann (ed.), *Optimization Techniques with Applications to Aerospace Systems*, Academic Press, New York, 1962.

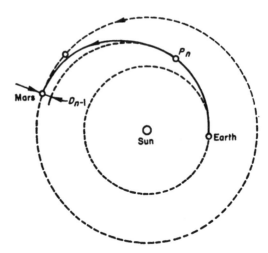

Fig. 1. Earth-Mars trajectory.

be with random errors in $\hat{D}_{n-1}$ due to random measurement errors.

The "control effectiveness" $\partial D(t)/\partial \mathbf{v}$ certainly decreases toward zero as the spaceship moves from Earth to Mars. Consequently if D_{n-1} were estimated correctly, the economical practice in terms of fuel expended on velocity correction would be to correct as soon as possible. On the average, however, D_{n-1} is not estimated correctly. This means that a correction at P_n still leaves us with a miss-distance D_n which may have to be reduced by further corrections. This supposedly reduced miss-distance is

$$D_n = D_{n-1} - \hat{D}_{n-1} + \frac{\partial D}{\partial \mathbf{v}}(t_n) \cdot \mathbf{v}_n', \qquad (1.2)$$

where the last term is due to a possible velocity mechanization error $\mathbf{v}_n'$. Moreover, we may expect that the error in estimating D_{n-1}, like the control effectiveness, decreases toward zero as we approach the target planet. The problem we face is that of choosing the correction points $P_1, P_2, \cdots, P_N$ so as to achieve in some average sense a required terminal accuracy with a minimum total velocity correction and hence a minimum expenditure of fuel for corrective thrusts.

2. Discussion

The problem as we have described it so far is two-dimensional. Actually, the true situation is three-dimensional, even if the "nominal" transfer trajectory is *coplanar* with the orbit of Mars, since it will be necessary to consider an "out-of-plane" miss-distance component related to out-of-plane position and velocity components. The out-of-plane one-dimensional correction problem is independent of the "in-plane" correction problem, except that both kinds of correction are made simultaneously so as to economize on

$$\sum_{n=1}^{N} \left| \mathbf{v}_n^{(c)} \right| .$$

Mention should be made at this point of a related problem treated in one dimension by Arnold Rosenbloom [1]. Instead of considering an average expenditure of fuel, Rosenbloom sets an upper limit on fuel available and inquires as to what to do to minimize some average terminal error. Mathematically this is a more difficult problem and one that will not be discussed further here.

Returning to our two-dimensional problem, we note that the velocity correction $\mathbf{v}_n^{(c)}$ is not uniquely determined by the relation (1.1). Naturally we resolve this choice by minimizing the individual velocity correction magnitude $\left| \mathbf{v}_n^{(c)} \right|$. It is easy to see that this amounts to choosing $\mathbf{v}_n^{(c)}$ either parallel or antiparallel to the vector $\partial D(t)/\partial \mathbf{v}$, depending on the sign of $\hat{D}_{n-1}$. It then follows that

$$\left| \mathbf{v}_n^{(c)} \right| = \frac{\left| \hat{D}_{n-1} \right|}{\left| \dfrac{\partial D}{\partial \mathbf{v}}(t_n) \right|} . \tag{2.1}$$

The control-effectiveness vector $\partial D/\partial \mathbf{v}(t)$ is to be evaluated along the nominal correction-free transfer orbit. Its magnitude and direction are indicated in Figure 2 for the case of a "Hohmann transfer" from Earth to Mars, that is, a 180° transfer along an ellipse cotangential to the Earth and Mars orbits, treated as coplanar heliocentric circles. The gravitational fields of Earth and Mars themselves were ignored in the calculation of $\partial D(t)/\partial \mathbf{v}$. The abscissa in Figure 2, the so-called mean anomaly, increases uniformly with time from 0° to 180°. It appears, then, that the effectiveness magnitude $\left| \partial D(t)/\partial \mathbf{v} \right|$ decreases to zero essentially linearly with time, while the direction of $\partial D(t)/\partial \mathbf{v}$, measured from the "transverse" direction perpendicular to the radius from the Sun, increases essentially linearly with time from 0° to 90°. Thus, as might be anticipated, an early correction is made forward or backward along the transfer orbit and the last correction is made perpendicular to the motion.

We may use (2.1) with n increased by 1, together with (1.2) to express $\left| \mathbf{v}_{n+1}^{(c)} \right|$ in terms of errors at P_n and P_{n+1}. In particular, if we

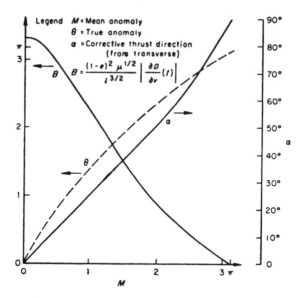

Fig. 2. Guidance chart for Earth-Mars Hohmann transfer.

neglect the mechanization errors $\mathbf{v}'$, we obtain:

$$\left| \mathbf{v}_{n+1}^{(c)} \right| = \frac{1}{\left| \dfrac{\partial D}{\partial \mathbf{v}}(t_{n+1}) \right|} \left| (\hat{D}_n - D_n) - (\hat{D}_{n-1} - D_{n-1}) \right|. \quad (2.2)$$

This tells us, for example, that if the latest miss-distance is correctly estimated ($\hat{D}_n = D_n$), the corrective velocity depends only on the last previous error in miss-distance estimation, regardless of how many corrections have been made. On the other hand, if the last previous miss-distance estimate were correct, the new correction would be due only to an incorrect new estimate of a miss-distance that is really zero.

If we now disregard any biases in the estimates $\hat{D}$, we may assume that the differences ($\hat{D} - D$) are normally distributed with zero mean and with variances and covariances that may be obtained in a straightforward manner from assumed variances in the various independent angular measurements involved, the errors of which are presumably normal with zero bias. The in-plane velocity correction magnitude $\left| \mathbf{v}_{n+1}^{(c)} \right|$ is thus expressed by means of (2.2) as the absolute value of a normal random variable with zero mean and computable variance.

Meanwhile we may use (1.2) to establish a time t_N for the last correction such that any earlier time would lead to an expected miss $E\{|D_N|\}$ in excess of some allowed terminal error. The "launching" error D_0 before the first correction may be presumed to be, on the average, far greater in absolute value than the allowed terminal error. What we would like to do, given t_N and a root-mean-square (rms) launching error σ_{D_0}, is to choose a sequence of times $t_1, t_2, \cdots, t_N$, with the integer N *not* specified, so that the 90 percentile, say, of the distribution of

$$\sum_{n=1}^{N} |\mathbf{v}_n^{(c)}|$$

is as small as possible. This, however, is an awkward quantity to compute because of the correlation between successive terms in the sum. A more workable, and closely related, criterion is the minimization of the sum

$$S_N = \sum_{n=1}^{N} E\{ |\mathbf{v}_n^{(c)}| \}. \tag{2.3}$$

We may take advantage here of the fact that the expected magnitude $E\{|x|\}$ of a normal random variable x with zero mean and variance σ_x^2 is just $\sigma_x \sqrt{2/\pi}$.

3. The Three-Dimensional Problem

The three-dimensional situation has been discussed by the author [2], [3]. If we denote the out-of-plane miss by D' and choose the z-direction perpendicular to the plane of motion, the out-of-plane velocity correction at P_{n+1} is

$$|\dot{z}_{n+1}^{(c)}| = \left| \left[\frac{\hat{D}_n' - D_n'}{\dfrac{\partial D'}{\partial \dot{z}}(t_{n+1})} - \dot{z}_{n+1}' \right] - \frac{\dfrac{\partial D'}{\partial \dot{z}}(t_n)}{\dfrac{\partial D'}{\partial \dot{z}}(t_{n+1})} \left[\frac{\hat{D}_{n-1}' - D_{n-1}'}{\dfrac{\partial D'}{\partial \dot{z}}(t_n)} - \dot{z}_n' \right] \right|, \tag{3.1}$$

where small mechanization errors $\dot{z}'$ are now included. The in-plane velocity correction becomes

$$\sqrt{\left(\dot{x}_{n+1}^{(c)}\right)^2 + \left(\dot{y}_{n+1}^{(c)}\right)^2} = \left|\left[\dfrac{\hat{D}_n - D_n}{\dfrac{\partial D}{\partial v}(t_{n+1})} - w_{n+1}'\right]\right.$$

$$\left.- \dfrac{\dfrac{\partial D}{\partial v}(t_n)}{\dfrac{\partial D}{\partial v}(t_{n+1})}\left[\dfrac{\hat{D}_{n-1} - D_{n-1}}{\dfrac{\partial D}{\partial v}(t_n)} - w_n'\right]\right|, \quad (3.2)$$

where $\partial D/\partial v(t)$ denotes

$$\sqrt{\left[\dfrac{\partial D}{\partial \dot{x}}(t)\right]^2 + \left[\dfrac{\partial D}{\partial \dot{y}}(t)\right]^2},$$

and where w' denotes the component of the mechanization error $\mathbf{v}'$ in the direction of the in-plane vector $(\partial D/\partial \dot{x}, \partial D/\partial \dot{y})$. Since

$$\sqrt{\left(\dot{x}_{n+1}^{(c)}\right)^2 + \left(\dot{y}_{n+1}^{(c)}\right)^2 + \left(\dot{z}_{n+1}^{(c)}\right)^2}$$

no longer has a simple statistical distribution, in spite of the assumed normality of the mechanization errors as well as the measurement errors, the criterion (2.3) is replaced by a related criterion, namely, that of minimizing

$$S_N = \sum_{n=1}^{N} \sqrt{\left(E\left\{\sqrt{[\dot{x}_n^{(c)}]^2 + [\dot{y}_n^{(c)}]^2}\right\}\right)^2 + \left(E\{|\dot{z}_n^{(c)}|\}\right)^2}, \quad (3.3)$$

which is more easily computable.

4. Examples

To carry out a minimization of S_N without a digital computer, we must make some simplifying assumptions relative to the miss-distance estimates D. We shall consider two examples. As our first example, in two dimensions, let us suppose the estimate D_{n-1} is based on measurements rather close to the position P_n so that they effectively measure position and velocity at P_n with an uncertainty in velocity that has a substantially greater effect on miss-distance than has the uncertainty in position. In this case, we have

$$\sigma_{\hat{D}_{n-1}} \cong \frac{\partial D}{\partial v}(t_n)\sigma_v, \tag{4.1}$$

where σ_v is the rms velocity measurement uncertainty in the direction of $\partial D/\partial v(t_n)$. We shall further suppose that this uncertainty is independent of time, as is the rms value σ' of w'.

The sum S_N now simplifies to

$$S_N = \sqrt{\frac{2}{\pi}} \left\{ \sqrt{\sigma_v^2 + \sigma'^2 + \frac{\sigma_{D_0}^2}{\left[\dfrac{\partial D}{\partial v}(t_1)\right]^2}} \right. $$

$$\left. + \sqrt{\sigma_v^2 + \sigma'^2} \sum_{n=2}^{N} \sqrt{1 + \left[\frac{\dfrac{\partial D}{\partial v}(t_{n-1})}{\dfrac{\partial D}{\partial v}(t_n)}\right]^2} \right\}. \tag{4.2}$$

It was shown in [3] that for sufficiently large rms launching error, in fact if

$$\sigma_{D_0} > 3\frac{\partial D}{\partial v}(0)\sqrt{\sigma_v^2 + \sigma'^2},$$

the optimum choice of correction times t_n must be such that $t_1 = 0$ (or as soon as feasible) and

$$\frac{\dfrac{\partial D}{\partial v}(t_{n-1})}{\dfrac{\partial D}{\partial v}(t_n)} = \rho,$$

a value independent of n, and that the optimum integer N is determined approximately by the condition $\rho = 3.0$. In the case of the Hohmann transfer from Earth to Mars, the approximate linearity of $\partial D/\partial v$ as a function of time leads to the following rough description of the optimum spacing of corrections: Make the first correction as soon as feasible; after any correction proceed two-thirds of the way to the target (i.e., wait for two-thirds of the remaining time) before the next correction. A similar result was given by Lawden [4]. Figure 3 shows

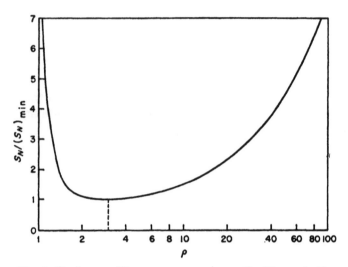

Fig. 3. Fuel expenditure versus spacing ratio (Example 1).

that the behavior of S_N as a function of ρ is not very sensitive to changes in "spacing ratio" from the optimum value 3. This "cascading" of corrections is surprisingly effective in reducing errors. Indeed, it was shown in [2], [5] that if we restrict our terminal error by the stringent requirement that in aiming to "bounce off" Mars in a particular direction (e.g., en route to Venus) along the "other asymptote" (see Fig. 4), our error in the subsequent velocity vector will be no greater than the velocity errors incurred at the correction points on the way to Mars; then 8 corrections suffice for the Earth-Mars leg, the last correction occurring about $3\frac{3}{4}$ hr before passing Mars.

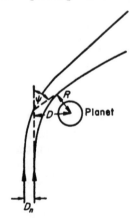

Fig. 4. Trajectory near planet.

The weak assumption in this first example is that D_{n-1} is estimated on the basis of observations close to P_n. It is certainly more plausible to assume that D_{n-1} is estimated on measurements at least as far back as P_{n-1}, if not all the way back to the start—the estimation in the latter case taking account of the previous corrective velocities.

To see the effect this might have on the optimum spacing, we choose for our second example a rather different "one-dimensional" situation (see Fig. 5). Suppose that a vehicle has a nominal straight-line motion

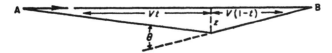

Fig. 5. Position determination by subtended angle.

from A to B with constant speed V in unit time, but that its actual position at time t has a small lateral component z perpendicular to AB in a fixed plane. Suppose further that lateral position z at any time is measured by means of the exterior subtended angle θ of which the standard deviation σ_θ is assumed to be independent of time. It is then easy to show that the standard deviation of lateral position determination z is

$$\sigma_z = Vt(1 - t)\sigma_\theta,$$

which is, of course, largest midway from A to B. Perfectly mechanized corrective thrusts are to be applied perpendicular to AB. The author has shown [3], [5] that if the miss-distance is estimated at any time from closely spaced measurements of z all the way back to A, the optimum choice of correction times t_n is such that

$$\frac{1 - t_{n-1}}{1 - t_n} \to 2.62 \quad \text{as} \quad t_n \to 1.$$

Figure 6, which is analogous to Figure 3, shows again that the expected fuel consumption for corrections is not very sensitive to a change of the spacing ratio ρ from the optimum value, in this case 2.62. Also included in Figure 6 is a curve representing the relative fuel expenditure, computed for a general constant spacing ratio ρ, for the case in which miss-distance is estimated only on the basis of closely spaced measurements since the last correction.

It is interesting to note here that because of the closeness of the last observations, and in spite of the assumed improvement in measurement as we approach B, the neglect of position information prior to the

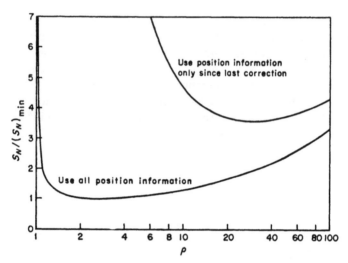

Fig. 6. Fuel expenditure versus spacing ratio (Example 2).

previous correction is costly. It is also interesting that the optimum spacing is not substantially different from that in the first example.

References

1. Rosenbloom, A., "Final-Value Systems with Total Effort Constraints," presented at the First International Congress of the International Federation on Automatic Control, Moscow, June, 1960.
2. Breakwell, J. V., "Fuel Requirements for Crude Interplanetary Guidance," presented at Second Western National Meeting of the American Astronautical Society, Los Angeles, Calif., August, 1959; published in *Advances in Astronautical Sciences*, Vol. 5, Plenum Press, New York, 1960.
3. Breakwell, J. V., "The Spacing of Corrective Thrusts in Interplanetary Navigation," presented at Third Annual West Coast Meeting of the American Astronautical Society, Seattle, Wash., August, 1960; published in *Advances in Astronautical Sciences*, Vol. 7, Plenum Press, New York, 1961.
4. Lawden, D. F., "Optimal Programme for Correctional Manoeuvres," *Astronaut. Acta*, Vol. 6, 1960, pp. 195–205.
5. Breakwell, J. V., "The Optimum Spacing of Corrective Thrusts in Interplanetary Navigation," in George Leitmann (ed.), *Optimization Techniques with Application to Aerospace Systems*, Academic Press, New York, 1962.

Chapter 4

The Analysis and Solution of Optimum Trajectory Problems

STUART E. DREYFUS

1. Introduction

There are extremely attractive alternatives to both the customary analytic approach (see [1]) and the method of numerical solution (see [2]) usually adopted in the analysis of trajectory problems. This chapter will indicate the nature of these alternatives and provide some references to fuller discussions.

2. Analytic Approach

Although the conventional variational approach characterizes optimality *globally* by means of comparison functions, it leads to a *local* theory consisting of the Euler–Lagrange differential equations and other local conditions.

On the other hand, dynamic programming [3] and the theory of Carathéodory [4] begin with a *local* characterization of optimality. It is shown in reference [5] how the classical results follow easily and intuitively from this approach.

The fundamental difference in approaches stems from the definition of an optimal solution. In the classical approach an optimal solution is a control function (or set of functions) of an independent variable, usually time, that yields a trajectory starting in a *specified initial state* configuration and satisfying certain terminal conditions.

In dynamic programming a solution is a mapping from state (that is, configuration, phase) space to a control space, so that each possible physical state has associated with it an optimal control action. This space of controls is so constructed that the trajectory thus determined from *any feasible initial point* satisfies any specified terminal conditions.

This new approach also allows a simple characterization of a relative extremal for the synthesis problem when the domain of the solution function is bounded and the solution consists of Euler curves and boundary curves [6].

3. Numerical Solution

An attractive method of numerical solution stems from the technique of successive approximations, where a nominal curve is guessed and then successively improved via a linearized theory. This idea is not new, of course, but it has recently found a number of successful applications [7], [8], [9]. Experimentation seems to indicate that this approach avoids the instabilities inherent in straightforward integration of the Euler equations.

The ordinary differential equations furnished by the Euler equation can be thought of as characteristics of the Hamilton–Jacobi partial differential equation derived directly via dynamic programming. Use of these equations in the approximation method mentioned above is preferable to the direct solution of the partial differential equation, or recurrence relation, of dynamic programming. The dynamic-programming method of solution, however, though time and space consuming, does guarantee the determination of the absolute extremum, and also finds applications in the study of stochastic and adaptive variational problems, where, as yet, no other general methods exist.

4. Optimal Guidance

Deviation from a preprogrammed optimal trajectory often occurs during flight, due to unpredictable forces such as wind, and occasionally as a result of mechanical malfunctions. Much attention has recently been concentrated on this problem. Earlier guidance schemes attempted to return to the old trajectory in the event of deviation, or to match the old terminal conditions. Current research recognizes that after deviation, particularly one of some magnitude, some new trajectory is optimal for the new problem at hand. If the optimal decision were known as a function of the coordinate in state space—rather than as a function of time as in the classical theory—optimal decisions could be easily rendered, despite disturbances, no matter how large.

Theories are currently being developed that determine the optimal decision *in the state-space neighborhood* of the optimal trajectory. These involve second-order analyses, including, in the classical case, the second variation [10], [11], and in the dynamic-programming approach, second partial derivatives [12].

5. Stochastic and Adaptive Variational Problems

Optimal guidance theories recognize that deviations may occur, but do not consider the probability of these deviations *in advance*. Stochastic variational theory seeks an optimal trajectory, usually in an expected-value sense, taking account of probable disturbances, the statistics of which are assumed known. Either the current state or the future forces may be statistical [13], [14].

If, initially, even the statistical description of the unknown forces, or of components of the state vector, is lacking, then the problem is called adaptive [15].

6. Conclusion

New problems, approaches, and results are appearing in the optimal-trajectory and control area. It is becoming obvious that a complete solution of a variational problem should consist of a mapping from state-variable space to the control space, so that each possible physical state has associated with it an optimal control action. If such a mapping can be found, both the optimal guidance and stochastic control problems are solved.

This mapping can generally be characterized as the solution of a partial differential equation. Since computation of the solution is often made prohibitive by the dimension of the state description, successive-approximation techniques are necessary. These schemes, often called gradient techniques, exist for deterministic systems.

While some analytic results exist for special stochastic problems, a practical method of numerical solution for general stochastic problems is yet to be developed.

References

1. Bliss, G. A., *Lectures on the Calculus of Variations*, University of Chicago Press, Chicago, Ill., 1946.
2. Breakwell, J. V., "The Optimization of Trajectories," *J. Soc. Indust. Appl. Math.*, Vol. 7, No. 2, 1959, pp. 215–247.
3. Bellman, R., *Dynamic Programming*, Princeton University Press, Princeton, N.J., 1957.

4. Carathéodory, C., *Variationsrechnung und Partielle Differential-gleichungen erster Ordnung*, B. G. Teubner, Leipzig, 1935, pp. 249–251.
5. Dreyfus, S. E., "Dynamic Programming and the Calculus of Varitions," *J. Math. Anal. and Appl.*, Vol. 1, No. 2, 1960, pp. 228–239.
6. Dreyfus, S. E., "Variational Problems with Inequality Constraints," *J. Math. Anal. and Appl.*, Vol. 4, No. 2, 1962, pp. 297–308.
7. Dreyfus, S. E., "Numerical Solution of Variational Problems," *J. Math. Anal. and Appl.*, Vol. 5, No. 1, 1962, pp. 30–45.
8. Kelley, H. J., "Gradient Theory of Optimal Flight Paths," *ARS Journal*, Vol. 30, No. 10, 1960, pp. 947–954.
9. Bryson, A. E., and W. Denham, "A Steepest Ascent Method for Solving Optimum Programming Problems," *J. Appl. Mech.*, Vol. 29, No. 2, 1962, pp. 247–257.
10. Kelley, H. J., "Guidance Theory and Extremal Fields," presented at National Aerospace Electronics Convention, May 14–16, 1962, Dayton, Ohio.
11. Breakwell, J. V., and A. E. Bryson, "Neighboring Optimum Terminal Control for Multivariable Nonlinear Systems," to appear in *Trans. ASME, Ser. D*.
12. Dreyfus, S. E., "Optimal Correction for Deviations from an Optimal Trajectory In-Flight," privately circulated, November, 1961.
13. Kalman, R. E., and R. S. Bucy, "New Results in Linear Filtering and Prediction Theory," *Trans. ASME, Ser. D*, Vol. 83D, 1961, pp. 95–108.
14. Florentin, J., "Optimal Control of Continuous Time, Markov, Stochastic Systems," *J. Electronics Control*, Vol. X, No. 6, 1961, pp. 473–488.
15. Bellman, R., *Adaptive Control Processes: A Guided Tour*, Princeton University Press, Princeton, N.J., 1961.

Part Two

COMMUNICATION, PREDICTION, AND DECISION

Chapter 5

A New Approach to the Synthesis of Optimal Smoothing and Prediction Systems[†]

EMANUEL PARZEN

1. Introduction

This chapter describes a new approach to a wide class of smoothing and prediction problems. The method can be applied to either stationary or nonstationary time series, with discrete or continuous parameters. It can easily be extended to time series observed in space-time and also to multiple time series, that is, those for which the observed value at each point of space-time is not a real number but a vector of real numbers.

Over the past few years I have been studying relationships between the theory of second-order stationary random functions, time series analysis, the theory of optimum design of communications and control systems, and classical regression analysis and analysis of variance. In the spring of 1957 I observed that reproducing-kernel Hilbert spaces provide a unified framework for these varied problems. The results obtained in 1957–1958 were theoretical elaborations of this idea, and were stated in a lengthy Stanford technical report [1] completed in the fall of 1958. Since then I have been concerned with developing examples and applications, well aware that the reproducing-kernel Hilbert space approach would be of no value unless it could provide new answers as well as old ones. It is hoped that the results presented here provide evidence that this approach is of value.

It may be of interest to relate this approach to one that is being

† Prepared with partial support of the Office of Naval Research.

developed in the Soviet Union by V. S. Pugachev ([2]–[5]). Pugachev has in recent years advanced a point of view that he calls the *method of canonic representations of random functions*, for which in a recent article [5] he makes the following claim: "The results of this article, together with the results of [previous] papers, permit us to state that the method of canonic representations of random functions is the foundation of the modern statistical theory of optimum systems." The methods to be presented in this chapter appear to provide a more powerful and elegant means of achieving in a unified manner the results that Pugachev has sought to unify by the method of canonic representations.

It may also be of interest to describe the standard approach to prediction and smoothing problems. The pioneering work of Wiener [6] and Kolmogorov [7] on prediction theory was concerned with a stationary time series observed over a semi-infinite interval of time, and sought predictors having minimum mean square over all possible linear predictors. Wiener showed how the solution of the prediction problem could be reduced to the solution of the so-called Wiener–Hopf integral equation, and gave a method (spectral factorization) for the solution of this integral equation. Simplified methods for solution of this equation in the practically important, special case of rational spectral density functions were given by Zadeh and Ragazzini [8] and Bode and Shannon [9]. Zadeh and Ragazzini [10] also treated the problem of regression analysis of time series with stationary fluctuation function by reducing the problem to one involving the solution of a Wiener-Hopf equation. There then developed an extensive literature treating prediction and smoothing problems involving a finite time of observation and nonstationary time series. The methods employed were either to reduce the solution of the problem to the solution of a suitable integral equation (generalization of the Wiener-Hopf equation) or to employ expansions (of Karhunen-Loève type) of the time series involved. In this chapter, we describe an approach to smoothing and prediction problems that may be called *coordinate free*, which, by the introduction of suitable coordinate systems, contains these previous approaches as special cases.

Finally, let us briefly outline the class of problems for which we shall give a unified, rigorous, and general treatment. A wide variety of problems concerning communication and control, or both (involving such diverse problems as the automatic tracking of moving objects, the reception of radio signals in the presence of natural and artificial disturbances, the reproduction of sound and images, the design of guidance systems, the design of control systems for industrial processes, fore-casting, the analysis of economic fluctuations, and the analysis of any kind of record representing observation over time), may be regarded as special cases of the following problem:

Let T denote a set of points on a time axis such that at each point t in T an observation has been made of a random variable $X(t)$. Given the observations $\{X(t), t \in T\}$, and a quantity Z related to the observation in a manner to be specified, one desires to form in an optimum manner estimates and tests of hypotheses about Z and various functions $\psi(Z)$.

This imprecisely formulated problem provides the general context in which to pose the following usual problems of communication and control.

Prediction or extrapolation: Observe the stochastic process $X(t)$ over the interval $s - T \leq t \leq s$; then predict $X(s + \alpha)$ for any $\alpha > 0$. The length T of interval of observation may be finite or infinite. The optimum system yielding the predicted value of $X(s + \alpha)$ is referred to as an optimum dynamic system if it provides estimates of $X(s + \alpha)$ for all $\alpha > 0$.

Smoothing or filtering: Over the interval $s - T \leq t \leq s$, observe the sum $X(t) = S(t) + N(t)$ of two stochastic processes or time series $S(t)$ and $N(t)$, representing signal and noise respectively; then estimate $S(t)$ for any value of t in $s - T \leq t \leq s$. The terminology "smoothing" derives from the fact that often the noise $N(t)$ consists of very high-frequency components compared with the signal $S(t)$; predicting $S(t)$ can then be regarded as attempting to pass a smooth curve through a very wiggly record.

Smoothing and prediction: Observe $S(t) + N(t)$ over $s - T \leq t \leq s$; then predict $S(s + \alpha)$ for any $\alpha > 0$.

Parameter estimation: Over an interval $0 \leq t \leq T$, observe $S(t) + N(t)$, where $S(t)$ represents the trajectory (given by $S(t) = x_0 + vt + at^2/2$, say) of a moving object and $N(t)$ represents errors of measurement; then estimate the velocity v and acceleration a of the object. More generally, estimate such quantities as $S(t)$ and $dS(t)/dt$ at any time t in $0 \leq t \leq T$, when the signal is of the form $S(t) = \beta_1 w_1(t) + \cdots + \beta_q w_q(t)$.

Signal extraction and detection: Observe $X(t) = A \cos \omega(t - \tau) + N(t)$ over an interval $0 \leq t \leq T$; then estimate the parameters A and τ, or test the hypothesis that $A = 0$ against the hypothesis that $A > 0$. This problem is not explicitly treated in this chapter, although it could be handled by means of the tools described here.

2. A New Approach to Prediction Problems

Let us consider a stochastic process or time series $\{X(t), t \in T\}$, which is a family of random variables indexed by a parameter t varying in some index set T. Assume that each random variable has a finite second moment. Let

$$K(s, t) = E[X(s)X(t)] \tag{2.1}$$

be the *covariance kernel* of the time series. It might be thought more logical to call the function defined by (2.1) the *product moment kernel*, and reserve the name covariance kernel for the function defined by

$$K(s, t) = \text{Cov}\,[X(s),\ X(t)] = E[X(s)X(t)] - E[X(s)]E[X(t)]. \tag{2.2}$$

This terminology seems cumbersome, however, and is not adopted. We shall call the function defined by (2.2) the *proper covariance kernel*.

Let Z be a random variable with finite second moment for which one knows the cross-covariance function $\rho_Z(\cdot)$, defined by

$$\rho_Z(t) = E[ZX(t)], \qquad t \text{ in } T. \tag{2.3}$$

A basic problem in statistical communication theory—which, as we shall see, is also basic to the study of the structure of time series—is that of *minimum mean-square error linear prediction*: Given a random variable Z with finite second moment, and a time series $\{X(t), t \in T\}$, find that random variable, linear in the observations, with smallest mean-square distance from Z. In other words, if we desire to predict the value of Z on the basis of having observed the values of the time series $\{X(t), t \in T\}$, one method might be to take that linear functional in the observations, denoted by $E^*[Z \mid X(t), t \in T]$, of which the mean-square error as a predictor is least.†

The existence and uniqueness of, and conditions characterizing, the best linear predictor are provided by the projection theorem of abstract Hilbert-space theory. (For proofs of the following assertions concerning Hilbert-space theory, see any suitable text, such as Halmos [13].)

By an abstract Hilbert space is meant a set H (with members u, v, $\cdots$, that are usually called vectors or points) that possesses the following properties:

i. H is a linear space. Roughly speaking, this means that for any vector u and v in H, and real numbers a, there exist vectors, denoted by

---

† The symbol E^* is used to denote a predictor because, in the case of jointly normally distributed random variables, the least linear predictor $E^*[Z \mid X(t), t \in T]$ coincides with the conditional expectation $E[Z \mid X(t), t \in T]$. For an elementary discussion of this fact, see Parzen [11], p. 387, or [12], Chap. 2.

$u + v$ and au, respectively, that satisfy the usual algebraic properties of addition and multiplication; also there exists a zero vector 0 with the usual properties under addition.

ii. H is an inner product space. That is, to every pair of points, u and v, in H there corresponds a real number, written (u, v) and called the inner product of u and v, possessing the following properties: for all points u, v, and w in H, and for every real number a,

   a. $(au, v) = a(u, v)$
   b. $(u + v, w) = (u, w) + (v, w)$,
   c. $(v, u) = (u, v)$,
   d. $(u, u) > 0$ if $u \neq 0$.

iii. H is a complete metric space under the norm $\|u\| = (u, u)^{1/2}$. That is, if $\{u_n\}$ is a sequence of points such that $\|u_m - u_n\| \to 0$ as $m, n \to \infty$, then there is a vector u in H such that $\|u_n - u\|^2 \to 0$ as $n \to \infty$.

The Hilbert space spanned by a time series $\{X(t), t \in T\}$ is denoted by $L_2(X(t), t \in T)$ and is defined as consisting of all random variables U that are either finite linear combinations of the random variables $\{X(t), t \in T\}$, or are limits of such finite linear combinations in the norm corresponding to the inner product defined on the space of square-integrable random variables by

$$(U, V) = E[UV]. \tag{2.4}$$

In words, $L_2(X(t), t \in T)$ consists of all linear functionals in the time series.

We next state without proof the projection theorem for an abstract Hilbert space.

PROJECTION THEOREM. *Let H be an abstract Hilbert space, let M be a Hilbert subspace of H, let v be a vector in H, and let v^* be a vector in M. A necessary and sufficient condition that v^* be the unique vector in M satisfying*

$$\|v^* - v\| = \min_{u \in M} \|u - v\| \tag{2.5}$$

is that

$$(v^*, u) = (v, u) \text{ for every } u \text{ in } M. \tag{2.6}$$

The vector v^* satisfying (2.5) is called the *projection of v onto M*, and is also written $E^*[v \mid M]$.

In the case that M is the Hilbert space spanned by a family of vectors $\{x(t), t \in T\}$ in H, we write $E^*[v \,|\, x(t), t \in T]$ to denote the projection of v onto M. In this case, a necessary and sufficient condition that v^* satisfy (2.5) is that

$$(v^*, x(t)) = (v, x(t)) \text{ for every } t \in T. \tag{2.7}$$

We are now in a position to solve the problem of obtaining an explicit expression for the minimum mean-square error linear prediction $E^*[Z \,|\, X(t), t \in T]$. From (2.7) it follows that the optimum linear predictor is the unique random variable in $L_2(X(t), t \in T)$ satisfying, for all t in T,

$$E[E^*[Z \,|\, X(t), \ t \in T] X(t)] = E[ZX(t)]. \tag{2.8}$$

Equation (2.8) may look more familiar if we consider the special case of an interval $T = \{t : a \leq t \leq b\}$. If one writes, heuristically,

$$\int_a^b X(s) w(s) \, ds \tag{2.9}$$

to represent a random variable in $L_2(X(t), t \in T)$, then (2.8) states that the weighting function $w^*(t)$ of the best linear predictor

$$E^*[Z \,|\, X(t), \ t \in T] = \int_a^b w^*(s) X(s) \, ds, \tag{2.10}$$

must satisfy the generalized Wiener–Hopf equation

$$\int_a^b w^*(s) K(s, t) \, ds = \rho_Z(t), \qquad a \leq t \leq b. \tag{2.11}$$

There is an extensive literature [14], [15], [16] concerning the solution of the integral equation in (2.11). However, this literature is concerned with an unnecessarily difficult problem—one in which the very formulation of the problem makes it difficult to be rigorous. The integral equation in (2.11) has a solution only if one interprets $w^*(s)$ as a generalized function including terms that are Dirac delta functions and derivatives of delta functions.

A simple reinterpretation of (2.11) avoids all of these difficulties. Let us not regard (2.11) as an integral equation for the weighting function $w^*(s)$; rather, let us compare (2.10) and (2.11). These equations say that *if one can find a representation for the function $\rho_Z(t)$ in terms of linear operations on the functions $\{K(s, \cdot), s \in T\}$, then the minimum mean-square error linear predictor $E^*[Z \,|\, X(t), t \in T]$ can be written in terms of the corresponding linear operations on the time series*

$\{X(s),\ s \in T\}$. It should be emphasized that the most important linear operations are integration and differentiation. Consequently, the problem of finding the best linear predictor is not one of solving an integral equation but rather one of hunting for a linear representation of $\rho_Z(t)$ in terms of the covariance kernel $K(s,\ t)$. A general method of finding such representations will be discussed in the following sections. In this section we illustrate the ideas involved by considering several examples.

Example 2A. Consider a stationary time series $X(t)$, with covariance kernel

$$K(s,\ t)\ =\ Ce^{-\beta|t-s|}, \tag{2.12}$$

which we have observed over a finite interval of time, $a \leq t \leq b$. Suppose that we desire to predict $X(b+c)$ for $c > 0$. Now, for $a \leq t \leq b$, we have

$$\rho(t)\ =\ E[X(t)X(b+c)]\ =\ Ce^{-\beta(b+c-t)}\ =\ e^{-\beta c}K(b,\ t). \tag{2.13}$$

In view of (2.13), by the interpretation of (2.10) and (2.11) just stated, it follows that

$$E^*[X(b+c)\,|\,X(t),\ a \leq t \leq b]\ =\ e^{-\beta c}X(b). \tag{2.14}$$

The present methods yield a simple proof of a widely quoted fact. Define a stationary time series $X(t)$ with a continuous covariance function $R(s-t)\ =\ E[X(s)X(t)]$ to be *Markov* if, for any real numbers $a < b$ and $c > 0$, the least linear predictor of $X(b+c)$, given $X(t)$ over the interval $a \leq t \leq b$, is a linear function of the most recent value $X(b)$; in symbols, $X(t)$ is Markov if

$$E^*[X(b+c)\,|\,X(t),\ a \leq t \leq b]\ =\ A(c)X(b) \tag{2.15}$$

for some constant $A(c)$ depending only on c.

Let us now establish the following result:

Doob's Theorem: *Equation* (2.15) *holds if and only if, for some constants C and β,*

$$R(u)\ =\ Ce^{-\beta|u|}. \tag{2.16}$$

Proof: From the fact that

$$\rho(t)\ =\ E[X(b+c)X(t)]\ =\ R(b-t+c),$$

it follows by the projection theorem that (2.15) holds if and only if, for every $a < b$, $c > 0$, and t in $a \leq t \leq b$, we have

$$R(b-t+c)\ =\ A(c)R(b-t). \tag{2.17}$$

By (2.17) it follows that for every $d \geq 0$ and $c \geq 0$ we have

$$R(d + c) = A(c)R(d). \qquad (2.18)$$

Letting $d = 0$, we obtain $A(c) = R(c)$; consequently, for every $c \geq 0$ and $d \geq 0$, $R(u)$ satisfies the equation

$$R(d + c) = R(d)R(c). \qquad (2.19)$$

It is well known (see Parzen [11], p. 263) that a continuous even function $R(u)$ satisfying (2.19) is of the form of (2.16).

Example 2B. (Reinterpretation of the Karhunen–Loève expansion.) Many writers on statistical communication theory (see [17], pp. 96, 244, 338–352, [18], and [19]) have made use of what is often called the Karhunen–Loève representation of a random function $X(t)$ of second order. The results obtained are clarified when looked at from the present point of view.

The fundamental fact underlying the Karhunen–Loève expansion may be stated as follows:

MERCER'S THEOREM. *If* $\{\varphi_n(t), n = 1, 2, \cdots\}$ *denotes the sequence of normalized eigenfunctions and* $\{\lambda_n, n = 1, 2, \cdots\}$ *the sequence of corresponding nonnegative eigenvalues satisfying the relations*

$$\int_a^b K(s, t)\varphi_n(s)\, ds = \lambda_n \varphi_n(t), \qquad a \leq t \leq b, \qquad (2.20)$$

$$\int_a^b \varphi_m(t)\varphi_n(t)\, dt = \delta(m, n), \qquad (2.21)$$

where $\delta(m, n)$ *is the Kronecker delta function, equal to 1 or 0 depending on whether* $m = n$ *or* $m \neq n$, *then the kernel* $K(s, t)$ *may be represented by the series*

$$K(s, t) = \sum_{n=1}^{\infty} \lambda_n \varphi_n(s)\varphi_n(t), \qquad (2.22)$$

and this series converges absolutely and uniformly for $a \leq s, t \leq b$.

If we wish to predict the value of a random variable Z on the basis of the observed values $X(t)$, $a \leq t \leq b$, we may write an explicit expression for the minimum mean-square error linear predictor as follows:

$$E^*[Z \mid X(t), a \leq t \leq b] = \sum_{n=1}^{\infty} \frac{1}{\lambda_n} \int_a^b \rho_Z(t)\varphi_n(t)\, dt$$

$$\times \int_a^b X(s)\varphi_n(s)\, ds. \qquad (2.23)$$

In order to prove the validity of (2.23), we need to prove that the infinite series is well defined and that it satisfies (2.8). Now

$$E\left[\int_a^b X(s)\varphi_m(s)\,ds \int_a^b X(t)\varphi_n(t)\,dt\right]$$

$$= \int_a^b \int_a^b K(s,t)\varphi_m(s)\varphi_n(t)\,ds\,dt = \lambda_n \delta(m,n). \qquad (2.24)$$

Therefore the mean square of the infinite series in (2.23) is equal to

$$\sum_{n=1}^{\infty} \frac{1}{\lambda_n}\left|\int_a^b \rho_Z(t)\varphi_n(t)\,dt\right|^2. \qquad (2.25)$$

Consequently, a necessary and sufficient condition that the infinite series in (2.23) be well defined is that the infinite series in (2.25) be finite, which may be shown always to be the case. Next, we can show that (2.8) is satisfied by verifying that, for any t in $a \le t \le b$,

$$E\left[X(t)\left\{\sum_{n=1}^{\infty} \frac{1}{\lambda_n}\int_a^b \rho_Z(s)\varphi_n(s)\,ds \int_a^b X(u)\varphi_n(u)\,ds\right\}\right]$$

$$= \sum_{n=1}^{\infty} \varphi_n(t)\int_a^b \rho_Z(s)\varphi_n(s)\,ds = \rho_Z(t). \qquad (2.26)$$

If it is permissible to interchange the processes of summation and integration in (2.23), then we may write

$$E^*[Z \mid X(t),\ a \le t \le b] = \int_a^b w^*(s)X(s)\,ds, \qquad (2.27)$$

where

$$w^*(s) = \sum_{n=1}^{\infty} \varphi_n(s)\frac{1}{\lambda_n}\int_a^b \rho_Z(t)\varphi_n(t)\,dt. \qquad (2.28)$$

The condition for the infinite series in (2.28) to be well defined is that

$$\sum_{n=1}^{\infty} \frac{1}{\lambda_n^2}\left|\int_a^b \rho_Z(t)\varphi_n(t)\,dt\right|^2 < \infty. \qquad (2.29)$$

It can be shown that if (2.29) holds, then (2.27) is valid. Although (2.25) is always finite, however, (2.29) rarely holds. The optimal predictor is not usually of the form of (2.27). Thus we again see that it is not desirable to reduce prediction problems to the solution of the integral equation in (2.11).

Example 2C. (*The method of shaping filters.*) Another technique employed in statistical communication theory is the method of *shaping filters* (see Lanning and Battin [14]). Let $X(t)$ be a stochastic process with covariance kernel $K(s, t)$. Let $\eta(t)$ be a white-noise process, and let $W(t, s)$ be a weighting function such that for every t we have

$$X(t) = \int_{-\infty}^{t} W(t, s)\eta(s)\, ds. \qquad (2.30)$$

In words, the time series $X(t)$ is represented as the response to a white-noise input of a system ("filter") described by a time-varying impulse-response function $W(t, s)$. If (2.30) holds, then $W(t, s)$ is called a *shaping filter* for the time series $X(t)$. We now show how to use shaping filters to solve the prediction problem, given a time series $X(t)$ that has been observed over a semi-infinite range, $-\infty < t < b$.

If (2.30) holds, and if the cross-covariance function $\rho_Z(t)$ may be written, for a square-integrable function $r(s)$, as

$$\rho_Z(t) = \int_{-\infty}^{t} W(t, s)r(s)\, ds, \qquad -\infty < t < b, \qquad (2.31)$$

then

$$E^*[Z \mid X(t),\ -\infty < t \leq b] = \int_{-\infty}^{b} r(s)\eta(s)\, ds. \qquad (2.32)$$

To prove (2.32), note that, for $-\infty \leq t \leq b$, we have

$$E\left[\int_{-\infty}^{t} W(t, s)\eta(s)\, ds \int_{-\infty}^{b} r(s)\eta(s)\, ds \right]$$

$$= \int_{-\infty}^{t} W(t, s)r(s)\, ds = \rho_Z(t). \qquad (2.33)$$

The expression given by (2.32) can be further simplified if we make the following reasonable assumptions about the shaping filter. Let L_t and M_t be differential operators of orders n and m respectively:

$$L_t = \sum_{k=0}^{n} a_k(t)\, \frac{d^k}{dt^k},$$

$$M_t = \sum_{k=0}^{m} b_k(t)\, \frac{d^k}{dt^k}. \qquad (2.34)$$

Let $H_L(t, s)$ and $H_M(t, s)$ be the respective one-sided Green's functions characterized by the property that any sufficiently differentiable function f is given by

$$f(t) = \int_{-\infty}^{t} H_L(t, s) L_s f(s) \, ds$$

$$= \int_{-\infty}^{t} H_M(t, s) M_s f(s) \, ds. \tag{2.35}$$

Suppose that the covariance kernel of $X(t)$ may be written

$$K(t, s) = \int_{-\infty}^{\min(t,s)} M_t H_L(t, u) M_s H_L(s, u) \, du, \tag{2.36}$$

or, equivalently, that

$$X(t) = \int_{-\infty}^{t} M_t H_L(t, s) \eta(s) \, ds. \tag{2.37}$$

For an interesting discussion of how to find differential operators satisfying (2.36), see Batkov [20]. It may be shown that if (2.36) holds, then the right-hand side of (2.32) may be written in the form

$$\int_{-\infty}^{b} dt \int_{-\infty}^{t} L_t H_M(t, u) \rho_Z(u) \, du \int_{-\infty}^{t} L_t H_M(t, u) X(u) \, du. \tag{2.38}$$

In the particular case $M_t \equiv 1$, (2.38) reduces to

$$\int_{-\infty}^{b} dt \{ L_t \rho_Z(t) \} \{ L_t X(t) \}. \tag{2.39}$$

For the sake of rigor, it should be noted that in (2.38) and (2.39) the highest-order derivative of the observed time series $X(t)$ may not exist, and we must then write $dX^{(n-1)}(t)$ for $X^{(n)}(t) \, dt$.

3. General Solution of the Problems of Linear Prediction

It is possible to give a treatment of problems of prediction and smoothing that distinguishes between the statistical and analytical aspects of the problem. Such methods as that of expansions in eigenfunctions used in example 2B and that of shaping filters used in example 2C are merely analytical means of evaluating certain abstract quantities that can be defined without reference to these methods. The statistical problems of prediction and smoothing may be solved in terms of these abstract quantities once and for all. Indeed, the theory we shall now describe underlies the solution of many optimization problems; for example, it includes as a special case the theory of generalized inverses of matrices (see Greville [21] for references to the history of the notion).

The basic tool in our theory is the notion of the *reproducing-kernel space* corresponding to a covariance kernel K.

THEOREM 3.1. (*Existence and uniqueness of the reproducing-kernel Hilbert space corresponding to a covariance function.*) *Let* $\{X(t), t \in T\}$ *be a time series with covariance kernel* $K(s, t)$ *given by (2.1). Let* $H(K)$ *consist of all functions* $h(\cdot)$ *defined on* T *and of the form, for some* U *in* $L_2(X(t), t \in T)$,

$$h(t) = E[X(t)U], \quad \text{for all } t \in T. \tag{3.1}$$

On $H(K)$ *define an inner product by*

$$(h, h)_K = E \,|\, U \,|^2. \tag{3.2}$$

Then $H(K)$ *is a Hilbert space. Further,* $H(K)$ *possesses the following two properties:* (a) *for every* $t \in T$,

$$K(\cdot, t) \quad \text{belongs to } H(K), \tag{3.3}$$

where $K(\cdot, t)$ *is the function defined on* T *with value at* s *equal to* $K(s, t)$; (b) *for every* t *in* T *and* $h(\cdot)$ *in* $H(K)$,

$$h(t) = (h, K(\cdot, t))_K. \tag{3.4}$$

One calls (3.4) the *reproducing property* of the kernel $K(s, t)$. Since (3.4) holds, we call $H(K)$ a reproducing-kernel Hilbert space, with reproducing kernel K (for the theory of such spaces, see [22]). The reproducing-kernel Hilbert space $H(K)$ is uniquely determined by the conditions (3.3) and (3.4).

Intuitively, a reproducing-kernel Hilbert space is a Hilbert space that contains a function playing the role of the Dirac delta function $\delta(t)$. It should be recalled that, for square-integrable functions $f(\cdot)$,

$$\int_{-\infty}^{\infty} f(s)\delta(s - t) \, ds = f(t). \tag{3.5}$$

Consequently, the kernel $K(s, t) = \delta(s - t)$ satisfies (3.4). It does not satisfy (3.3), however, and therefore it is not truly a reproducing kernel.

THEOREM 3.2. (*General solution of the prediction problem.*) *Let* $\{X(t), t \in T\}$ *be a time series with covariance kernel* $K(s, t)$, *and let* $H(K)$ *be the corresponding reproducing-kernel Hilbert space. Between* $L_2(X(t), t \in T)$ *and* $H(K)$ *there exists a one-to-one inner product pre-*

serving linear mapping under which $X(t)$ and $K(\cdot, t)$ are mapped into one another. Denote by $(h, X)_K$ the random variable in $L_2(X(t), t \in T)$ that corresponds under the mapping to the function $h(\cdot)$ in $H(K)$. Then the general solution to the prediction problem may be written as follows. If Z is a random variable with finite second moment, and if

$$\rho_Z(t) = E[ZX(t)],$$

then

$$E^*[Z \mid X(t), t \in T] = (\rho_Z, X)_K, \tag{3.6}$$

with mean-square error of prediction given by

$$E[\mid Z - E^*[Z \mid X(t), t \in T]\mid^2] = E\mid Z\mid^2 - (\rho_Z, \rho_Z)_K. \tag{3.7}$$

PROOF. The validity of Theorem 3.2 follows immediately from the definition of the concepts involved. However, it may be instructive to give a proof of the theorem, using the following properties of the mapping $(h, X)_K$. For any functions g and h in $H(K)$ and random variables Z with finite second moment, we have

$$E[(h, X)_K(g, X)_K] = (h, g)_K, \tag{3.8}$$

$$E[Z(h, X)_K] = (\rho_Z, h)_K, \tag{3.9}$$

where $\rho_Z(t) = E[ZX(t)]$. Now a random variable in $L_2(X(t), t \in T)$ may be written $(h, X)_K$ for some h in $H(K)$. Consequently, the mean-square error between any linear functional $(h, X)_K$ and Z may be written thus:

$$
\begin{aligned}
E[\mid (h, X)_K - Z\mid^2] &= E[(h, X)_K^2] + E[Z^2] - 2E[Z(h, X)_K] \\
&= E[Z^2] + (h, h)_K - 2(\rho_Z, h)_K \\
&= E[Z^2] - (\rho_Z, \rho_Z)_K + (h - \rho_Z, h - \rho_Z)_K. \tag{3.10}
\end{aligned}
$$

From (3.10) it is immediately seen that $(\rho_Z, X)_K$ is the minimum mean-square error linear predictor of Z, with mean-square prediction error equal to $E[Z^2] - (\rho_Z, \rho_Z)_K$. The proof of Theorem 3.2 is thus complete.

Theorem 3.2 represents a coordinate-free solution of the prediction problem. The usual methods of explicitly writing optimum predictors, using either eigenfunction expansions, Green's functions (impulse response functions), or (power) spectral density functions, are merely methods of writing down the reproducing-kernel inner product corresponding to the covariance kernel $K(s, t)$ of the observed time series.

Example 3A. (*Eigenfunction expansions.*) Let $X(t)$, $a \leq t \leq b$, be a time series of which the covariance kernel $K(s, t)$ has the eigenfunction expansion (2.22). The corresponding reproducing-kernel Hilbert space consists of all square-integrable functions $h(t)$ on the interval $a \leq t \leq b$ such that

$$\int_a^b |h(t)|^2 dt = \sum_{n=1}^{\infty} \left| \int_a^b h(t)\varphi_n(t) dt \right|^2$$

and

$$\sum_{n=1}^{\infty} \frac{1}{\lambda_n} \left| \int_a^b h(t)\varphi_n(t) dt \right|^2 < \infty. \qquad (3.11)$$

The reproducing-kernel inner product between two such functions is given by

$$(h, g)_K = \sum_{n=1}^{\infty} \frac{1}{\lambda_n} \int_a^b h(t)\varphi_n(t) dt \int_a^b g(t)\varphi_n(t) dt. \qquad (3.12)$$

The random variable $(h, X)_K$ in $L_2(X(t), a \leq t \leq b)$ corresponding to $h(\cdot)$ in $H(K)$ under the mapping described in Theorem 3.2 is given by (3.12) with g replaced by X.

Example 3B. (*Autoregressive schemes.*) The reproducing-kernel Hilbert space and inner product corresponding to time series of the type described in example 2C can be determined; the reader may easily infer them from (2.32) and (2.38). Here let us consider a stationary time series $X(t)$, observed over a finite interval $a \leq t \leq b$, of the type that statisticians call an *autoregressive scheme*.

A continuous-parameter stationary time series $X(t)$ is said to be an autoregressive scheme of order m if its covariance function may be written (see Doob [23], p. 542) as

$$R(s - t) = E[X(s)X(t)] = \int_{-\infty}^{\infty} \frac{e^{i(s-t)\omega}}{2\pi \left| \sum_{k=0}^{m} a_k(i\omega)^{m-k} \right|^2} d\omega, \qquad (3.13)$$

where the polynomial

$$\sum_{k=0}^{m} a_k z^{m-k}$$

has no zeros in the right-hand half of the complex z-plane. It can be shown that, given observations of such a time series over a finite interval $a \leq t \leq b$, the corresponding reproducing-kernel Hilbert space contains all functions $h(t)$ on $a \leq t \leq b$ that are continuously differentiable of order n. The reproducing-kernel inner product is given by

$$(h, g)_K = \int_a^b (L_t h)(L_t g)\, dt + \sum_{j,k=0}^m d_{j,k} h^{(j-1)}(a) g^{(k-1)}(a), \quad (3.14)$$

where

$$L_t h = \sum_{k=0}^m a_k h^{(m-k)}(t), \quad (3.15)$$

$$\{d_{j,k}\}^{-1} = \left\{ \frac{\partial^{j+k-2}}{\partial t^{j-1} \partial u^{k-1}} R(t - u) \Big|_{t=a, u=a} \right\}. \quad (3.16)$$

The first- and second-order autoregressive schemes are of particular importance.

A time series $X(t)$ is said to satisfy a *first-order autoregressive scheme* if it is the solution of a first-order linear differential equation with input a white noise $\eta'(t)$ (the symbolic derivative of a process $\eta(t)$ with independent stationary increments):

$$\frac{dX}{dt} + \beta X = \eta'(t). \quad (3.17)$$

It should be remarked that, from a mathematical point of view, (3.17) should be written as

$$dX(t) + \beta X(t)\, dt = d\eta(t). \quad (3.18)$$

Even then, in saying that $X(t)$ satisfies (3.17) or (3.18) we mean that

$$X(t) = \int_{-\infty}^t H(t - s)\, d\eta(s), \quad (3.19)$$

where $H(t - s) = e^{-\beta(t-s)}$ is the one-sided Green's function of the differential operator

$$L_t f = f'(t) + \beta f(t).$$

The covariance function of the time series $X(t)$ is

$$R(t - u) = \frac{1}{2\beta} e^{-\beta|u-t|}. \quad (3.20)$$

The corresponding reproducing-kernel Hilbert space $H(K)$ contains all differentiable functions. The inner product is given by

$$(f, g) = \int_a^b (f' + \beta f)(g' + \beta g)\, dt + 2\beta f(a) g(a). \quad (3.21)$$

More generally, corresponding to the covariance function

$$K(s, t) = Ce^{-\beta|s-t|},\tag{3.22}$$

the reproducing-kernel inner product is

$$(h, g)_K = \frac{1}{2\beta C}\left\{\int_a^b (h' + \beta h)(g' + \beta g)\, dt + 2\beta h(a)g(a)\right\}$$

$$= \frac{1}{2\beta C}\int_a^b (h'g' + \beta^2 hg)\, dt + \frac{1}{2C}\{h(a)g(a) + h(b)g(b)\}.\tag{3.23}$$

The random variable $(h, X)_K$ in $L_2(X(t), a \le t \le b)$, corresponding to $h(\cdot)$ in $H(K)$, may be written as

$$(h, X)_K = \frac{1}{2\beta C}\left\{\beta^2 \int_a^b h(t)X(t)\, dt + \int_a^b h'(t)\, dX(t)\right\}$$

$$+ \frac{1}{2C}\{h(a)X(a) + h(b)X(b)\}.\tag{3.24}$$

Note that $X'(t)$ does not exist in any rigorous sense; consequently, we write $dX(t)$ where $X'(t)\, dt$ seems to be called for. It can be shown that (3.24) makes sense. In the case that $h(\cdot)$ is twice differentiable, one may integrate by parts and write

$$\int_a^b h'(t)\, dX(t) = h'(b)X(b) - h'(a)X(a) - \int_a^b X(t)h''(t)\, dt.\tag{3.25}$$

A time series $X(t)$ is said to satisfy a *second-order autoregressive scheme* if it is the solution of a second-order linear differential equation with input a white noise $\eta'(t)$:

$$\frac{d^2X}{dt^2} + 2\alpha \frac{dX}{dt} + \gamma^2 X = \eta'(t).\tag{3.26}$$

If $\omega^2 = \gamma^2 - \alpha^2 > 0$, the covariance function of the time series is

$$R(t - u) = \frac{e^{-\alpha|u-t|}}{4\alpha\gamma^2}\left\{\cos \omega\,(u - t) + \frac{\alpha}{\omega}\sin \omega\,|u - t|\right\}.\tag{3.27}$$

The corresponding reproducing-kernel Hilbert space contains all twice-differentiable functions on the interval $a \le t \le b$ with inner product

$$(h, g)_K = \int_a^b (h'' + 2\alpha h' + \gamma^2 h)(g'' + 2\alpha g' + \gamma^2 g)\, dt$$

$$+ 4\alpha\gamma^2 h(a)g(a) + 4\alpha h'(a)g'(a).\tag{3.28}$$

To write $(h, X)_K$, we use the same considerations as those in (3.24).

4. General Solution of the Problem of Linear Smoothing (Regression Analysis)

Let $\{X(t), t \in T\}$ be a time series of which the proper covariance kernel

$$K(s, t) = \text{Cov}\,[X(s), X(t)] \tag{4.1}$$

is known. The mean-value function,

$$m(t) = E[X(t)], \tag{4.2}$$

is only assumed to belong to a known class M. One case of particular importance is that in which M consists of all finite linear combinations of q known functions $w_1(t), \cdots, w_q(t)$, so that the mean-value function is of the form

$$m(t) = \beta_1 w_1(t) + \cdots + \beta_q w_q(t) \tag{4.3}$$

for unknowns $\beta_1, \cdots, \beta_q$ that are to be estimated.

In this section we consider the problem of estimating various functionals of the true mean-value function $m(\cdot)$; in statistical theory, this is known as the problem of regression analysis of time series (see Parzen [24]). We seek estimates that (a) are linear in the observations $\{X(t), t \in T\}$ in the sense that they belong to $L_2(X(t), t \in T)$, (b) are *unbiased*, in a sense to be defined, and (c) have *minimum* variance among all linear unbiased estimates.

THEOREM 4.1. (*General solution of the problem of minimum variance unbiased linear estimation.*) *Let* $\{X(t), t \in T\}$ *be a time series with known proper covariance kernel* $K(s, t)$, *and unknown mean-value function* $m(t)$ *belonging to a known class* M *of functions. Let* $H(K)$ *be the corresponding reproducing-kernel Hilbert space, and assume that* M *is a subset of* $H(K)$.

i. *Between* $L_2(X(t), t \in T)$ *and* $H(K)$ *there exists a one-to-one linear mapping with the following properties: for every* t *in* T, *and* h *and* g *in* $H(K)$,

$$(K(\cdot, t), X)_K = X(t), \tag{4.4}$$

$$E_m[(h, X)_K] = (h, m)_K \qquad \text{for all } m \text{ in } M, \tag{4.5}$$

$$\text{Cov}\,[(h, X)_K, (g, X)_K] = (h, g)_K, \tag{4.6}$$

where $(h, X)_K$ *denotes the random variable in* $L_2(X(t), t \in T)$ *that cor-*

responds under the mapping to the function $h(\cdot)$ in $H(K)$. The subscript m on an expectation operator is written to indicate that the expectation is computed under the assumption that $m(\cdot)$ is the true mean-value function.

ii. *A random variable $(h, X)_K$ in $L_2(X(t), t \in T)$ is said to be an unbiased linear estimate of the value $m(t)$ at a particular time t of the mean-value function $m(\cdot)$ if*

$$E_m[(h, X)_K] = (h, m)_K = m(t) \qquad \text{for all m in M.} \qquad (4.7)$$

The uniformly minimum variance unbiased linear estimate $m^*(t)$ of $m(t)$ is given by

$$m^*(t) = (E^*[K(\cdot, t) \mid \overline{M}], X)_K, \qquad (4.8)$$

in which $\overline{M}$ is the smallest Hilbert subspace of $H(K)$ containing M, and $E^*[K(\cdot, t)|\overline{M}]$ is the projection onto $\overline{M}$ of $K(\cdot, t)$.

iii. *In the special case that $\overline{M}$ is finite dimensional and is spanned by q functions $w_1, \cdots, w_q$ that are linearly independent as functions in $H(K)$, we can write explicitly*

$$W m^*(t) = - \begin{vmatrix} (w_1, w_1)_K & \cdots & (w_1, w_q)_K & (X, w_1)_K \\ \vdots & & \vdots & \vdots \\ (w_q, w_1)_K & \cdots & (w_q, w_q)_K & (X, w_q)_K \\ w_1(t) & \cdots & w_q(t) & 0 \end{vmatrix}, \qquad (4.9)$$

$$W \operatorname{Var}[m^*(t)] = - \begin{vmatrix} (w_1, w_1)_K & \cdots & (w_1, w_q)_K & w_1(t) \\ \vdots & & \vdots & \vdots \\ (w_q, w_1)_K & \cdots & (w_q, w_q)_K & w_q(t) \\ w_1(t) & \cdots & w_q(t) & 0 \end{vmatrix}, \qquad (4.10)$$

where

$$W = \begin{vmatrix} (w_1, w_1)_K & \cdots & (w_1, w_q)_K \\ \vdots & & \vdots \\ (w_q, w_1)_K & \cdots & (w_q, w_q)_K \end{vmatrix}. \qquad (4.11)$$

More generally, for any linear function $\psi(\beta)$ of the parameters $\beta_1, \cdots, \beta_q$,

$$\psi(\beta) = \psi_1 \beta_1 + \cdots + \psi_q \beta_q, \qquad (4.12)$$

where the constants $\psi_1, \cdots, \psi_q$ are known, the minimum variance unbiased linear estimate of $\psi(\cdot)$ is

$$\psi^* = \psi_1 \beta_1^* + \cdots + \psi_q \beta_q^*, \qquad (4.13)$$

where $\beta_1^*, \cdots, \beta_q^*$ are any solution of the set of normal equations

$$\begin{bmatrix} (w_1, w_1)_K & \cdots & (w_1, w_q)_K \\ & \vdots & \\ (w_q, w_1)_K & \cdots & (w_q, w_q)_K \end{bmatrix} \begin{bmatrix} \beta_1^* \\ \vdots \\ \beta_q^* \end{bmatrix} = \begin{bmatrix} (w_1, X)_K \\ \vdots \\ (w_q, X)_K \end{bmatrix}. \qquad (4.14)$$

In particular, if the true mean-value function $m(\cdot)$ is of the form

$$m(t) = \beta w(t), \qquad (4.15)$$

where $w(\cdot)$ is known and β is a constant to be estimated, then

$$m^*(t) = \beta^* w(t), \qquad \beta^* = \frac{(w, X)_K}{(w, w)_K}, \qquad (4.16)$$

$$\text{Var}\,[m^*(t)] = \frac{1}{(w, w)_K}. \qquad (4.17)$$

If the true mean-value function is of the form

$$m(t) = \beta_1 w_1(t) + \beta_2 w_2(t), \qquad (4.18)$$

where $w_i(\cdot)$ are known functions and β_1 and β_2 are constants to be estimated, then

$$m^*(t) = \beta_1^* w_1(t) + \beta_2^* w_2(t), \qquad (4.19)$$

$$\text{Var}\,[m^*(t)] = W^{11} w_1^2(t) + 2W^{12} w_1(t)w_2(t) + W^{22} w_2^2(t). \qquad (4.20)$$

In (4.19), we have

$$\begin{aligned} \beta_1^* &= W^{11}(w_1, X)_K + W^{12}(w_2, X)_K, \\ \beta_2^* &= W^{21}(w_1, X)_K + W^{22}(w_2, X)_K, \end{aligned} \qquad (4.21)$$

where

$$W^{11} = \frac{(w_2, w_2)_K}{W},$$

$$W^{22} = \frac{(w_1, w_1)_K}{W},$$

$$W^{12} = W^{21} = -\frac{(w_1, w_2)_K}{W}, \qquad (4.22)$$

$$W = (w_1, w_1)_K (w_2, w_2)_K - |(w_1, w_2)_K|^2.$$

To establish Theorem 4.1 there is no need to employ the method of Lagrange multipliers as so many writers do (see, for example, Lanning and Battin [14], pp. 300–302); rather, we use the *projection* theorem. The minimum-variance unbiased linear estimate of $m(t)$ may be characterized as the linear functional $(h, X)_K$ that, among all linear functionals satisfying

$$E_m[(h, X)_K] = (h, m) = m(t) = (K(\cdot, t), m)_K \qquad (4.23)$$

for all m in M, has minimum norm square

$$\|h\|_K^2 = \mathrm{Var}\,[(h, X)_K]. \qquad (4.24)$$

By the projection theorem, the function in $H(K)$ having minimum norm among all functions satisfying the restraints (4.23) is $E^*[K(\cdot, t)\,|\,\overline{M}]$. Consequently, (4.8) has been proved. For a complete proof of Theorem 4.1, the reader is referred to [24].

Example 4A. To illustrate the use of the foregoing formulas, let us consider an example that has been treated by many authors. The statement of this problem is given by Lanning and Battin ([14], pp. 294, 303, 307): "Consider the problem of predicting a future position of a moving target by a system which receives target data, in the presence of noise, starting at $t = 0$." Its position $S(t)$ is an unknown linear function of time t,

$$S(t) = \beta_1 + \beta_2 t, \qquad (4.25)$$

where β_1 and β_2 are unknown constants; in Section 6 we consider the case in which β_1 and β_2 are random variables. The observed $X(t)$ is assumed to be the sum of $S(t)$ and a stationary random noise $N(t)$, with covariance function

$$R(u) = E[N(t)N(t + u)] = Ce^{-\beta|u|}. \qquad (4.26)$$

It is desired to use observations $X(t)$, $0 \le t \le T$, to estimate the particle's position $S(t)$ at any given time t. Since $S(t) = E[X(t)]$, the problem of estimating $S(t)$ is equivalent to the problem of estimating the mean-value function of an observed time series. Consequently, the minimum-variance unbiased linear estimate $S^*(t)$ of the value of $S(t)$ at a particular time t is given by the right-hand side of (4.19), with $w_1(t) = 1$ and $w_2(t) = t$. The inner products appearing in (4.22) are explicitly given by means of (3.23) as follows:

$$(1, 1)_K = \frac{\beta T + 2}{2C},$$

$$(1, t)_K = \frac{\beta^2 T^2 + 2\beta T}{4C\beta},$$

$$(t, t)_K = \frac{\beta^3 T^3 + 3\beta^2 T^2 + 3\beta T}{6C\beta^2},$$ (4.27)

$$W = (1, 1)_K (t, t)_K - (1, t)_K^2 = \frac{(\beta T)^4 + 8(\beta T)^3 + 24(\beta T)^2 + 24(\beta T)}{48C^2\beta^2}.$$

The variance of the estimate $S^*(t)$ is given by the right-hand side of (4.20).

If the time series $X(t)$ is assumed to be normal (or Gaussian), or if linear functionals $(h, X)_K$ may be assumed to be approximately normally distributed, then one may state a confidence band for the entire mean-value function $m(t)$ as follows. Given a confidence level α, let $K_q(\alpha)$ denote the α percentile of the χ^2 distribution with q degrees of freedom; in symbols,

$$P[\chi_q^2 \geq K_q(\alpha)] = \alpha.$$ (4.28)

In particular, for $q = 2$ and $\alpha = 95$ per cent, $K_q(\alpha)$ is approximately 6.

It can be shown that if the space M of possible mean-value functions has finite dimension q, then the interval

$$m^*(t) - \sqrt{K_q(\alpha)}\, \sigma[m^*(t)] \leq m(t)$$
$$\leq m^*(t) + \sqrt{K_q(\alpha)}\, \sigma[m^*(t)],$$ (4.29)

for all t in $-\infty < t < \infty$, is a simultaneous confidence band for all values of the mean-value function with a level of significance not less than α; that is, if $m(\cdot)$ is the true mean-value function, then (4.29) holds with a probability greater than or equal to α.

5. Iterative Evaluation of Reproducing-Kernel Inner Products

In this section we give an iterative method of evaluating the reproducing-kernel inner product $(h, h)_K$ and corresponding random variable $(h, X)_K$ that makes possible the approximate synthesis of an optimum linear communication or control system in the presence of noise for which the covariance kernel K can be of any form and can be known either analytically or numerically. The method to be described is a gradient method related to the method of steepest descent. For a general discussion of the role of such methods in solving integral equations, see Kantorovich ([25], Chap. III), and in solving partial differential equations and algebraic linear equations, see Forsythe and Wasow ([26], Sec. 2).

Let $K(s, t)$ be a covariance kernel, defined for $a \leq s,\ t \leq b$. Let $H(K)$ be the corresponding reproducing-kernel Hilbert space. Let $C(a, b)$ be the space of continuous functions on the interval a to b.

For a given function h in $H(K)$, it is of interest to develop methods of generating sequences $\{H_n\}$ of functions in $C(a, b)$ having the properties that

$$\lim_{n \to \infty} E\big[\,|\,(X, h)_K - \int_a^b H_n(t)X(t)\ dt\,|^2\big] = 0, \tag{5.1}$$

$$(h, h)_K = \lim_{n \to \infty} \int_a^b \int_a^b H_n(s)K(s, t)H_n(t)\ ds\ dt. \tag{5.2}$$

It is easily shown that sequences $\{H_n\}$ satisfying (5.1) and (5.2) exist. As in example 2B, let values λ_n be the eigenvalues (arranged in decreasing order, $\lambda_1 \geq \lambda_2 \geq \cdots$) and let $\varphi_n(\cdot)$ be the corresponding eigenfunctions of the kernel $K(s, t)$. Then a function h belongs to $H(K)$ if and only if

$$\int_a^b |\,h(t)\,|^2 = \sum_{n=1}^{\infty} \Big|\int_a^b h(t)\varphi_n(t)\ dt\Big|^2$$

and

$$(h, h)_K = \sum_{n=1}^{\infty} \frac{1}{\lambda_n} \Big|\int_a^b h(t)\varphi_n(t)\ dt\Big|^2 < \infty. \tag{5.3}$$

Consequently, define

$$H_n(t) = \sum_{k=1}^{n} \varphi_k(s) \frac{1}{\lambda_k} \int_a^b h(s)\varphi_k(s)\ ds. \tag{5.4}$$

Clearly $H_n(\cdot)$ belongs to $C(a, b)$.

It may be verified that

$$\int_a^b \int_a^b H_n(s)K(s, t)H_n(t)\ ds\ dt = \sum_{k=1}^{n} \frac{1}{\lambda_k} \Big|\int_a^b h(t)\varphi_k(t)\ dt\Big|^2 \tag{5.5}$$

and

$$\int_a^b H_n(t)X(t)\ dt = \sum_{k=1}^{n} \frac{1}{\lambda_k} \int_a^b h(s)\varphi_k(s)\ ds \int_a^b X(t)\varphi_k(t)\ dt. \tag{5.6}$$

Therefore the sequence defined by (5.4) satisfies (5.1) and (5.2). It is not computationally convenient, however, to use (5.4), inasmuch as it involves the calculation of eigenvalues and eigenfunctions.

Define a transformation T on functions H in $C(a, b)$ as follows:

$$TH(t) = \int_a^b H(s)K(s, t)\, ds, \qquad a \le t \le b. \tag{5.7}$$

It can be proved that

$$\int_a^b H(t)X(t)\, dt = (TH, X)_K, \tag{5.8}$$

$$\int_a^b \int_a^b H(s)K(s, t)H(t)\, ds\, dt = (TH, TH)_K. \tag{5.9}$$

Next, define a sequence of functions H_n as follows: Let α be a constant to be specified. Let $H_0(t) = 1$, or some other function in $C(a, b)$. For $n \ge 1$, let†

$$H_{n+1} = H_n - \alpha(TH_n - h). \tag{5.10}$$

We claim that if α is chosen in an interval specified by (5.18) or (5.21), then the sequence H_n defined by (5.10) satisfies (5.1) and (5.2). To prove this assertion it suffices to show that

$$E[|\,(h, X)_K - (TH_n, X)_K\,|^2] = \|(h - TH_n)\|_K^2 \to 0 \quad \text{as } n \to \infty. \tag{5.11}$$

From (5.10) we may write

$$TH_{n+1} - h = (TH_n - h) - \alpha T(TH_n - h)$$
$$= (I - \alpha T)(TH_n - h), \tag{5.12}$$

where I is the identity operator, $Ih(t) = h(t)$. From (5.12) it follows that, for $n \ge 0$,

$$TH_n - h = (I - \alpha T)^n(TH_0 - h). \tag{5.13}$$

We next note that for any function g in $H(K)$,

$$g(t) = \sum_{n=1}^{\infty} \varphi_n(t) \int_a^b \varphi_n(s)g(s)\, dt, \tag{5.14}$$

$$Tg(t) = \sum_{n=1}^{\infty} \varphi_n(t)\lambda_n \int_a^b \varphi_n(s)g(s)\, ds, \tag{5.15}$$

$$\|(I - \alpha T)g\|_K^2 = \sum_{n=1}^{\infty} \frac{1}{\lambda_n}\left\{\int_a^b \varphi_n(s)g(s)\, ds\right\}^2 \{1 - \alpha\lambda_n\}^2. \tag{5.16}$$

---

† Leonov gives an iterative procedure similar to the one given here in his very interesting paper [27], which he correctly describes as the first application of the methods of functional analysis to the problem of determining the weight function of an optimal system. Although he mentions the problem of establishing the convergence of the procedure, the proof he sketches does not seem to be satisfactory.

Defining $g = TH_0 - h$ and $\gamma_n = \int_a^b \varphi_n(s)g(s)\,ds$, from (5.13) and (5.16) we have

$$\| TH_n - h \|_K^2 = \sum_{m=1}^{\infty} \frac{1}{\lambda_m} \gamma_m^2 \{1 - \alpha_m\}^{2n}. \tag{5.17}$$

Let α be chosen so that, for every integer m,

$$-1 < 1 - \alpha\lambda_m < 1 \quad \text{or} \quad 0 < \alpha < 2/\lambda_m. \tag{5.18}$$

If (5.18) holds, then for any integer M

$$\| TH_n - h \|_K^2 \leq \sum_{m=1}^{M} \frac{1}{\lambda_m} \gamma_m^2 \{1 - \alpha\lambda_m\}^{2n} + \sum_{m>M} \frac{1}{\lambda_m} \gamma_m^2, \tag{5.19}$$

which tends to 0 as we first let n tend to ∞, and then let M tend to ∞ [note that the last term in (5.19) is the remainder term of a convergent series]. We have thus shown that if (5.18) is satisfied, then (5.11) holds. Further, the procedure converges monotonically, in the sense that

$$\| TH_{n+1} - h \|_K \leq \| TH_n - h \|_K. \tag{5.20}$$

If M is a constant such that $\max_m \lambda_m < M$, then (5.18) is satisfied if we choose α so that

$$0 < \alpha \leq 2/M. \tag{5.21}$$

A convenient choice for M is

$$M = \sum_{m=1}^{\infty} \lambda_m = \int_a^b K(t, t)\,dt. \tag{5.22}$$

It should be remarked that (5.19) implies that

$$\lim_{n \to \infty} \int_a^b |\,(TH_n - h)(t)\,|^2\,dt = 0, \tag{5.23}$$

since, for any g in $H(K)$,

$$|\,g(t)\,|^2 \leq \|g\|_K^2 K(t, t),$$

$$\int_a^b |\,g(t)\,|^2 \leq \|g\|_K^2 \int_a^b K(t, t)\,dt. \tag{5.24}$$

The iterative method given by (5.10) undoubtedly does not converge very quickly. Other iterative methods (such as an analogue of the conjugate gradient method [28]) can be developed and should be studied.

6. Random Regression Coefficients

Let $\{X(t),\, t \in T\}$ be a time series of the form

$$X(t) = m(t) + Y(t). \tag{6.1}$$

It is assumed that $Y(t)$ is a time series with known mean-value and covariance functions:

$$E[Y(t)] = 0, \qquad E[Y(s)Y(t)] = R_Y(s,\, t). \tag{6.2}$$

It is assumed that $m(t)$ is of the form

$$m(t) = \beta_1 w_1(t) + \cdots + \beta_q w_q(t), \tag{6.3}$$

where the functions $w_1, \cdots, w_q$ are known, and $\beta_1, \cdots, \beta_q$ are *random variables* independent of $\{Y(t),\, t \in T\}$ with known means

$$\mu_j = E[\beta_j], \qquad j = 1, \cdots, q, \tag{6.4}$$

and covariance matrix $\Gamma = \{\Gamma_{ij}\}$, where, for $i, j = 1, \cdots, q$,

$$\Gamma_{ij} = \mathrm{Cov}\,[\beta_i,\, \beta_j]. \tag{6.5}$$

We call the foregoing set of assumptions *the case of random regression coefficients*.

The problem of estimating (or predicting) the value of $m(t)$ under the assumption of random regression coefficients has been considered by Lanning and Battin ([14], pp. 305–309) and Bendat ([29], Chap. 9). We here consider the more general problem of estimating a parametric function

$$\psi(\beta) = \psi_1 \beta_1 + \cdots + \psi_q \beta_q. \tag{6.6}$$

Strictly speaking, the problem before us is one of pure prediction. The minimum mean-square error predictor of the random variable $\psi(\beta)$, given the observations $X(t),\, t \in T$, is the projection $E^*[\psi(\beta)\,|\,X(t),\, t \in T]$. Consequently, our aim in this section is to give an explicit formula for the projection.

One answer to this problem was given in Section 2, namely

$$E^*[\psi(\beta)\,|\,X(t),\, t \in T] = (\rho,\, X)_{R_X}, \tag{6.7}$$

where

$$R_X(s,\, t) = E[X(s)X(t)], \quad \rho(t) = E[\psi(\beta)X(t)]. \tag{6.8}$$

We easily verify that

$$R_X(s, t) = E[m(s)m(t)] + E[Y(s)Y(t)]$$

$$= \sum_{j,k=1}^{q} w_j(s)(\Gamma_{jk} + \mu_j\mu_k)w_k(t) + R_Y(s, t), \qquad (6.9)$$

$$\rho(t) = E[\psi(\beta)X(t)] = \sum_{j,k=1}^{q} \psi_j(\Gamma_{jk} + \mu_j\mu_k)w_k(t). \qquad (6.10)$$

We now propose to obtain an expression for the best estimate of $\psi(\beta)$ in terms of the reproducing-kernel inner product corresponding to R_Y, and the matrices

$$\Gamma = \{\Gamma_{ij}\}, \qquad K = \{K_{ij}\}, \qquad K_{ij} = (w_i, w_j)_{R_Y}. \qquad (6.11)$$

THEOREM 6.1. *The minimum mean-square error linear predictor of*

$$\psi(\beta) = \psi_1\beta_1 + \cdots + \psi_q\beta_q, \qquad (6.12)$$

given the observations $\{X(t), t \in T\}$, *is*

$$\psi(\beta^*) = \psi_1\beta_1^* + \cdots + \psi_q\beta_q^*, \qquad (6.13)$$

where

$$\begin{bmatrix} \beta_1^* \\ \vdots \\ \beta_q^* \end{bmatrix} = (\Gamma^{-1} + K)^{-1}\left\{ \begin{bmatrix} (w_1, X)_{R_Y} \\ \vdots \\ (w_q, X)_{R_Y} \end{bmatrix} + \Gamma^{-1}\begin{bmatrix} \mu_1 \\ \vdots \\ \mu_q \end{bmatrix} \right\}. \qquad (6.14)$$

The estimates $\beta_1^*, \cdots, \beta_q^*$ *have covariance matrix*

$$\{\mathrm{Cov}\,[\beta_j^*, \beta_k^*]\} = \Gamma K(\Gamma^{-1} + K)^{-1}, \qquad (6.15)$$

and mean-square error matrix

$$\{E[(\beta_j^* - \beta_j)(\beta_k^* - \beta_k)]\} = (\Gamma^{-1} + K)^{-1}. \qquad (6.16)$$

Application. To understand the meaning of Theorem 6.1, let us consider the case $q = 1$. We then observe that $X(t) = \beta w(t) + Y(t)$, where $Y(t)$ satisfies (6.2), $w(t)$ is a known function, and β is a random variable (independent of $Y(t)$, $t \in T$) with mean μ and variance σ^2.

The minimum mean-square error linear predictor of β is

$$\beta^* = \frac{\dfrac{\mu}{\sigma^2} + (w, X)_R}{\dfrac{1}{\sigma^2} + (w, w)_R}, \tag{6.17}$$

$$\text{Var } [\beta^*] = \frac{\sigma^2(w, w)_R}{\dfrac{1}{\sigma^2} + (w, w)_R}, \tag{6.18}$$

$$E[\,|\,\beta^* - \beta\,|^2] = \text{Var } [\beta] - \text{Var } [\beta^*] = \left\{\frac{1}{\sigma^2} + (w, w)_R\right\}^{-1}. \tag{6.19}$$

On the other hand, if β is assumed to be an unknown constant rather than a random variable, then the minimum mean-square error unbiased linear estimate of β is

$$\beta^* = \frac{(w, X)_R}{(w, w)_R}, \tag{6.20}$$

$$E_\beta\,|\,\beta^* - \beta\,|^2 = \text{Var}_\beta\,[\beta^*] = \frac{1}{(w, w)_R}. \tag{6.21}$$

One sees that for $\mu = 0$ and σ very large, (6.17) and (6.20) yield approximately the same expression for β^*. This result was previously obtained by Lanning and Battin ([14], p. 309).

PROOF OF THEOREM 6.1. Let us write tr to denote *transpose*, and define vectors μ, β, β^*, $w(t)$ in the obvious manner; for example, $\psi^{tr} = (\psi_1, \cdots, \psi_q)$. To prove (6.13), it suffices to prove that for every t in T we have

$$E[\beta X(t)] = E[\beta^* X(t)]. \tag{6.22}$$

Let A be the second-moment matrix of β, defined by $A = \Gamma + \mu\mu^{tr}$. Clearly we have

$$E[\beta X(t)] = Aw(t).$$

To evaluate the right-hand side of (6.22), let us write

$$\beta^* = (\Gamma^{-1} + K)^{-1}V + \mu,$$

where $V^{tr} = (V_1, \cdots, V_q)$, $V_j = (w_j, X - E[X]_{R_Y})$, and $E[X]$ is the function of t defined by $E[X](t) = \mu^{tr}w(t)$. It may be verified that

$$E[VX(t)] = (K\Gamma + I)w(t) = (\Gamma^{-1} + K)\Gamma w(t),$$

$$E[\beta^* X(t)] = \Gamma w(t) + \mu\mu^{tr}w(t) = Aw(t).$$

The proof of (6.22) is complete. To prove (6.15), verify that

$$\{\mathrm{Cov}\,[\beta_j^*, \beta_k^*]\} = (\Gamma^{-1} + K)^{-1}E[VV^{tr}](\Gamma^{-1} + K)^{-1},$$

$$E[VV^{tr}] = (K\Gamma + I)K = (\Gamma^{-1} + K)\Gamma K.$$

To prove (6.16), verify that

$$\begin{aligned}
\{E[(\beta_j^* - \beta_j)(\beta_k^* - \beta_k)]\} &= \{\mathrm{Cov}\,[\beta_j, \beta_k]\} - \{\mathrm{Cov}\,[\beta_j^*, \beta_k^*]\} \\
&= \Gamma - \Gamma K(\Gamma^{-1} + K)^{-1} \\
&= \{\Gamma(\Gamma^{-1} + K) - \Gamma K\}(\Gamma^{-1} + K)^{-1} \\
&= (\Gamma^{-1} + K)^{-1}.
\end{aligned}$$

7. Minimum-Variance Linear Unbiased Prediction

Let $\{X(t), t \in T\}$ be a time series of which the proper covariance function,

$$K(s, t) = \mathrm{Cov}\,[X(s), X(t)], \tag{7.1}$$

is known. The mean-value function $m(t) = E[X(t)]$ is known only to be a member of a class M of possible mean-value functions, where M is a subset of the reproducing-kernel Hilbert space $H(K)$ corresponding to K. To make the discussion concrete we assume that M consists of all functions $m(t)$ of the form

$$m(t) = \beta_1 w_1(t) + \cdots + \beta_q w_q(t), \tag{7.2}$$

where the functions $w_1, \cdots, w_q$ are known.

Let Z be a random variable for which we know the variance $\mathrm{Var}\,[Z]$ and the covariance

$$\rho_Z(t) = \mathrm{Cov}\,[Z, X(t)]. \tag{7.3}$$

The mean of Z depends on the true mean-value function as follows:

$$E[Z] = (h, m)_K \qquad \text{for every } m \text{ in } M, \tag{7.4}$$

for some h in $H(K)$. If M consists of all functions of the form (7.2), then

$$E_\beta[Z] = \psi_1\beta_1 + \cdots + \psi_q\beta_q \qquad (7.5)$$

for some known constants $\psi_1, \cdots, \psi_q$.

One case of particular importance is $Z = X(t_0)$, where t_0 does not belong to T; then $\psi_j = w_j(t_0)$ for $j = 1, \cdots, q$.

It is desired to predict Z, given the observations $\{X(t), t \in T\}$. Now if the means $E[X(t)]$ and $E[Z]$ were known, then the minimum variance linear predictor Z^* of Z would satisfy

$$Z^* - E[Z] = (\rho_Z, X - m)_K, \qquad (7.6)$$

from which it follows that

$$Z^* = (\rho_Z, X)_K + \sum_{i=1}^{q} \beta_i\psi_i - \sum_{i=1}^{q} \beta_i(\rho_Z, w_i)_K. \qquad (7.7)$$

One might think it plausible in the case of unknown means that the minimum-variance unbiased linear predictor is given by

$$Z^* = (\rho_Z, X)_K + \sum_{i=1}^{q} \beta_i^*\{\psi_i - (\rho_Z, w_i)_K\}, \qquad (7.8)$$

where $\beta_i^*, \cdots, \beta_q^*$ are any solution of the "normal equations" given in (4.14). We now show that this conjecture is correct.

THEOREM 7.1. *Let* $\{X(t), t \in T\}$ *have known proper covariance kernel* K, *and unknown mean-value function* m *belonging to a subspace* M *of* $H(K)$. *Let* Z *be a random variable with cross-covariance function* $\rho_Z(t) = \text{Cov}\,[Z, X(t)]$; *and let its mean, for each* m *in* M, *be given by* $E_m[Z] = (h, m)_K$, *where* h *belongs to* $H(K)$. *The minimum-variance linear unbiased predictor* Z^* *of* Z, *given the observations* $\{X(t), t \in T\}$, *is*

$$Z^* = (X, \rho_Z)_K + (X, E^*[h - \rho_Z \,|\, M])_K, \qquad (7.9)$$

with mean-square error of prediction

$$E\,|Z^* - Z|^2 = \text{Var}\,[Z] - \|\rho_Z\|_K^2 + \|E^*[h - \rho_Z \,|\, M]\|_K^2. \qquad (7.10)$$

REMARK. A linear estimate $(X, g)_K$ is said to be an *unbiased linear predictor* of Z, if for all m in M we have

$$E_m[(X, g)_K] = (m, g)_K = (m, h)_K = E_m[Z]. \qquad (7.11)$$

The notion of unbiased linear prediction was first considered by Dolph and Woodbury [30].

PROOF. The mean-square error of prediction of an unbiased linear estimate of Z is given, independently of m, by

$$E\,|\,Z - (X, g)_K\,|^2 = \mathrm{Var}\,[Z - (X, g)_K]$$
$$= \mathrm{Var}\,[Z] + \mathrm{Var}\,[(X, g)_K]$$
$$- 2\,\mathrm{Cov}\,[Z, (X, g)_K]. \qquad (7.12)$$

It may be shown that ρ_Z belongs to $H(K)$ and that

$$\mathrm{Cov}\,[Z, (X, g)_K] = (\rho_Z, g)_K. \qquad (7.13)$$

In view of (7.13), we can write

$$E\,|\,Z - (X, g)_K\,|^2 = \mathrm{Var}\,[Z] + (g, g)_K - 2(\rho_Z, g)_K$$
$$= \mathrm{Var}\,[Z] - \|\rho_Z\|_K^2 + \|g - \rho_Z\|_K^2. \qquad (7.14)$$

Letting $g = \rho_Z + f$, we see that the best predictor is given by $Z^* = (X, \rho_Z + f)_K$, where f is the function of minimum norm $\|f\|_K$ satisfying the constraints

$$(m, f)_K = (m, h - \rho_Z)_K \qquad \text{for all } m \text{ in } M. \qquad (7.15)$$

It is clear that $f = E^*[h - \rho_Z\,|\,M]$. The proof of Theorem 7.1 is now complete.

Let us now exhibit an explicit formula for the best predictor $X^*(t)$ of $X(t)$, for t not in T. From Theorem 7.1, it follows that if $m(t)$ is of the form of (7.2), then

$$WX^*(t) = - \begin{vmatrix} (w_1, w_1)_K & \cdots & (w_1, w_q)_K & (X, w_1)_K \\ \vdots & & \vdots & \vdots \\ (w_q, w_1)_K & \cdots & (w_q, w_q)_K & (X, w_q)_K \\ d_1(t) & \cdots & d_q(t) & (X, K(\cdot, t))_K \end{vmatrix}, \qquad (7.16)$$

$$WE\,|\,X^*(t) - X(t)\,|^2 = \begin{vmatrix} (w_1, w_1)_K & \cdots & (w_1, w_q)_K & d_1(t) \\ \vdots & & \vdots & \vdots \\ (w_q, w_1)_K & \cdots & (w_q, w_q)_K & d_q(t) \\ d_1(t) & \cdots & d_q(t) & d(t) \end{vmatrix}, \qquad (7.17)$$

where

$$d_j(t) = w_j(t) - (w_j, K(\cdot, t))_K, \qquad (7.18)$$
$$d(t) = K(t, t) - (K(\cdot, t), K(\cdot, t))_K, \qquad (7.19)$$

$$W = \begin{vmatrix} (w_1, w_1)_K & \cdots & (w_1, w_q)_K \\ \vdots & & \vdots \\ (w_q, w_1)_K & \cdots & (w_q, w_q)_K \end{vmatrix}. \qquad (7.20)$$

8. Decision Theoretic Extensions

The problems considered in the foregoing discussion have all involved linear estimates chosen according to a criterion expressed in terms of mean-square error. Nevertheless the mathematical tools developed continue to play an important role if one desires to develop communication theory from the viewpoint of statistical decision theory or any other theory of statistical inference (see [31], [32], [33]). All modern theories of statistical inference take as their starting point the idea of the probability density function of the observations. Thus in order to apply any principle of statistical inference to communication problems, it is first necessary to develop the notion of the probability density function (or functional) of a stochastic process. In this section we state a result showing how one can write a formula for the probability density functional of a stochastic process that is normal (Gaussian).

Given a normal time series $\{X(t),\ t \in T\}$ with known covariance function

$$K(s, t) = \mathrm{Cov}\ [X(s), X(t)] \tag{8.1}$$

and mean-value function $m(t) = E[X(t)]$, let P_m be the probability measure induced on the space of sample functions of the time series. Next, let m_1 and m_2 be two functions, and let P_1 and P_2 be the probability measure induced by normal time series with the same covariance kernel K and with mean-value functions equal to m_1 and m_2, respectively. By the Lebesgue decomposition theorem it follows that there is a set N of P_1-measure 0 and a nonnegative P_1-integrable function, denoted by dP_2 / dP_1, such that, for every measurable set B of sample functions,

$$P_2(B) = \int_B \frac{dP_2}{dP_1} dP_1 + P_2(BN). \tag{8.2}$$

If $P_2(N) = 0$, then P_2 is absolutely continuous with respect to P_1, and dP_2/dP_1 is called the *probability density function* of P_2 with respect to P_1. Two measures that are absolutely continuous with respect to one another are called *equivalent*. Two measures P_1 and P_2 are said to be *orthogonal* if there is a set N such that $P_1(N) = 0$ and $P_2(N) = 1$.

It has been proved, independently by various authors under various hypotheses (for references, see [24], Sec. 6), that two normal probability measures are either equivalent or orthogonal. From the point of view of obtaining an explicit formula for the probability density function, the following formulation of this theorem is useful.

THEOREM (*Parzen* [24]). *Let P_m be the probability measure induced on the space of sample functions of a time series $\{X(t),\ t \in T\}$ with covariance kernel K and mean-value function m. Assume that either* (a) *T is countable or* (b) *T is a separable metric space, K is continuous, and the stochastic process $\{X(t),\ t \in T\}$ is separable. Let P_0 be the probability measure corresponding to the normal process with covariance kernel K and mean-value function $m(t) = 0$. Then P_m and P_0 are equivalent or orthogonal, depending on whether m does or does not belong to the reproducing-kernel Hilbert space $H(K)$. If $m \in H(K)$, then the probability density functional of P_m with respect to P_0 is given by*

$$f(X, m) = \frac{dP_m}{dP_0} = \exp\left\{(X, m)_K - \frac{1}{2}(m, m)_K\right\}. \tag{8.3}$$

Using the concrete formula for the probability density functional of a normal process provided by (8.3), we have no difficulty in applying the concepts of classical statistical methodology to problems of inference on normal time series.

References

1. Parzen, E., *Statistical Inference on Time Series by Hilbert Space Methods, I*, Department of Statistics, Stanford University, Technical Report No. 23, January 2, 1959.
2. Pugachev, V. S., *Theory of Random Functions and Its Application to Automatic Control Problems* (Russian), State Publishing House of Theoretical Technical Literature (Gostekhizdat), Moscow, 1957.
3. Pugachev, V. S., "Application of Canonic Expansions of Random Functions in Determining an Optimum Linear System" (Russian), *Automation and Remote Control*, Vol. 17, 1956, pp. 489–499.
4. Pugachev, V. S., "Integral Canonic Representations of Random Functions and Their Application in Determining Optimum Linear Systems" (Russian), *Automation and Remote Control*, Vol. 18, 1957, pp. 971–984; English translation, pp. 1017–1031.
5. Pugachev, V. S., "A Method of Solving the Basic Integral Equation of Statistical Theory of Optimum Systems in Finite Form" (Russian), *J. Appl. Math. Mech.*, Vol. 23, 1959, pp. 3–14; English translation, pp. 1–16.
6. Wiener, N., *The Extrapolation, Interpolation, and Smoothing of Stationary Time Series*, John Wiley & Sons, Inc., New York, 1949.
7. Kolmogorov, A., "Interpolation and Extrapolation," *Bull. Acad. Sci. URSS. Sér. Math.*, Vol. 5, 1941, pp. 3–14.
8. Zadeh, L. A., and J. R. Ragazzini, "An Extension of Wiener's Theory of Prediction," *J. Appl. Phys.*, Vol. 21, 1950, pp. 645–655.
9. Bode, H. W., and C. E. Shannon, "A Simplified Derivation of Linear Least-Squares Smoothing and Prediction Theory," *Proc. IRE*, Vol. 38, 1950, pp. 417–425.

10. Zadeh, L. A., and J. R. Ragazzini, "Optimum Filters for the Detection of Signals in Noise," *Proc. IRE*, Vol. 40, 1952, pp. 1223–1231.

11. Parzen, E., *Modern Probability Theory and Its Applications*, John Wiley & Sons, Inc., New York, 1960.

12. Parzen, E., *Stochastic Processes*, Holden-Day, San Francisco, 1962.

13. Halmos, P., *Introduction to Hilbert Space*, Chelsea Publishing Co., New York, 1951.

14. Lanning, J. H., and R. H. Battin, *Random Processes in Automatic Control*, McGraw-Hill Book Company, Inc., New York, 1956.

15. Miller, K. S., and L. A. Zadeh, "Solution of an Integral Equation Occurring in the Theories of Prediction and Detection," *Trans. IRE*, PGIT-2, No. 2, June, 1956, pp. 72–76.

16. Shinbrot, M., "Optimization of Time-varying Linear Systems with Non-stationary Inputs," *Trans. ASME*, Vol. 80, 1958, pp. 457–462.

17. Davenport, W. B., and W. L. Root, *Introduction to the Theory of Random Signals and Noise*, McGraw-Hill Book Company, Inc., New York, 1958.

18. Davis, R. C., "On the Theory of Prediction of Non-stationary Stochastic Processes," *J. Appl. Phys.*, Vol. 23, 1952, pp. 1047–1053.

19. Grenander, U., "Stochastic Processes and Statistical Inference," *Ark. Mat.*, Vol. 1, 1950, pp. 195–277.

20. Batkov, A. M., "Generalization of the Shaping Filter Method To Include Non-stationary Random Processes" (Russian), *Automation and Remote Control* (Russian), Vol. 20, 1959, pp. 1081–1094; English translation, pp. 1049–1062.

21. Greville, T. N. E., "Some Applications of the Pseudoinverse of a Matrix," *SIAM Rev.*, Vol. 2, 1960, pp. 15–22.

22. Aronszajn, N., "Theory of Reproducing Kernels," *Trans. Amer. Math. Soc.*, Vol. 68, 1950, pp. 337–404.

23. Doob, J. L., *Stochastic Processes*, John Wiley & Sons, Inc., New York, 1953.

24. Parzen, E., "Regression Analysis of Continuous Parameter Time Series," *in* Jerzy Newman (ed.), *Proceedings of the Fourth Berkeley Symposium on Probability and Mathematical Statistics*, Vol. I (*Theory of Statistics*), University of California Press, Berkeley and Los Angeles, California, 1961, pp. 469–489.

25. Kantorovich, L. W., "Functional Analysis and Applied Mathematics" (Russian), *Uspehi Mat. Nauk*, Vol. 3, 1948, pp. 89–189; English translation issued by National Bureau of Standards, Los Angeles, 1952.

26. Forsythe, G. E., and W. Wasow, *Finite Difference Methods for Partial Differential Equations*, John Wiley & Sons, Inc., New York, 1960.

27. Leonov, P., "On an Approximate Method for Synthesizing Optimal Linear Systems for Separating Signal from Noise" (Russian), *Automation and Remote Control*, Vol. 20, 1959, pp. 1071–1080; English translation, pp. 1039–1048.

28. Hestenes, M. R., and E. Stiefel, "Method of Conjugate Gradients for Solving Linear Systems," *J. Res. Nat. Bur. Standards*, Vol. 49, 1952, pp. 409–436.

29. Bendat, J. S., *Principles and Applications of Random Noise Theory*, John Wiley & Sons, Inc., New York, 1958.

30. Dolph, C. L., and M. A. Woodbury, "On the Relation between Green's Functions and Covariances of Certain Stochastic Processes and Its Application to Unbiased Linear Prediction," *Trans. Amer. Math. Soc.*, Vol. 72, 1952, pp. 519–550.

31. Middleton, D., *Introduction to Statistical Communication Theory*, Part IV, McGraw-Hill Book Company, Inc., New York, 1960.

32. Pugachev, V. S., "Determination of an Optimum System Using a General Criterion" (Russian), *Automation and Remote Control*, Vol. 19, 1958, pp. 519–539; English translation, pp. 513–532.

33. Pugachev, V. S., "Optimum System Theory Using a General Bayes Criterion," *Trans. IRE*, PGIT-6 No. 1, March, 1960, pp. 4–7.

Chapter 6

Adaptive Matched Filters †

THOMAS KAILATH

1. Introduction

There has been a considerable interest in the field of adaptive systems in the past few years [1], [2]. Since the subject is still relatively new, few optimally adaptive systems have been found. In fact, there is still discussion [3] of the characteristics and properties that entitle a system to be called "adaptive."

We have encountered, as a result of some studies in communication through randomly varying media, an "optimum" receiver that we feel qualifies as an adaptive system. The adaptive features of this system materialized from direct calculation of the optimum receiver and were not inserted on the basis of any intuitive or heuristic arguments. Nevertheless, our directly, if somewhat fortuitously, obtained adaptive receiver enables us to compare some of its characteristics with intuitive ideas and guesses regarding its detailed adaptive and optimal nature. This leads us to some interesting conclusions; not surprisingly, we find that intuitive extrapolations and guesses about optimal adaptive procedures are not always confirmed mathematically.

The problem we shall examine is depicted in Figure 1, where one of a set of known signals $x^{(k)}(t)$ of finite duration is transmitted through a random linear time-variant channel operator, or filter, A. The result is a waveform, $z^{(k)}(t)$, which is further corrupted by additive noise, $n(t)$,

† This work was sponsored in part by the U.S. Army Signal Corps, the Air Force Office of Scientific Research, and the Office of Naval Research. The paper is based on work being done in partial fulfillment of the requirements for the degree of Doctor of Science in the Department of Electrical Engineering at the Massachusetts Institute of Technology. The author wishes to express his gratitude to Dr. R. Price of Lincoln Laboratory, M.I.T., and to Professors J. M. Wozencraft and W. M. Siebert of the Department of Electrical Engineering, M.I.T., for helpful discussions.

before becoming available to the receiver. The final received signal is called $y(t)$. Let T denote the duration, or the interval of observation, of $y(t)$. We then define the optimum receiver, in the sense of Woodward [4], as being one that computes the set of a posteriori probabilities $[p(x^{(k)}(t)\,|\,y(t))]$, or functions that are monotonically related to these probabilities.

The term "optimum" merits some explanation. Woodward has shown that all the information concerning the transmitted signals that is present in the received signal $y(t)$ is contained in the set of a posteriori probabilities $[p(x^{(k)}(t)\,|\,y(t))]$. These probabilities can then be weighted and combined according to different criteria [5], [6]—for example, Neyman–Pearson, ideal observer, and minimum average risk—to make the final decision as to which signal $x^{(k)}(t)$ was actually present. In

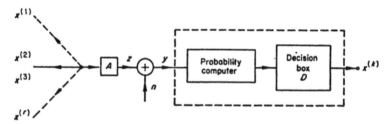

Fig. 1. The communication system.

Figure 1, the box marked D denotes this latter processing. We shall consider only how to obtain a posteriori probabilities, and this will define our optimum receiver.

Our first task in the following discussion will be to set up a model for the channel and signals. After setting up this model, we are able to state our assumptions more definitely and then to proceed to the computation of the optimum receiver. A discussion of some of the adaptive features of this receiver is given in Section 4. This leads to the derivation of the "Rake" system, which employs a practically demonstrated adaptive receiver that is designed on the basis of the previous mathematical theory, tempered by some heuristic ideas combined with engineering balance and judgment. In Sections 1 through 5, we are concerned only with detection in a single time interval. In Section 6 we shall discuss some questions connected with the step-by-step detection of successive signals transmitted through a channel having a long memory, that is, a channel with statistical dependencies extended in time for periods considerably greater than a single-waveform duration. In the concluding Section 7 we give a detailed summary of the paper. This section may profitably be read before beginning Section 2. Finally, in Appendix A we derive the Wiener minimum-variance estimator, and in Appendix B we give results for a threshold or weak (defined more precisely by Equation (5.4)) signal case.

Because of the general nature of this symposium, most of the participants not being communication theorists, we have included much tutorial material, which has had to be drawn from previously published papers by the author. However, the material in the latter half of Section 4, in Sections 5 and 6, and in Appendix B is largely new.

2. A Model for the Channel

The channel operator A is assumed to be a linear time-variant filter. No restriction is placed on its memory or rate of variation. We shall replace the continuous channel by a discrete approximation, thus converting it into a sampled-data channel. This is done chiefly for convenience in analysis and interpretation; the same results can be obtained by using Grenander's method of observable coordinates [7], [8], as has already been done to some extent by Davis [9] and Helstrom [10]. We might also point out that in these days of increasing digital-computer usage, not only are such sampled-data channels becoming increasingly important but it is often almost mandatory to replace continuous channels by their discrete approximations. The solution for the continuous case, which we shall not discuss here, can usually be found as the limit in the mean, as the sampling density becomes infinite, of the finite discrete solution. A discussion of some of the mathematical problems involved is given in [7] and [8].

In setting up the discrete model, the first step is to obtain a discrete analogue of the convolution integral

$$z(t) = \int_0^t a(t - \tau, t)x(\tau) \, d\tau, \qquad (2.1)$$

where $a(\tau, t)$ is the impulse response of the filter, that is, the response of the filter measured at time t to an impulse input τ seconds ago. We shall assume that $a(\tau, t) = 0$ for $\tau < 0$, so that the filter can be physically realized. The input to the filter is $x(t)$, with $x(t) = 0$ for $t < 0$. A discrete approximation to (2.1) can be written

$$z(m) = \sum_{k=0}^m a(m - k, m)x(k), \qquad (2.2)$$

where we choose the samples on a suitable time scale, one unit apart.

A convenient interpretation of (2.2) is given by the sampled-data delay line filter shown in Figure 2. This "multipath" model serves as a convenient discrete approximation to the actual channel. We should note that this model, with its uniform tap arrangement, does not need

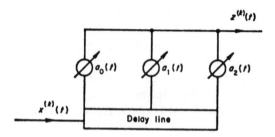

Fig. 2. A simple delay-line channel model.

to bear any direct structural relationship to the actual channel configuration, though in some cases—for example, channels with paths at known delays—it is convenient to make them coincide.

We now introduce matrix notation and rewrite (2.2) in the form

$$\mathbf{z} = \mathbf{Ax}, \tag{2.3}$$

where $\mathbf{x}$ and $\mathbf{z}$ are column matrices of the sample values of $x(t)$ and $z(t)$, and $\mathbf{A}$ represents the channel. (Boldface symbols will be used throughout this chapter to denote matrices.) As an illustration, let us consider a three-tap channel. Then we have

$$\begin{bmatrix} z_0 \\ z_1 \\ z_2 \\ z_3 \end{bmatrix} = \begin{bmatrix} a_0(0) & 0 \\ a_1(1) & a_0(1) \\ a_2(2) & a_1(2) \\ 0 & a_2(3) \end{bmatrix} \begin{bmatrix} x_0 \\ x_1 \end{bmatrix}. \tag{2.4}$$

We shall write $a_k(m) = a(m, k)$ as a_{km} for convenience. Notice that the matrix has all zeros above the main diagonal. This reflects the realizability condition: no output before an input.

We can rewrite (2.4) with the roles of channel and signal interchanged [11],

$$\begin{bmatrix} z_0 \\ z_1 \\ z_2 \\ z_3 \end{bmatrix} = \begin{bmatrix} x_0 & 0 & 0 & 0 & 0 & 0 \\ 0 & x_1 & x_0 & 0 & 0 & 0 \\ 0 & 0 & 0 & x_1 & x_0 & 0 \\ 0 & 0 & 0 & 0 & 0 & x_1 \end{bmatrix} \begin{bmatrix} a_{00} \\ a_{01} \\ \cdots \\ a_{11} \\ a_{12} \\ \cdots \\ a_{22} \\ a_{23} \end{bmatrix}, \tag{2.5}$$

$$z = Xa, \tag{2.6}$$

where X is a matrix in which the elements are sample values of $x(t)$, and z is a matrix in which the sample values belonging to each tap are arranged sequentially in a column. The advantage of rewriting the convolution formula in this way is that it enables us to compute the covariance matrix of z, Φ_{zz}, in a straightforward fashion. Thus, if we assume that the tap functions are composed of a mean component $a(t)$ and a zero mean random component $a^r(t)$, then we can write, with an obvious notation,

$$z = z_r + \bar{z} = A^r x + \bar{A}x = Xa^r + X\bar{a}. \tag{2.7}$$

Now the covariance matrix of z is given by

$$\Phi_{zz} = \overline{(z - \bar{z})(z - \bar{z})}_t = \overline{(z_r)(z_r)}_t \tag{2.8}$$

$$= \overline{Xa^r a^r_t X}_t = X\Phi_{AA} X_t. \tag{2.9}$$

The bar denotes an ensemble average over the random processes controlling the taps, and t denotes the transpose of the matrix. To illustrate again, using for notational simplicity only two taps and deleting the superscript r, we have

$$
\Phi_{zz} =
\begin{bmatrix}
x_0 & 0 & 0 & 0 \\
0 & x_1 & x_0 & 0 \\
0 & 0 & 0 & x_1
\end{bmatrix}
\begin{bmatrix}
\overline{a_{00}^2} & \overline{a_{00}a_{01}} & \overline{a_{00}a_{11}} & \overline{a_{00}a_{12}} \\
\overline{a_{01}a_{00}} & \overline{a_{01}^2} & \overline{a_{01}a_{11}} & \overline{a_{01}a_{12}} \\
\overline{a_{11}a_{00}} & \overline{a_{11}a_{01}} & \overline{a_{11}^2} & \overline{a_{11}a_{12}} \\
\overline{a_{12}a_{00}} & \overline{a_{12}a_{01}} & \overline{a_{12}a_{11}} & \overline{a_{12}^2}
\end{bmatrix}
$$

$$
\cdot
\begin{bmatrix}
x_0 & 0 & 0 \\
0 & x_1 & 0 \\
0 & x_0 & 0 \\
0 & 0 & x_1
\end{bmatrix}
\tag{2.10}
$$

Notice that Φ_{AA} can be conveniently partitioned, as shown, into blocks representing the self- and crossvariances of the sample values of the different tap functions. Thus, if the tap functions were statistically independent, all "off-diagonal" blocks would be zero. The covariance matrices Φ_{AA} and Φ_{zz} are either positive semidefinite or positive definite. If they are positive semidefinite, we have singular covariance matrices; examples are provided by random time-invariant channels [11].

A more detailed discussion of the channel model—including a discussion of time-invariant channels, narrow-band channels, multi-link (for example, diversity) and/or multidimensional (for example, optical) channels is given in [11] and [12].

3. Assumptions for the Problem

Using the discrete model and the matrix notation, we can write (see Fig. 1)

$$\mathbf{y} = \mathbf{z}^{(k)} + \mathbf{n} = \mathbf{A}\mathbf{x}^{(k)} + \mathbf{n} = \mathbf{A}^r\mathbf{x}^{(k)} + \overline{\mathbf{A}}\mathbf{x}^{(k)} + \mathbf{n}, \qquad (3.1)$$

where $\mathbf{x}^{(k)}$ represents the kth transmitted signal and $\mathbf{y}$ represents the received signal. Our major assumptions will be the following: (a) The output signal $\mathbf{z}^{(k)}$ has a Gaussian distribution for each k. (b) The noise $\mathbf{n}$ is also Gaussian, not necessarily white for the present discussion.† (c) The noise has zero mean and a nonsingular covariance matrix $\boldsymbol{\Phi}_{nn}$. (d) The noise $\mathbf{n}$ and the output signal $\mathbf{z}^{(k)}$ are statistically independent for all k. (e) The statistics of $\mathbf{z}^{(k)}$ and $\mathbf{n}$ are known a priori.

Under our assumptions, for each $\mathbf{x}^{(k)}$, $\mathbf{y}$ is a Gaussian signal with mean $\bar{\mathbf{z}}^{(k)} = \overline{\mathbf{A}}\mathbf{x}^{(k)}$ and covariance matrix $\boldsymbol{\Phi}_{yy}^{(k)} = \boldsymbol{\Phi}_{zz}^{(k)} + \boldsymbol{\Phi}_{nn}$. Since the sum of a positive definite and a positive semidefinite quadratic form is always positive, the covariance matrix $\boldsymbol{\Phi}_{yy}^{(k)}$ is nonsingular even when $\boldsymbol{\Phi}_{zz}^{(k)}$ is singular. We may therefore (see [13]) write

$$p(\mathbf{y} \mid \mathbf{x}^{(k)}) = \left(\frac{1}{2\pi}\right)^{N/2} \frac{1}{\mid \boldsymbol{\Phi}_{yy}^{(k)} \mid^{1/2}}$$

$$\cdot \exp\left[-\frac{1}{2}\left\{(\mathbf{y} - \bar{\mathbf{z}}^{(k)})_t[\boldsymbol{\Phi}_{yy}^{(k)}]^{-1}(\mathbf{y} - \bar{\mathbf{z}}^{(k)})\right\}\right], \quad (3.2)$$

where N, the number of samples in the vector $\mathbf{y}$, is related to the duration, say T seconds, of the observation $y(t)$. $\mid \boldsymbol{\Phi}_{yy} \mid$ denotes the determinant of $\boldsymbol{\Phi}_{yy}$. Thus far, no assumptions as to the channel or the noise being stationary are required. These specifications cover a wide variety of channels, particularly in scatter-multipath communications. Of course, our assumptions also exclude several types of channels, especially those with paths for which the delays fluctuate randomly, because then the $\mathbf{z}^{(k)}$ no longer have Gaussian statistics. We can show, however, that for threshold conditions our results hold for arbitrary channel statistics, provided the noise is Gaussian [11], [14]. (See also

---

† In the present discrete formulation, noise will be called "white" if samples of it have equal variance and are uncorrelated.

Appendix B.) We should also mention that many of the results obtained apply to the detection of Gaussian signals in Gaussian noise.

In fact, for these results the filter **A** need not be linear as long as it yields a Gaussian random function $z^{(k)}$ when $x^{(k)}$ is the input. However, the assumption of linearity enables us to obtain more explicit results for communication channels (see the following and [11]).

4. The Optimum Receiver

In [11] a fairly comprehensive discussion of optimum receiver structures obtained directly by using (3.2) has been given. Here we are interested only in one particular structure, namely, that having the estimator-correlator property mentioned in Section 1. We shall therefore give the estimator-correlator derivation here only for the restricted case of additive white Gaussian noise,† and for variety we shall use a different method of proof from that given earlier [15].

Computation of A Posteriori Probabilities

As stated in Section 1, the optimum receiver essentially computes the set of a posteriori probabilities $[p(x^{(k)}|y)]$. We shall show how to obtain one of these, say $p(x^{(k)}|y)$. If we use Bayes' rule and assume that the a priori probabilities $p(x^{(k)})$ are known, then what we essentially have to compute is the "forward" probability $p(y|x^{(k)})$. For this we have‡

$$p(y\,|\,x) = \int_{-\infty}^{\infty} p(y\,|\,Xa)p(a)\,da. \tag{4.1}$$

Since $y = Xa + n$, where **a** has a Gaussian distribution with mean $\bar{a}$ and covariance matrix Φ_{AA}, and where **n** has zero mean and covariance matrix $N_0 I$ (**I** is the identity matrix), (4.1) can be rewritten

$$p(y\,|\,X) = \frac{1}{(2\pi)^{N/2}N_0^{1/2}}\,\frac{1}{(2\pi)^{M/2}\,|\,\Phi_{AA}\,|^{1/2}}\,Q(y\,|\,X),$$

where

$$Q(y\,|\,X) = \int_{-\infty}^{\infty} \exp\left(-\frac{1}{2}\left\{(y - Xa^r - X\bar{a})_t\,\frac{1}{N_0}\,(y - Xa^r - X\bar{a})\right.\right.$$
$$\left.\left. + a_t^r\Phi_{AA}^{-1}a^r\right\}\right)da^r. \tag{4.2}$$

---

† The colored-noise case may be reduced easily to the white-noise case by means of whitening filters [15].

‡ The superscript k has been dropped for convenience but will be restored whenever necessary to avoid confusion.

The letters N and M denote the number of samples in the $\mathbf{y}$ and $\mathbf{a}$ vectors, respectively. We have here assumed that $\mathbf{\Phi}_{AA}$ is nonsingular; we shall show later how to handle the situation in which $\mathbf{\Phi}_{AA}$ is singular. The integral can be evaluated using a result given in Cramér [13], and we finally get

$$p(\mathbf{y}\mid \mathbf{X}) = K_1 \exp\left\{-\left(\frac{\mathbf{y}_t + \mathbf{X}\bar{\mathbf{a}}}{N_0}\right)\right\}$$

$$\cdot \exp\left\{\frac{1}{2N_0^2}(\mathbf{y}-\mathbf{X}\bar{\mathbf{a}})_t\mathbf{X}\left(\mathbf{\Phi}_{AA}^{-1}+\frac{\mathbf{X}_t\mathbf{X}}{N_0}\right)^{-1}\mathbf{X}_t(\mathbf{y}-\mathbf{X}\bar{\mathbf{a}})\right\} \quad (4.3)$$

$$= K_1 \exp\left\{-\left(\frac{\mathbf{y}_t\mathbf{X}\bar{\mathbf{a}}}{N_0}\right)\right\}$$

$$\cdot \exp\left\{\frac{-1}{2N_0^2}[(\mathbf{y}-\mathbf{X}\bar{\mathbf{a}})_t\mathbf{X}\mathbf{\Phi}_{AA}\left(\mathbf{I}+\frac{\mathbf{X}_t\mathbf{X}\mathbf{\Phi}_{AA}}{N_0}\right)^{-1}\mathbf{X}_t(\mathbf{y}-\mathbf{X}\bar{\mathbf{a}})]\right\},$$

$$(4.4)$$

where

$$K_1 = \exp\left\{-\frac{\mathbf{y}_t\mathbf{y}+\bar{\mathbf{a}}_t\mathbf{X}_t\mathbf{X}\bar{\mathbf{a}}}{2N_0}\ \frac{1}{(2\pi)^{M/2}}\left|\frac{(N_0^{-1}\mathbf{I}+\mathbf{X}_t\mathbf{X}\mathbf{\Phi}_{AA})|^{1/2}}{N_0}\right|\right\}.$$

This can be further rewritten as

$$K_1 = \frac{1}{(2\pi)^{M/2}}\exp\left[\frac{-\mathbf{y}_t\mathbf{y}}{2N_0}\right]K_2^{(k)}, \quad (4.5)$$

where the superscript has been reintroduced to show that K_2 depends on (k). Now, restoring superscripts everywhere, we have

$$p(\mathbf{X}^{(k)}\mid \mathbf{y}) = \frac{p(\mathbf{y}\mid \mathbf{X}^{(k)})p(\mathbf{X}^{(k)})}{\sum_k p(\mathbf{y}\mid \mathbf{X}^{(k)})p(\mathbf{X}^{(k)})}. \quad (4.6)$$

In comparing the a posteriori probabilities, it is convenient first to take logarithms of these quantities, so that we have

$$\Lambda'^{(k)} = \frac{1}{2N_0}\left[\frac{1}{N_0}(\mathbf{y}-\mathbf{X}^{(k)}\bar{\mathbf{a}})_t\mathbf{X}^{(k)}\mathbf{\Phi}_{AA}\left(\mathbf{I}+\frac{\mathbf{X}_t^{(k)}\mathbf{X}^{(k)}}{N_0}\mathbf{\Phi}_{AA}\right)^{-1}\right.$$

$$\left.\cdot\mathbf{X}_t^{(k)}(\mathbf{y}-\mathbf{X}^{(k)}\bar{\mathbf{a}})+2\mathbf{y}_t\mathbf{X}\bar{\mathbf{a}}\right]$$

$$+[\ln K_2^{(k)}+\ln p(\mathbf{x}^{(k)})]+\ln\left(\sum \cdots\right). \quad (4.7)$$

The first term on the right-hand side gives the receiver structure; that is, it determines the operations to be performed on the received data in testing the kth hypotheses. *We shall henceforth denote this term by* $\Lambda^{(k)}$. The second term is a "bias," or weighting, term. The last term is the same for all hypotheses and can be omitted in comparisons of the $\Lambda'^{(k)}$.

Before proceeding to examine the receiver structure, we must clarify the question of the singularity of Φ_{AA}. Clearly in (4.3) and (4.4) the proof is based on a nonsingularity of Φ_{AA}; in (4.4), however, Φ_{AA}^{-1} does not appear, and we might therefore suspect that this equation is valid even when Φ_{AA} is singular. This is in fact so, and can be proved in several ways. Two methods are given in [15] and [16]; here we shall indicate another argument. Since the sum of a positive definite matrix and a positive semidefinite matrix is always a positive definite (and therefore nonsingular) matrix, the matrix $[\Phi_{AA} + \epsilon I]$, where $\epsilon > 0$, is positive definite. We can use this perturbed matrix in all the steps of our preceding proof, and at the end—in (4.4)—we may let $\epsilon \to 0$. This sort of "continuity" argument is often used in matrix analysis [17].

The Receiver Structure

Let us now return to the study of the receiver, for which we have

$$\Lambda^{(k)} = \frac{1}{N_0^2}\,(\mathbf{y} - \mathbf{X}^{(k)}\bar{\mathbf{a}})_t \mathbf{X}^{(k)}\Phi_{AA}\left(\mathbf{I} + \frac{\mathbf{X}_t^{(k)}\mathbf{X}^{(k)}\Phi_{AA}}{N_0}\right)^{-1}$$

$$\cdot \mathbf{X}_t^{(k)}(\mathbf{y} - \mathbf{X}^{(k)}\bar{\mathbf{a}}) + \frac{2\mathbf{y}_t \mathbf{X}^{(k)}\bar{\mathbf{a}}}{N_0}\,. \tag{4.8}$$

It is instructive first to consider the case in which the random component of the channel, $\mathbf{a}^r$, is zero and in which the channel is therefore completely known to the receiver as $\bar{\mathbf{a}}$. In this case, we have

$$N_0\Lambda^{(k)} = 2\mathbf{y}_t\mathbf{X}\bar{\mathbf{a}} = 2\mathbf{y}_t(\bar{\mathbf{A}}\mathbf{x}^{(k)}) = 2\mathbf{y}_t\bar{\mathbf{z}}^{(k)}. \tag{4.9}$$

Thus the essential receiver operation is the formation of the dot, or inner, product of $\mathbf{y}$ and $\bar{\mathbf{z}}^{(k)}$; this is equivalent, for continuous signals, to the crosscorrelation of y and $z^{(k)}$, and therefore the $\Lambda^{(k)}$ computer can be represented by the block diagram of Figure 3. This conclusion is of course a trivial extension of the well-known result for the detection of known signals $\mathbf{x}^{(k)}$ in white Gaussian noise. The presence of the known channel $\bar{\mathbf{A}}$ is taken into account by modifying $\mathbf{x}^{(k)}$ by $\bar{\mathbf{A}}$ to produce $\bar{\mathbf{z}}^{(k)}$ at the receiver, which treats the $\bar{\mathbf{z}}^{(k)}$ as known signals in additive Gaussian noise.

Now, however, suppose that channel **A** is purely random; in this case, the $\mathbf{z}^{(k)}(=\mathbf{A}^r\mathbf{x}^{(k)})$ are still signals corrupted by additive noise, but the receiver cannot reconstruct the $\mathbf{z}^{(k)}$ since the channel $\mathbf{A}^r$ is random and not completely known at the receiver. It is not easy to say offhand what the optimum operation in such a situation should be. One suggestion might be to use a known sounding signal to estimate the channel, and then, if we assume that the channel is not varying too rapidly, to use this estimate of the channel as the actual channel during the transmission of a succeeding information-bearing signal [18]. Another suggestion might be first to consider the signal being tested as having actually been sent, then to make a maximum-like-

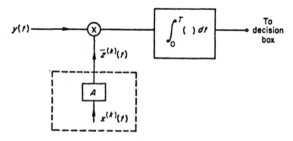

Fig. 3. An element of the optimum receiver for the case of a signal perturbed by a known channel A.

lihood† estimate of the unknown channel parameters, and finally to use these estimated values as the actual parameter values in computing the likelihood function [8], [19]. A third suggestion might be to form an "average" estimate of the channel and use this average channel to generate signals $\mathbf{z}^{(k)}$ at the receiver, to be correlated with the received waveform **y**.

All of these suggestions seem reasonable and in fact are often used in adaptive systems of many kinds [1], [2] that operate in the face of changing and incompletely known system parameters. In our case of Gaussian statistics, however, we can find the ideal receiver explicitly and thereby rigorously test those intuitive notions in at least one concrete case.

For the purely random channel, with $\bar{\mathbf{a}} = 0$, from (4.8) we have

$$N_0\Lambda^{(k)} = \mathbf{y}_t\mathbf{X}\boldsymbol{\Phi}_{AA}(N_0\mathbf{I} + \mathbf{X}_t\mathbf{X}\boldsymbol{\Phi}_{AA})^{-1}\mathbf{X}_t\mathbf{y}. \tag{4.10}$$

This may be rewritten as

$$N_0\Lambda^{(k)} = \mathbf{y}_t\mathbf{H}^{(k)}\mathbf{y} = \mathbf{y}_t\mathbf{z}_e^{(k)}, \tag{4.11}$$

---

† By maximum likelihood estimator, we mean here the value of $\mathbf{a}^{(k)}$ that maximizes $p(\mathbf{y}|\mathbf{X}^{(k)}\mathbf{a})$; see [13].

where

$$z_e^{(k)} = \mathbf{H}^{(k)}\mathbf{y} = \mathbf{X}\mathbf{\Phi}_{AA}(N_0\mathbf{I} + \mathbf{X}_t\mathbf{X}\mathbf{\Phi}_{AA})^{-1}\mathbf{X}_t\mathbf{y}. \tag{4.12}$$

A block diagram for the receiver structure implied by (4.11) is shown in Figure 4.

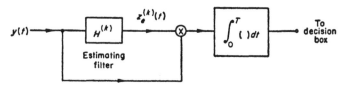

Fig. 4. An element of the optimum receiver for a purely random channel A for white Gaussian noise.

Equation (4.11) is of the same form as (4.9); that is, the optimum receiver crosscorrelates the received signal $\mathbf{y}$ against a waveform $z_e^{(k)}$. In this case, however, $z_e^{(k)}$ is not known a priori at the receiver but is computed from the received data $\mathbf{y}$ by means of the operations represented by $\mathbf{H}^{(k)}$, and $\mathbf{H}^{(k)}$ depends on the channel and noise covariances $\mathbf{\Phi}_{AA}$ and $N_0\mathbf{I}$, and also on the signal $\mathbf{X}^{(k)}$. On closer examination it is found that the series of operations (4.12) on $\mathbf{y}$ that give $z_e^{(k)}$ are equivalent to the optimum extraction of $z^{(k)}$ from $\mathbf{y} = z^{(k)} + \mathbf{n}$ on a minimum-error-variance criterion. This is perhaps more easily seen (cf. Appendix A) if we recast (4.12), with the aid of some matrix algebra, in the form

$$\begin{aligned} z_e^{(k)} &= \mathbf{X}^{(k)}\mathbf{\Phi}_{AA}\mathbf{X}_t^{(k)}(N_0\mathbf{I} + \mathbf{X}^{(k)}\mathbf{\Phi}_{AA}\mathbf{X}_t^{(k)})^{-1}\mathbf{y} \\ &= \mathbf{\Phi}_{zz}^{(k)}(N_0\mathbf{I} + \mathbf{\Phi}_{zz}^{(k)})^{-1}\mathbf{y} \\ &= \mathbf{H}^{(k)}\mathbf{y}. \end{aligned} \tag{4.13}$$

The expression for $\mathbf{H}^{(k)}$ is reminiscent of the formula for the unrealizable Wiener filter in the frequency domain [20], namely,

$$H(\omega) = \phi_{zz}(\omega)[N_0 + \phi_{zz}(\omega)]^{-1}. \tag{4.14}$$

A proof that $\mathbf{H}^{(k)}$ is indeed the minimum-variance estimator is given in Appendix A. We should also note that since we are dealing with Gaussian statistics, this estimator is also optimum for a fairly general class of criteria [21]. We should point out that $\mathbf{H}^{(k)}$ is a symmetric matrix, and therefore represents an unrealizable filter. By a simple artifice [22], [15], however, it can be replaced by a realizable filter if desired; such a filter is obtained by deleting all terms in $\mathbf{H}^{(k)}$ above the main diagonal and by doubling all terms below it. Although $z_e^{(k)}$ is then no longer a minimum-variance estimate, receiver output is unchanged.

This interpretation of $z_e^{(k)}$ fits quite happily with our intuition. With the knowledge of the solution in the deterministic case, in which $z^{(k)}$ can be computed exactly at the receiver, it seems eminently reasonable that in the random-channel case, in which $z^{(k)}$ cannot be computed exactly, we should say that $z^{(k)}$ should be estimated from the received data and this estimate $z_e^{(k)}$ should be used in place of the un-available exact $z^{(k)}$. This interpretation was first recognized by Price [22] for the single-path channel and at low signal-to-noise ratios for a more general channel, and was later extended by Kailath [15].

This rather satisfying interpretation of the receiver action leads us to believe that this form of receiver, which may be described as an estimator-correlator receiver (see Fig. 4), is effective even in situations that do not conform exactly to our assumption of Gaussian statistics. This interpretation also enables us to make engineering approximations to the operations demanded by (4.11). Thus the Wiener–Hopf equation for $H^{(k)}$ can be solved only in a few special cases, but in our equipment we may for convenience use simpler estimating filters, for example, narrow-band RC filters, and simpler estimating operations, for example, crosscorrelation. In fact, such simplifications were made in constructing the Rake antimultipath receiver [23] that grew out of the above interpretation.

We should also point out that this estimator-correlator can be re-garded as an "adaptive" matched filter. This point of view depends on the fact that the crosscorrelation

$$ y_i z_e^{(k)} \quad \text{or} \quad \int_0^T y(t) z_e^{(k)}(t)\, dt $$

can be alternatively performed by a filter matched to $z_e^{(k)}(t)$, that is, by a filter with impulse response $z_e^{(k)}(-t)$. In our case, however, the specification of the matched filter is not completely determined a priori, but its impulse response is calculated from the received data. The re-ceiver may therefore be regarded as adapting, or matching, itself to the state of the channel. Figure 5 is a block diagram reflecting this point of view.

Finally, let us study the receiver for the case where random and deterministic components are both present. For physical reasons, in communication channels the deterministic (or mean value) component is often called the specular component. The receiver formula is (4.8), which through (4.9), (4.11), and (4.12) can be rewritten

$$ N_0 \Lambda^{(k)} = 2 y_i z_e^{(k)} + (y - \bar{z}^{(k)})_i z_{er}^{(k)}, \tag{4.15} $$

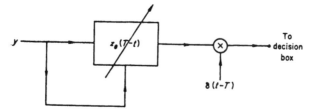

Fig. 5. The estimator-correlator receiver as an
adaptive matched filter.

where $z_{er}^{(k)} = H^{(k)} (y - \bar{z}^{(k)})$ = the minimum variance estimate of
the random component of Z. A block diagram for (4.15) is shown
in Figure 6. This receiver will reappear in Section 6, where the
specular component will arise by extrapolation from previous data and
decisions.

Discussion of the Receiver Operation

Having now determined the general structure of the optimum receiver, we can compare it with our earlier intuitive ideas of what the
receiver should do.

We see that the optimum receiver does not first use the received
signal to make a maximum-likelihood estimate of **a** that is then used
as if it were exact. This type of operation was suggested by Root and
Pitcher [19], [8] for the case in which the statistics of the channel
were not known. It is readily shown that this type of test—called a
generalized maximum-likelihood test by Davenport and Root [8]—
also leads to an estimator-correlator receiver. The estimator, however,
turns out to be a maximum-likelihood, or least-squares, estimator.
Such estimators do not take advantage of any a priori knowledge of
the channel statistics and hence lead to a relatively weaker type of
receiver than the one we have found. As a matter of fact, we can show
that the least-squares estimate of z in our case of additive Gaussian
noise is also a minimax [6] estimate. It is therefore based on the most
pessimistic view of the channel statistics; if we can obtain any information about the actual channel statistics, it is worth trying to do so.
We might mention, however, that the generalized maximum-likelihood
test has often been suggested and has been used to advantage in
statistics [10].

Whether we use a minimum-variance or a minimax estimator, however, we see that the receiver does adjust its parameters to take account
of the channel conditions as they are reflected in the received signal.

It is important to note that, contrary to intuition, the mathematics
shows that the optimum receiver for Gaussian statistics does not di-

rectly estimate the channel; that is, the receiver does not somehow obtain an estimate, say $\mathbf{a}_e$, averaged over all possible transmitted signals, of the channel, from which it obtains the $\mathbf{z}_e^{(k)}$ by the operation $\mathbf{z}_e^{(k)} = \mathbf{X}^{(k)}\mathbf{a}_e$. Consequently, from (4.11) we see that $\mathbf{z}_e^{(k)}$ may be written

$$\mathbf{z}_e^{(k)} = \mathbf{X}^{(k)}\mathbf{a}_e^{(k)},$$

where

$$\mathbf{a}_e^{(k)} = \mathbf{\Phi}_{AA}(N_0\mathbf{I} + \mathbf{X}_t^{(k)}\mathbf{X}^{(k)}\mathbf{\Phi}_{AA})^{-1}\mathbf{X}_t^{(k)}\mathbf{y} = \mathbf{F}^{(k)}\mathbf{X}_t^{(k)}\mathbf{y}, \text{ say.}\dagger \quad (4.16)$$

Therefore $\mathbf{a}_e^{(k)}$ may be regarded as an estimate of the channel vector $\mathbf{a}$ under the hypothesis that the signal $\mathbf{x}^{(k)}$ was sent. The last clause is

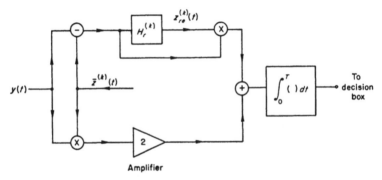

Fig. 6. An element of the optimum receiver for a channel A having a specular and a random component and white Gaussian noise.

important: $\mathbf{a}_e^{(k)}$ is not an estimate of the channel $\mathbf{a}$ itself, unless $\mathbf{x}^{(k)}$ was the signal actually transmitted.

Equation (4.14) shows, moreover, that it is impossible to have an optimum receiver consisting of a single filter matched to the channel, followed by a bank of filters matched to the transmitted signals $\mathbf{X}^{(k)}$. Such a receiver is possible (see (4.9)) only when the channel is known to the receiver. We should point out, however, that this does not mean that such receivers should not be considered. Although they do not conform to the a posteriori probability criterion, they might still be valuable in practice and, in fact, in theory also. Thus, using this receiver, Green gives a provocative discussion of a communication system

$\dagger$ We should remark that $\mathbf{a}^{(k)}$ as given by (4.15) is the discrete version (for time-variant or time-invariant channels) of Turin's [24] frequency domain estimate of time-invariant channels. We have already noticed such similarity— (4.13) and (4.14)—for the Wiener filter. Our remark suggests, in view of the close relation between $\mathbf{a}_e$ and $\mathbf{z}_e$, that Turin's result can be directly obtained from (4.14).

in which the transmitter, but not the receiver, knows the instantaneous channel behavior [25].

As our last point in this section, let us study more closely the estimator-correlator feature for the receiver of (4.8), where we consider random as well as specular components. We see that (4.15) may be rewritten

$$N_0\Lambda^{(k)} = (\mathbf{y} - \bar{\mathbf{z}}^{(k)})_t\bar{\mathbf{z}}_{re}^{(k)} + 2\mathbf{y}_t\bar{\mathbf{z}}^{(k)}$$

$$= \mathbf{y}_t(\mathbf{z}_{re}^{(k)} + \bar{\mathbf{z}}^{(k)}) + (\mathbf{y} - \mathbf{z}_{re}^{(k)})_t\bar{\mathbf{z}}^{(k)}. \qquad (4.17)$$

The first term on the right-hand side may be regarded as the cross-correlation of $\mathbf{y}$, and a total estimate of $\mathbf{z}^{(k)}$, viz., the sum of the known mean component of $\bar{\mathbf{z}}^{(k)}$ and the estimated random component of $\mathbf{z}^{(k)}$. This term corresponds to an estimator-correlator operation. This, however, does not completely describe the receiver because of the term $(\mathbf{y} - \mathbf{z}_{re}^{(k)})_t\bar{\mathbf{z}}^{(k)}$. Therefore, in the case of a channel with specular as well as random components, the receiver *cannot* be considered an estimator-correlator receiver and in this sense is not a natural extension of the deterministic-channel case. The factor 2 for the specular component intervenes and prevents such an interpretation. Hence, care must be exercised in using the estimator-correlator concept. However, even though the general receiver of Equation (4.15) or Figure 6 does not directly correspond to an estimator-correlator receiver, the form shown—which handles specular and random components separately— is in fact entirely satisfactory.

5. The Rake Receiver

We have shown in Section 4 that we cannot have an optimum receiver comprised of a single filter matched in some sense to an "estimated" channel, followed by a bank of filters matched to the signals $\mathbf{X}^{(k)}$. From (4.15) we see that the closest we can approach this type of receiver is to arrange for $\mathbf{X}_t^{(k)}\mathbf{X}^{(k)}$ to be the same for all k. This condition requires that the $\mathbf{X}^{(k)}$ differ by the orthogonal matrices $\mathbf{U}^{(k)}$; that is, we have $\mathbf{X}^{(k)} = \mathbf{U}^{(k)}\mathbf{X}$, where $\mathbf{U}_t^{(k)}\mathbf{U}^{(k)} = \mathbf{I}$, so that $\mathbf{X}_t^{(k)}\mathbf{X}^{(k)}$ is a constant. For a channel with a single time-variant path, this condition (see Sec. 2) is equivalent to the requirement that all signals have identical envelopes. This condition is met, for example, in frequency-shift keying. For a single time-invariant path, the condition implies that the signals have equal energies. For more general channels, the conditions are more complicated but are of the same nature as those above. With the condition that $\mathbf{X}_t^{(k)}\mathbf{X}^{(k)}$ be the same for all k, the filter $\mathbf{F}^{(k)}$, given (see (4.15)) by

$$\mathbf{F}^{(k)} = \mathbf{\Phi}_{AA}(N_0\mathbf{I} + \mathbf{X}_t^{(k)}\mathbf{X}^{(k)}\mathbf{\Phi}_{AA})^{-1},$$

will be the same for all hypotheses.

For the binary case, decision is based on the difference of the functions $\Lambda^{(k)}$; that is, the important operation is given by

$$\Lambda = \Lambda'^{(1)} - \Lambda'^{(2)} = \Lambda^{(1)} - \Lambda^{(2)} + \left(\ln\frac{K_2^{(1)}}{K_2^{(1)}}\right). \qquad (5.1)$$

As before, $\Lambda^{(1)} - \Lambda^{(2)}$ determines the receiver operations and the other term is a biasing, or weighting, term. Using (4.10) and (4.15),† we have

$$\begin{aligned}
\Lambda^{(1)} - \Lambda^{(2)} &= \mathbf{y}_t\mathbf{X}^{(1)}\mathbf{F}\mathbf{X}_t^{(1)}\mathbf{y} - \mathbf{y}_t\mathbf{X}^{(2)}\mathbf{F}\mathbf{X}_t^{(2)}\mathbf{y} \\
&= \mathbf{y}_t\mathbf{X}^{(1)}\mathbf{F}\mathbf{X}_t^{(1)}\mathbf{y} + (\mathbf{y}_t\mathbf{X}^{(2)}\mathbf{F}\mathbf{X}_t^{(1)}\mathbf{y} - \mathbf{y}_t\mathbf{X}^{(1)}\mathbf{F}\mathbf{X}_t^{(2)}\mathbf{y}) \\
&\quad + \mathbf{y}_t\mathbf{X}^{(2)}\mathbf{F}\mathbf{X}_t^{(2)}\mathbf{y} \\
&= \mathbf{y}_t[\mathbf{X}^{(1)} + \mathbf{X}^{(2)}]\mathbf{F}[\mathbf{X}^{(1)} - \mathbf{X}^{(2)}]_t\mathbf{y}. \qquad (5.2)
\end{aligned}$$

The two terms inside the parentheses in the middle equality are transposes of each other; since they are scalars, they are equal and thus cancel each other out.

A block diagram for the receiver implied by (5.2) is shown in Figure 7. In the single-path channel case, $\mathbf{X}^{(k)}$ represents a multiplication

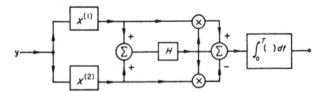

Fig. 7. The general Rake receiver.

operation, and this block diagram of Figure 7 is almost exactly that of the Price–Green Rake receiver (see Fig. 3 of [23]). The difference arises because—except in the case of a single time-invariant channel path— $\mathbf{F}$ is an unrealizable operator. To overcome this difficulty, we can either use a delay or break up $\mathbf{F}$ into two filters, $\mathbf{F}_+$ and $\mathbf{F}_-$. $\mathbf{F}_+$ ($\mathbf{F}_-$) is $\mathbf{F}$ with the diagonal elements halved and the elements above (below) the diagonal set equal to zero.

At this point, a host of speculations may be raised by the realization

---

† This technique was suggested by some unpublished work of Professor Siebert.

that we can regard the signals $X^{(1)}$ and $X^{(2)}$ as being composed of a common part, $(X^{(1)} + X^{(2)})/2$, and an information-bearing part, $\pm (X^{(1)} - X^{(2)})/2$, for example,

$$X^{(1)} = \frac{X^{(1)} + X^{(2)}}{2} + \frac{X^{(1)} - X^{(2)}}{2} . \qquad (5.3)$$

Thus, we might ask whether, in (5.2), $y_t [X^{(1)} + X^{(2)}]F$ provides an estimate of the channel by the known sounding signal $(X^{(1)} + X^{(2)})/2$, which is then multiplied by the information-bearing signal $\pm (X^{(1)} - X^{(2)})/2$ to provide $z_s^{(t)}$. The answer is no. This is disappointing, but, in view of the previous discussions of Section 4, not surprising. This is not to say, however, that receivers constructed on the above philosophy will not work; it just means that they will not be optimum from a decision-theory point of view. In practice, it might even be preferable to build a receiver based on the "erroneous" philosophy. It would be interesting to study how far from optimum such a receiver is.

But one conclusion must be drawn that is of relevance to studies in adaptive and learning systems: The intuitive choice is not always consistent with the requirements of some theoretically powerful concept such as the computation of a posteriori probabilities.

To return to the Rake structure, for the multipath case, the filters $X^{(k)}$ perform multiplications of the received signal with shifted replicas of the stored signal $x^{(k)}$. The filter $F^{(k)}$ does not break down in any simple way, except in the threshold case. The threshold case is defined by the condition (see Appendix B) that

$$\lambda_{\max}(\Phi_{zz}) < N_0, \qquad (5.4)$$

where $\lambda_{\max}$ is the largest eigenvalue of Φ_{zz}. This is a necessary and sufficient condition [26] for a Neumann series expansion of F in (5.2). If we retain only the first term in the expansion, even though retaining additional terms is not particularly troublesome, we get (see Appendix B)

$$N_0(\Lambda^{(1)} - \Lambda^{(2)}) = y_t [X^{(1)} + X^{(2)}]\Phi_{AA}[X^{(1)} - X^{(2)}]_t y. \qquad (5.5)$$

Let us also assume that the channel tap functions are uncorrelated, so that only the diagonal blocks, say Φ_i, of Φ_{AA} are nonzero. The covariance matrix for the ith tap is Φ_i. Now, closer study of (5.5) shows that it can be implemented as shown in Figure 8. The receiver is a cascade of units, one for each channel tap, strung out on the delay line. Notice that the units are arranged on the receiver delay line in the opposite order to their arrangement on the delay line, thus providing a generalized type of *matched filter*, matched both to the channel and to the transmitted signals. Figure 8 is identical with the receiver

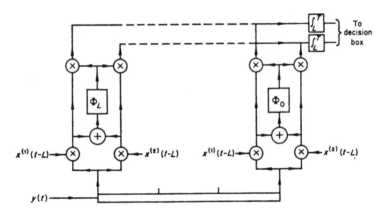

Fig. 8. The original Rake receiver.

originally obtained by Price and Green [23] by a brilliant combination of theoretical and physical ideas. Our derivation above is more compact and leads to the final result much more directly than in [23]. (For those more interested in the details of Rake, we remark that here we obtain the Type II Rake directly as the optimum system.) It shows under what conditions the Price–Green Rake is optimum and also yields the receiver structure (5.2) for the general case. However, this more complete theoretical analysis was only done long after the original Rake had been built and proved successful. It is a tribute to the deep theoretical understanding and fine engineering judgment of Price and Green that they converged exactly on the optimum system. One of the explanations they gave for the action of each unit on the receiver delay line is of interest:

Suppose $x^{(1)}$ was transmitted. The received signal y should be multiplied by $x^{(1)}$ to provide a preliminary weighting against the noise (suppressing portions of y where the signal x is weak and reinforcing portions where x is high). The product should then be smoothed in a filter Φ_i whose bandwidth is equal to the fluctuation rate of the corresponding tap to get an approximate estimate of the tap-gain function. This estimate can then be used to perturb the stored signals before the final correlation with y. We would not know, of course, whether $x^{(1)}$ was actually transmitted. But since either $x^{(1)}$ or $x^{(2)}$ was transmitted, the sum $y_t(x^{(1)} + x^{(2)})$ can be used in lieu of the proper but unknown $y_t x^{(1)}$. Thus the structure of each unit in Figure 8 is obtained. However, we should notice again that this intuitively very reasonable explanation of the adaptive features of Rake is not optimal in general from a decision-theoretic point of view.

Price and Green went a step further and took advantage of the fact that the communication extends over several time intervals in each of which either $\mathbf{x}^{(1)}$ or $\mathbf{x}^{(2)}$ is transmitted. They allow the filters Φ_i to ring (that is, have a memory) over several time intervals. This longer ring time provides a more accurate channel estimate than can be obtained by restricting the filter memory to a single interval. This has been a simple and reasonably successful method of accounting for channel memory. However, no attempts seem to have been made to justify it on a decision-theoretic basis. We shall turn now to such an attempt.

6. Sequential Operation

In the preceding pages, we have studied the optimum structure for the reception of signals in just a single time interval, namely, the observation time T. In a communication system, however, we would be sending a sequence of signals. For continuous operation, we could operate on an interval-to-interval basis, treating each interval independently of the others. If the channel has a statistical memory that is greater than the duration of the interval, such a procedure clearly throws away information. One method of compensating for this loss of information is to form longer intervals—perhaps sentences or paragraphs long—so that the channel is now essentially independent from interval to interval, though this procedure enlarges the set of possible messages and the bank of receiver filters by a prohibitive factor. Another method is to retain the original interval length but attempt to use information from other intervals in making a decision in any particular interval. We shall discuss some aspects of this latter procedure, particularly those dealing with the learning or adaptive nature of the scheme.

In the rest of this section, we discuss an approach to using interval-to-interval information; we show how to obtain the conditional probabilities required in any sequential scheme and also discuss the physical interpretation of the operations involved; and finally we give a more explicit formula for the computation of the conditional probabilities, which is obtained simply by partitioning the matrix formulas of (4.8) and (4.14). We also show how the formulas simplify in the threshold case and how in a simple case they provide a rationale for the Rake operation mentioned above.

Some Methods of Operation

Numerical subscripts will be used for values of corresponding quantities in the specified time intervals: Zero will refer to the interval about which a decision is being made; the negative integers, to previous intervals; the positive integers, to later intervals. We shall assume a sequential decoding scheme such that when a decision is to be made about a given interval (denoted by the subscript zero), all previous intervals (denoted by negative subscripts) are assumed to have been correctly decoded. Now all the relevant information pertaining to the decision in interval zero is contained in the a posteriori probability

$$p(\mathbf{x}_0^{(k)} \mid \mathbf{y}_0, \mathbf{y}_{-1}, \mathbf{x}_{-1}^{(\ )}, \mathbf{y}_{-2}, \mathbf{x}_{-2}^{(\ )}, \cdots, \mathbf{y}_1, \mathbf{x}_1^{(l)}, \mathbf{y}_2, \mathbf{x}_2^{(m)}, \cdots).$$

We leave the parentheses ( ) for $\mathbf{x}_{-1}, \mathbf{x}_{-2}, \cdots$ unfilled to indicate that these have been decided upon and are known without error. If we assume that the channel memory extends over M intervals, clearly the terms $\mathbf{y}_{-M-1}, \mathbf{x}_{-M-1}, \mathbf{y}_{-M-2}, \mathbf{x}_{-M-2}, \cdots$ can be omitted. This is not true, however, for $\mathbf{y}_{+M+1}, \mathbf{x}_{M+1}, \mathbf{y}_{M+2}, \mathbf{x}_{M+2}, \cdots$, even though it appears plausible at first glance. Clearly, $\mathbf{y}_1, \mathbf{x}_1, \cdots, \mathbf{y}_M, \mathbf{x}_M$ will also contribute information about the channel behavior in interval zero; if we had better information about $\mathbf{y}_1, \mathbf{x}_1$, we could make a better decision in this interval. But $\mathbf{y}_2, \mathbf{x}_2, \cdots, \mathbf{y}_{M+1}, \mathbf{x}_{M+1}$ all contribute information about interval one, and therefore also aid the decision in interval zero. Similarly, we can argue that $\mathbf{y}_{M+2}, \mathbf{x}_{M+2}, \cdots$ all provide useful information about interval zero. Thus, theoretically, an infinite delay is required if we are to conserve all the information. In practice, however, one might be content with looking only M intervals ahead. In any case, for a decision in interval zero, we would pick on a maximum a posteriori probability criterion (which corresponds to minimum probability of error), say the $\mathbf{x}_0^{(k)}$ that maximizes the sum

$$\sum{}' p(\mathbf{x}_0^{(k)} \mid \mathbf{y}_{-m}, \cdots, \mathbf{y}_{-1}, \mathbf{x}_{-M}^{(\ )}, \cdots, \mathbf{x}_{-1}^{(\ )}, \mathbf{y}_0, \mathbf{y}_1, \cdots, \mathbf{x}_1^{(l)}, \mathbf{x}_2^{(m)}, \cdots),$$

where the prime denotes the sum over all $l, m, \cdots$. If we assume that the $\mathbf{x}^{(k)}$ are equally likely and are picked independently in each interval, this reduces to the maximum-likelihood criterion: Pick $\mathbf{x}_0^{(k)}$ to maximize

$$\sum{}' p(\mathbf{y}_0 \mid \mathbf{x}_0^{(k)}, \mathbf{y}_{-M}, \cdots, \mathbf{x}_{-M}, \cdots, \mathbf{y}_1, \cdots, \mathbf{x}_1^{(l)}, \cdots).$$

In practice, even when we drop $\mathbf{y}_{M+1}, \cdots, \mathbf{x}_{M+1}^{(p)}, \cdots$, these schemes involve too much computation, and it is profitable to search for simpler nonoptimum methods. The convolutional-encoding and sequential-decoding concept developed by J. M. Wozencraft [27] is, as far as we know, the only feasible technique suggested so far. However, we shall not enter upon a discussion of this method here; since we observe that a necessary step in all methods is the computation of a posteriori probabilities and likelihood functions, we shall study this calculation more closely.

Calculation of Probabilities and Interpretation of Receiver

Dropping all superscripts for convenience, let us consider

$$\Gamma = p(\mathbf{y}_0 \,|\, \mathbf{y}_{-M}, \cdots, \mathbf{y}_1, \cdots, \mathbf{y}_N, \mathbf{x}_{-M}, \cdots, \mathbf{x}_{-1}, \mathbf{x}_0, \mathbf{x}_1, \cdots, \mathbf{x}_N).$$

Now

$$\Gamma = K \cdot p(\mathbf{y}_0, \mathbf{y}_1, \cdots, \mathbf{y}_N \,|\, \mathbf{y}_{-M}, \cdots, \mathbf{y}_{-1}, \mathbf{x}_{-M}, \cdots, \mathbf{x}_0, \cdots, \mathbf{x}_N),$$

where K is a constant, since we are assuming that a particular set $\mathbf{x}_0 \cdots \mathbf{x}_N$ and $\mathbf{y}_0 \cdots \mathbf{y}_N$ is given. Now we define two new matrices,

$$\mathbf{y}_{ut} = [\mathbf{y}_{0t}, \mathbf{y}_{it}, \cdots, \mathbf{y}_{Nt}] \quad \text{and} \quad \mathbf{y}_{dt} = [\mathbf{y}_{-Mt}, \cdots, \mathbf{y}_{1t}].$$

Similarly defining $\mathbf{x}_u$ and $\mathbf{x}_d$, we have

$$\Gamma = K \cdot p(\mathbf{y}_u \,|\, \mathbf{y}_d, \mathbf{x}_d, \mathbf{x}_u). \tag{6.1}$$

The subscripts u and d are chosen to represent the undecided intervals and decided intervals, respectively. Thus Γ tells us in what way our knowledge of the previously transmitted symbols affects our impressions of later intervals. We would expect that, from the previous intervals and $\mathbf{y}_d$, $\mathbf{x}_d$, we should be able to form an estimate, reliable to within the factor of uncertainty due to the additive noise, of the channel behavior in the past. Now, since there is a correlation between successive channel values, this knowledge should help to reduce our uncertainty about the present and future behavior of the channel. We would expect that, based on the past, we could make a prediction about the channel behavior in the future, thus reducing the randomness of future values. For our Gaussian channel, this is in fact so: If the covariance matrix

$$\mathbf{\Phi}_{yy} = \begin{bmatrix} \mathbf{\Phi}_{dd} & \mathbf{\Phi}_{du} \\ \hline \mathbf{\Phi}_{ud} & \mathbf{\Phi}_{uu} \end{bmatrix} \tag{6.2}$$

is known (note that this matrix depends on $\mathbf{x}_d$ and the assumed $\mathbf{x}_u$), and we assume a purely random channel (i.e., $\bar{\mathbf{A}} = 0$),† then we have

$$\ln \Gamma \sim (\mathbf{y}_u - \mathbf{\Phi}_{ud}\mathbf{\Phi}_{dd}^{-1}\mathbf{y}_d)_t (\mathbf{\Phi}_{uu} - \mathbf{\Phi}_{ud}\mathbf{\Phi}_{dd}^{-1}\mathbf{\Phi}_{du})^{-1} (\mathbf{y}_u - \mathbf{\Phi}_{ud}\mathbf{\Phi}_{dd}^{-1}\mathbf{y}_d). \tag{6.3}‡$$

---

† To avoid complications, we shall assume for the present discussion that we have no intersymbol interference from interval to interval. This will be the situation if we have a single-path channel or if we allow sufficient "dead-time" between symbols for the channel response to one signal to die out before the next signal is transmitted.

‡ The symbol $\sim$ is used to denote that $\ln \Gamma$ is directly related to the right-hand side.

Thus we now have a mean component, $\hat{\mathbf{y}}_u = \boldsymbol{\Phi}_{ud}\boldsymbol{\Phi}_{dd}^{-1}\mathbf{d}$, which is the minimum-variance prediction, or extrapolation, of $\mathbf{y}_u$ given $\mathbf{y}_d$, for assumed $\mathbf{x}_u$. This is evident because the conditional mean is the optimum predictor for the minimum-variance criterion [13].

The presence of the new mean component, $\hat{\mathbf{y}}_u$, reduces the variance of $\mathbf{y}_u$ from $\boldsymbol{\Phi}_{uu} = N_0\mathbf{I} + \boldsymbol{\Phi}_{z_u z_u}$ to

$$\overline{(\mathbf{y}_u - \hat{\mathbf{y}}_u)(\mathbf{y}_u - \mathbf{y}_u)_t} = \boldsymbol{\Phi}_{uu} - \boldsymbol{\Phi}_{ud}\boldsymbol{\Phi}_{dd}^{-1}\boldsymbol{\Phi}_{du}$$
$$= N_0\mathbf{I} + \boldsymbol{\Phi}_{z_u z_u}, \text{ say.} \qquad (6.4)$$

After some algebra (using the identity

$$[N_0\mathbf{I} + \mathbf{B}]^{-1} = N_0^{-1} - N_0^{-1}\mathbf{B}[N_0\mathbf{I} + \mathbf{B}]^{-1}),$$

an element of a receiver implementing (6.3) can be written

$$N_0\Lambda^{(k)} = 2\mathbf{y}_{ut}\hat{\mathbf{y}}_u^{(k)} + (\mathbf{y}_u - \hat{\mathbf{y}}_u)_t\hat{\mathbf{H}}^{(k)}(\mathbf{y}_u - \hat{\mathbf{y}}_u^{(k)}), \qquad (6.5)$$
$$N_0\Lambda^{(k)} = 2\mathbf{y}_{ut}\hat{\mathbf{y}}_u^{(k)} + (\mathbf{y}_u - \mathbf{y}_u)_t\hat{\mathbf{z}}_{ure}^{(k)}, \qquad (6.6)$$

say, where

$$\hat{\mathbf{z}}_{ure}^{(k)} = \hat{\mathbf{H}}^{(k)}(\mathbf{y}_u - \hat{\mathbf{y}}_u^{(k)}), \qquad (6.7)$$
$$\hat{\mathbf{H}}^{(k)} = \hat{\boldsymbol{\Phi}}_{z_u z_u}^{(k)}[N_0\mathbf{I} + \hat{\boldsymbol{\Phi}}_{z_u z_u}^{(k)}]^{-1}. \qquad (6.8)$$

We see that (6.6) is similar to (4.15) for the single interval specular plus random component case. The difference is that in sequential operation we replace the single interval mean value $\bar{\mathbf{y}} = \bar{\mathbf{z}}$ (assumed to be zero here for convenience) and covariance $N_0\mathbf{I} + \boldsymbol{\Phi}_{zz}$ by the *conditional* mean and covariance $\hat{\mathbf{y}}_u$ and $N_0\mathbf{I} + \hat{\boldsymbol{\Phi}}_{z_u z_u}$.

The previous data $\mathbf{y}_d$, $\mathbf{x}_d$ are used to make a minimum-error-variance extrapolation of the observed waveform in the present interval, and this prediction helps reduce the "randomness" in the present interval. Now the usual single interval estimator-correlator receiver can be used. This interesting generalization of the estimator-correlator receiver is what we would intuitively expect and it is satisfying to see the theoretical analysis bear this out.

We should notice that this solution applies not only to communication problems but also to Bayesian learning and pattern recognition problems [28] where the $\mathbf{y}_d$ are completely identified (as to source) observations that are used to classify a new observation $\mathbf{y}_u$. Thus the problem studied in [29] and [30] can be considered a special case of (6.5). We have pointed out there [30] that our interpretation of (6.5) as a generalized estimator-correlator receiver also applies to the pattern recognition problem. In [29] and [30] we have $\mathbf{y}_d = \mathbf{z}_d^{(t)} + n$, where the $\mathbf{z}_d^{(t)}$ is drawn from a Gaussian distribution with mean zero and covariance $\Phi_{zz}^{(k)}$ and the $\mathbf{z}_d^{(k)}$, $d = -1, -2, \cdots, -M$, are all identical. That is, the "pattern" $\mathbf{z}^{(k)}$ remains invariant during the learning and observation period. With this constancy assumption, the conditional mean and variance simplify to

$$\hat{\mathbf{y}}_u = \Phi_{zz}\left[\Phi_{zz} + \frac{N_0}{M}\mathbf{I}\right]^{-1}\mathbf{s}, \qquad (6.9)$$

$$\hat{\Phi}_{z_u z_u} = \Phi_{zz} - \Phi_{zz}\left[\Phi_{zz} + \frac{N_0}{M}\mathbf{I}\right]^{-1}\Phi_{zz}. \qquad (6.10)$$

where

$$\mathbf{s} = \frac{\mathbf{y}_{-1} + \mathbf{y}_{-2} + \cdots + \mathbf{y}_{-m}}{M}$$

$$= \text{sample mean of learning observations.} \qquad (6.11)$$

(Equations (6.9), (6.10), and (6.11) can be obtained by using the matrix identity

$$[\mathbf{I} + \mathbf{a}\mathbf{a}_t]^{-1} = \mathbf{I} - \mathbf{a}[\mathbf{I} + \mathbf{a}_t\mathbf{a}]^{-1}\mathbf{a}_t,$$

where $\mathbf{a}$ is a column vector whose elements may again be column vectors.)

In this case the sample mean is a simple sufficient statistic [13] for the learning observations. We expect that this result, first obtained in a much different way by Braverman [29], can be applied to the Rake system as follows. Consider a simple time-invariant single-path channel and let the signalling waveforms $\mathbf{x}^{(1)}$ and $\mathbf{x}^{(2)}$ be orthogonal. Then the $\mathbf{y}_{dt}$ ($\mathbf{x}_d^{(1)} + \mathbf{x}_d^{(2)}$) provides a completely identified learning observation for the channel tap gains. Averaging of these learning observations to obtain the mean according to (6.11) can be done by a narrow-band (integrating) filter, which is actually what is used in Rake [23]. This simple argument needs to be more completely examined and generalized. We have not yet done this. However, in conclusion we shall present a more explicit solution in terms of the signals $\mathbf{x}$ and the channel covariance.

More Explicit Formulas

Thus we have

$$\Gamma \sim p(\mathbf{y}_u \mid \mathbf{y}_d, \mathbf{x}_d, \mathbf{x}_u) \sim p(\mathbf{y}_u, \mathbf{y}_d \mid \mathbf{x}_u, \mathbf{x}_d) = p(\mathbf{y} \mid \mathbf{x}), \qquad (6.12)$$

where

$$\mathbf{y}_t = [\mathbf{y}_{ut} \vdots \mathbf{y}_{dt}], \qquad \mathbf{x}_t = [\mathbf{x}_{ut} \vdots \mathbf{x}_{dt}].$$

This, however, looks again like the one-shot problem, so we can immediately write, by (4.8),

$$\ln \Gamma \sim \mathbf{y}_t \mathbf{X} \mathbf{\Phi}_{AA} \left(\mathbf{I} + \frac{\mathbf{X}_t \mathbf{X}}{N_0} \mathbf{\Phi}_{AA} \right)^{-1} \mathbf{X}_t \mathbf{y}. \qquad (6.13)$$

Moreover, in a threshold situation (see Appendix B), by choosing the first term in the Neumann series expansion for the term in parentheses, we have

$$\ln \Gamma \sim \mathbf{y}_t \mathbf{X} \mathbf{\Phi}_{AA} \mathbf{X}_t \mathbf{y}$$
$$= \mathbf{y}_{ut} \mathbf{X}_u \mathbf{\Phi}_{a_u a_u} \mathbf{X}_{ut} \mathbf{y}_u + 2 \mathbf{y}_{ut} \mathbf{X}_u \mathbf{\Phi}_{a_u a_d} \mathbf{X}_{dt} \mathbf{y}_u + \mathbf{y}_{dt} \mathbf{X}_d \mathbf{\Phi}_{a_d a_d} \mathbf{X}_{dt} \mathbf{y}. \quad (6.14)$$

This expression has also been obtained by Wozencraft [31]. The first term on the right-hand side corresponds to the normal operations in interval u; the second term represents the information contributed by the data and decisions of interval d, and it depends on the correlation of the channel $\mathbf{\Phi}_{a_u a_d}$; the last term is a constant that has already been computed in interval d. Equation (6.14) provides a form that is convenient for realization, but we shall not discuss this here, since our chief purpose was to illustrate the adaptive features of this type of receiver.

7. Concluding Remarks

The application of adaptive concepts to the design of a communication system receiver offers attractive possibilities, just as it does in other areas in which random processes are encountered, such as in control systems. The incorporation of adaptive capability into any system is for the most part likely to remain based on intuition and heuristic argument, as well as on trial-and-error procedures, because a strictly mathematical approach usually fails to take proper account of engineering considerations. It is nonetheless important to seek guidelines from mathematical analysis wherever possible, if only to establish standards by which practical systems may be judged and expedient improvement sought.

There is already established a firm mathematical discipline in the field of communications, known as statistical detection (or decision) theory. This theory may be applied to optimize signal reception, virtually without invoking any preconceived notions of receiver structure such as that adaptivity is a "good thing" to have. In the application of this theory to the particular reception problem considered here, an adaptive network has materialized from the mathematics used to specify the optimum receiver, thus giving reassuring confirmation of the basic soundness of the idea of adaptivity. At the same time, however, close examination of this abstractly synthesized adaptive behavior —which can, in such an instance, justifiably be called optimal adaptive behavior—has provided the opportunity to see where intuition may perhaps go astray. This observation serves to indicate that adaptive system design should not be approached too dogmatically.

This evaluation of adaptivity has been based on a particular reception problem: that of deciding—from observation of a received waveform and complete knowledge of the statistical characteristics of a particular channel—which one of a finite set of possible transmitted waveforms (that are also known to the receiver) has actually been sent over the channel. For simplicity, the waveforms have been considered to be of a sampled-data type and the channel to consist of a time-varying, linear, sampled-data filter, followed by stationary, zero-mean additive Gaussian noise that is independent from sample to sample as well as independent of the transmitted waveform and of the channel-filter variations. The channel filter has been represented as a tapped delay line with tap outputs summed through time-varying gain controls.

From this model it has been shown that if the instantaneous time-varying filter behavior is at all times exactly known to the receiver, then on the basis of detection theory, the optimum receiver operation is to crosscorrelate, or form inner products of, the received data and the various possible waveforms that could exist at the channel-filter output just ahead of the noise. One way of performing the crosscorrelation is to pass the received signal into a set of special time-invariant linear filters called *matched filters*. Since we lack any direct knowledge, however, of the instantaneous channel-filter behavior, such operations cannot be performed. It would then seem plausible to estimate, from the received data and other available knowledge, the possible waveforms that could exist just ahead of the noise, and then to crosscorrelate the received data against these estimates. Thus the notion of an estimator-correlator has been introduced, or equivalently, an adaptive matched filter that is governed by a prenoise waveform estimate. It has been shown that such a conjectured mode of operation is precisely what detection theory specifies for the optimum receiver (or for its mathematical equivalent, since the structure of the optimum receiver is not unique), provided the set of channel-filter-tap gain variations—which can be negative as well as positive—form a zero mean, multidimensional, Gaussian random process with parameters known to the receiver. More general gain variations are allowed if the noise is large. The prenoise waveform estimates have been shown to be of the minimum-error-variance type.

To serve as a warning that intuition does not necessarily lead to the mathematically best adaptive systems, although it certainly must be used when encountering practical difficulties, two illustrations have been given that use the communication system model already described. In the first example, it has been shown that the reasonable idea of first forming a single estimate of the channel-filter behavior, using it to modify transmitted waveform replicas in the same way that the channel filter modifies the actual transmitted waveforms, and then of crosscorrelating the received signal against the modified replicas, is nonoptimal, although doubtlessly effective. The second contradiction of intuition occurs when the apparently slight modification is made of letting the Gaussian filter gain variations have nonzero mean—the optimum receiver is then no longer a pure estimator-correlator, although many of its basic features remain unchanged.

The Rake receiver has been cited as an example of a practical communications application of the adaptive-matched-filter concept. This development is based largely on detection theory but also includes heuristic concessions to practical requirements. However, some of the originally heuristic modifications have recently been shown to be supported by detection theory.

An important problem in communication systems is the proper use of previously received data and previously made decisions in improving the reliability of the current decision. This problem has been formulated as a Bayesian learning problem, and thus applies also to Bayesian pattern recognition schemes. Some general results have been obtained, but much work still remains to be done.

Appendix A: The Minimum-Variance Estimator

It is a well-known result [13] that for Gaussian signals in Gaussian noise the conditional mean provides the best estimate of the signal for a minimum-variance criterion. In fact, for Gaussian signals this estimate is optimum for a wide class of criteria [21]. In our problem, we have

$$\mathbf{y} = \mathbf{z} + \mathbf{n}$$

and wish to find an estimate $\mathbf{z}_e$ such that

$$\epsilon_i = \overline{(z_i - z_{ei})^2} \tag{A.1}$$

is minimized for each i. Now

$$p(\mathbf{z} \mid \mathbf{y}) = p(\mathbf{y} \mid \mathbf{z})p(\mathbf{z})/p(\mathbf{y})$$

$$= K \exp\left(-\frac{1}{2}\left\{\mathbf{z}_t\left[\boldsymbol{\Phi}_{zz}^{-1} + \frac{\mathbf{I}}{N_0}\right]\mathbf{z} - 2\mathbf{z}_t\mathbf{y}/N_0\right\}\right), \tag{A.2}$$

where K is a constant that is independent of $\mathbf{z}$. From (A.2) we see that

z has a Gaussian distribution. By comparison with the standard equation,

$$p(\mathbf{x}) = K' \exp\left(-\tfrac{1}{2}\{(\mathbf{x} - \bar{\mathbf{x}})_t \Phi_{zz}^{-1}(\mathbf{x} - \bar{\mathbf{x}})\}\right), \qquad (\text{A.3})$$

$$p(\mathbf{x}) = K'' \exp\left(-\tfrac{1}{2}\{(\mathbf{x}_t\Phi_{zz}^{-1}\mathbf{x} - 2\mathbf{x}_t\Phi_{zz}^{-1}\bar{\mathbf{x}}\}\right), \qquad (\text{A.4})$$

we see that

$$\frac{\mathbf{z}_t\mathbf{y}}{N_0} = \mathbf{z}_t\left[\Phi_{zz}^{-1} + \frac{\mathbf{I}}{N_0}\right]^{-1}\mathbf{z}_e, \qquad (\text{A.5})$$

and therefore

$$\mathbf{z}_e = \frac{1}{N_0}\left[\Phi_{zz}^{-1} + \frac{\mathbf{I}}{N_0}\right]^{-1}\mathbf{y} = \Phi_{zz}[\Phi_{zz} + N_0\mathbf{I}]^{-1}\mathbf{y}. \qquad (\text{A.6})$$

This is (4.11) of the text. We have assumed Φ_{zz} to be nonsingular in this proof; by a continuity argument similar to that used in Section 4, however, we see that (A.6) is valid also for singular Φ_{zz}.

We notice also, from (A.2), that the $\mathbf{z}$ that maximizes $p(\mathbf{z}|\mathbf{y})$ is the one making the argument of the exponent zero. Therefore, the maximum a posteriori probability estimate of z is also given by

$$\mathbf{z}_e = \frac{1}{N_0}\left[\Phi_{zz}^{-1} + \frac{\mathbf{I}}{N_0}\right]^{-1}\mathbf{y},$$

as in (A.6), for the minimum-variance estimate. Other proofs can be found in [15] and [16].

From the identity

$$[N_0\mathbf{I} + \mathbf{X}_t\mathbf{X}\Phi_{AA}]\mathbf{X}_t = \mathbf{X}_t[\mathbf{X}\Phi_{AA}\mathbf{X}_t + N_0\mathbf{I}],$$

assuming all inverses exist, we would obtain

$$\mathbf{X}\Phi_{AA}\mathbf{X}_t[\mathbf{X}\Phi_{AA}\mathbf{X}_t + N_0\mathbf{I}]^{-1} = \Phi_{zz}[\Phi_{zz} + N_0\mathbf{I}]^{-1}$$
$$= \mathbf{X}\Phi_{AA}[N_0\mathbf{I} + \mathbf{X}_t\mathbf{X}\Phi_{AA}]^{-1}\mathbf{X}_t, \qquad (\text{A.7})$$

thus proving the equivalence of (4.12) and (4.13) of the text. Now $[N_0\mathbf{I}+\mathbf{X}\Phi_{AA}\mathbf{X}_t]=\Phi_{yy}$ is always nonsingular and so is $[N_0\mathbf{I}+\mathbf{X}_t\mathbf{X}\Phi_{aa}]$. In fact, its inverse is given by

$$[N_0\mathbf{I} + \mathbf{X}_t\mathbf{X}\Phi_{AA}]^{-1} = N_0^{-1} - N_0^{-1}\mathbf{X}_t(N_0\mathbf{I} + \mathbf{X}\Phi_{AA}\mathbf{X}_t)^{-1}\mathbf{X}\Phi_{AA}.$$

Appendix B: The Threshold Case; More General Channel Statistics

In the threshold case, the matrix equations can be solved by iteration. Let us consider the formula

$$\Lambda = \mathbf{y}_t\frac{\mathbf{H}}{N_0}\mathbf{y}, \qquad (\text{B.1})$$

where

$$N_0 \mathbf{H} = \boldsymbol{\Phi}_{zz} \left(1 + \frac{\boldsymbol{\Phi}_{zz}}{N_0} \right)^{-1}. \tag{B.2}$$

We can make a Neumann series expansion [26] for the term in parentheses,

$$\left(\mathbf{I} + \frac{\boldsymbol{\Phi}_{zz}}{N_0} \right)^{-1} = \mathbf{I} - \frac{\boldsymbol{\Phi}_{zz}}{N_0} + \frac{(\boldsymbol{\Phi}_{zz})^2}{N_0} \cdots . \tag{B.3}$$

This expansion is valid if the norm of $\boldsymbol{\Phi}_{zz}$, regarded as a linear transformation, is less than N_0.

The norm $\|\boldsymbol{\Phi}_{zz}\|$ is defined by $\|\boldsymbol{\Phi}_{zz}\| = \min K$ such that $(\boldsymbol{\Phi}_{zz}\mathbf{x}, \boldsymbol{\Phi}_{zz}\mathbf{x}) \leq K(\mathbf{x}, \mathbf{x})$, where $(\ ,\ )$ stands for the inner, or scalar, product. This definition is equivalent to

$$\|\boldsymbol{\Phi}_{zz}\|^2 = \max_{\mathbf{x}} \ (\mathbf{x}_t [\boldsymbol{\Phi}_{zz}]_t \boldsymbol{\Phi}_{zz}\mathbf{x}; \ \mathbf{x}_t\mathbf{x} = 1),$$

which is the largest eigenvalue of

$$[\boldsymbol{\Phi}_{zz}]_t \boldsymbol{\Phi}_{zz} = [\boldsymbol{\Phi}_{zz}]^2,$$

because $\boldsymbol{\Phi}_{zz}$ is symmetric. Since an eigenvalue of $[\boldsymbol{\Phi}_{zz}]^2$ is the square of an eigenvalue of $\boldsymbol{\Phi}_{zz}$, we have $\|\boldsymbol{\Phi}_{zz}\| = \lambda_{\max}(\boldsymbol{\Phi}_{zz})$, the largest eigenvalue of $\boldsymbol{\Phi}_{zz}$. Since $\boldsymbol{\Phi}_{zz}$ is positive semidefinite, $\lambda_{\max}$ is always positive.

Therefore the condition for the validity of (B.3) is that

$$\lambda_{\max}(\boldsymbol{\Phi}_{zz}) < N_0. \tag{B.4}$$

Using this expansion, we have

$$\Lambda = \frac{1}{N_0^2} [\mathbf{y}_t \boldsymbol{\Phi}_{zz}\mathbf{y}] - \frac{1}{N_0^3} [\mathbf{y}_t \boldsymbol{\Phi}_{zz}^2 \mathbf{y}] + \cdots . \tag{B.5}$$

A simple iterative scheme can be set up to compute as many terms of the right-hand side as desired.

There is another instructive way of deriving this result in our particular case. Since $\boldsymbol{\Phi}_{zz}$ is a positive semidefinite matrix, it can be diagonalized by pre- and postmultiplication by a suitable orthogonal matrix and its transpose. Thus $\boldsymbol{\Phi}_{zz} = \mathbf{P} \boldsymbol{\Gamma} \mathbf{P}_t$, where $\mathbf{P}$ is an orthogonal matrix and $\boldsymbol{\Gamma}$ is a diagonal matrix with the eigenvalues, λ_i, of $\boldsymbol{\Phi}_{zz}$ on the diagonal. Then it is easily seen that

$$\boldsymbol{\Phi}_{zz}(\boldsymbol{\Phi}_{zz} + N_0\mathbf{I})^{-1} = \mathbf{P}\boldsymbol{\Gamma}(\boldsymbol{\Gamma} + N_0\mathbf{I})^{-1}\mathbf{P}_t. \tag{B.6}$$

Now our receiver formula is

$$N_0\Lambda = \mathbf{y}_t \boldsymbol{\Phi}_{zz}(\boldsymbol{\Phi}_{zz} + N_0\mathbf{I})^{-1}\mathbf{y} = \mathbf{y}_t \mathbf{P}\boldsymbol{\Gamma}(\boldsymbol{\Gamma} + N_0\mathbf{I})^{-1}\mathbf{P}_t\mathbf{y}$$

$$= \mathbf{y}_t'\boldsymbol{\Gamma}(\boldsymbol{\Gamma} + N_0\mathbf{I})^{-1}\mathbf{y}' = \sum_{i=1}^{N} \frac{\lambda_i}{\lambda_i + N_0} y_i'^2, \tag{B.7}$$

where y_i' is the ith component of $\mathbf{y}' = \mathbf{P}_t\mathbf{y}$. (We note that y_i' is the inner product of the ith eigenvector of $\mathbf{\Phi}_{zz}$ and $\mathbf{y}$.) If now the largest eigenvalue, say λ_1, is less than N_0, we can expand each term $\lambda_i/\lambda_i + N_0$ in a geometric series. If we do this and collect terms suitably, we get

$$N_0\Lambda = \sum_i \frac{\lambda_i}{N_0} y_i'^2 - \sum_i \frac{\lambda_i^2}{N_0^2} y_i'^2 + \cdots. \qquad (B.8)$$

This can now be rewritten as

$$N_0\Lambda = \mathbf{y}_t \frac{\mathbf{\Phi}_{zz}}{N_0}\mathbf{y} - \mathbf{y}_t \frac{[\mathbf{\Phi}_{zz}]^2}{N_0^2}\mathbf{y} + \cdots, \qquad (B.9)$$

which is the result we would have got from the Neumann series expansion. (Note that a parallel argument holds in the continuous case in which Mercer's theorem [8] can be directly used to expand the continuous analogue of the kernel $\mathbf{\Phi}_{zz}$.)

This method not only gives the same condition as before, $\lambda_{\max}(\mathbf{\Phi}_{zz}) < N_0$, but also reveals some other interesting facts. First, we need not appeal to the theory of the Neumann series to obtain the result (B.9) that we desire. Second, from (B.8) we see that since λ_i is less than N_0 for all i, each term on the right-hand side of the equation is smaller in magnitude than all the preceding terms. This method also suggests a very simple means of finding an upper bound on the norm of the error for a truncated Neumann series. Thus the error is

$$\mathbf{E}_N = \left(\mathbf{I} + \frac{\mathbf{\Phi}_{zz}}{N_0}\right)^{-1} - \sum_{i=0}^{N-1}\left(\frac{\mathbf{\Phi}_{zz}}{N_0}\right)^2 = \sum_{i=N}^{\infty}\left(\frac{\mathbf{\Phi}_{zz}}{N_0}\right)^2, \qquad (B.10)$$

and

$$\|\mathbf{E}_N\| \leq \sum_{i=N}^{\infty}\left\|\left(\frac{\mathbf{\Phi}_{zz}}{N_0}\right)^2\right\| = \frac{(\lambda_{\max}/N_0)^N}{1 - \lambda_{\max}/N_0}. \qquad (B.11)$$

In some cases, it is sufficient to take only the first term of the expansion (B.3), so that

$$N_0^2\Lambda = \mathbf{y}_t\mathbf{\Phi}_{zz}\mathbf{y}. \qquad (B.12)$$

This is usually known as the threshold receiver and is a form often used to simplify (error probability) calculations. We repeat, however, that by a simple iterative arrangement—though at the cost of a greater delay in obtaining the final answer—we can compute as many terms of the series for $N_0\Lambda$ as desired. *We should point out that condition*

(B.4) *is not synonymous with low detectability. The detectability has to do with sums of powers of eigenvalues (often, only squares of eigenvalues) and may be quite large, even if the largest eigenvalue is less than N_0.*

Various bounds on $\lambda_{\max}$ can be found in the literature [26], [17]. A particularly suggestive one is due to Szegö [32]: If z is stationary, $\lambda_{\max}$ is less than the peak value of the power spectrum of z.

Arbitrary Signal Statistics

In this situation, we obtain, for large N_0, a receiver structure corresponding to (B.12), above. Thus we have

$$\mathbf{y} = \mathbf{z}^{(k)} + \mathbf{n},$$

where the noise is white Gaussian. Then, for hypothesis k, we have

$$\Lambda \sim \int p(\mathbf{y} \mid \mathbf{z}) p(\mathbf{z})\, d\mathbf{z} = \overline{p_n(\mathbf{y} - \mathbf{z})},$$

where the bar denotes averaging over the random variable $\mathbf{z}$, whence

$$\Lambda = \overline{K \cdot \exp \left\{ -\frac{1}{2} (\mathbf{y} - \mathbf{z})_t \frac{1}{N_0} (\mathbf{y} - \mathbf{z}) \right\}}$$

$$= \overline{K \cdot \exp - \frac{1}{2N_0} \mathbf{y}_t \mathbf{y} \cdot \left[1 + \frac{1}{2N_0} (2\bar{\mathbf{z}}_t \mathbf{y} - \overline{\mathbf{z}_t \mathbf{z}}) + \frac{4}{4N_0^2} \overline{\mathbf{z}\mathbf{z}_t \mathbf{y}} + \overline{O(\mathbf{z}^3)} \right]}$$

$$= K' \cdot \left[1 + \frac{1}{N_0} \bar{\mathbf{z}}_t \mathbf{y} + \frac{1}{N_0^2} \mathbf{y}_t \mathbf{\Phi}_{zz} \mathbf{y} + \overline{O(\mathbf{z}^3)} \right].$$

Keeping only the first three terms, we get

$$\ln p(\mathbf{y} \mid \mathbf{x}) = \frac{1}{N_0} \bar{\mathbf{z}}_t \mathbf{y} + \frac{1}{N_0^2} \mathbf{y}_t \mathbf{\Phi}_{zz} \mathbf{y} + \text{"bias" term } (\ln K'). \quad (\text{B.13})$$

This agrees with (B.12), where the specular term is not included, as we had set out to show. This result has recently been obtained also by Rudnick [14].

References

1. Aseltine, J. A., A. R. Mancini, and C. Sarture, "A Survey of Adaptive Control Systems," *Trans. IRE*, PGAC-6, December, 1958, pp. 102–108.
2. Stromer, P. R., "Adaptive or Self-optimizing Control Systems—A Bibliography," *Trans. IRE*, PGAC-4, No. 1, May, 1959, pp. 65–68.

3. Eykhoff, P., "Adaptive and Optimalizing Control Systems," *Trans. IRE*, PGAC-5, No. 2, June, 1960, pp. 148–151.

4. Woodward, P. M., *Probability and Information Theory with Applications to Radar*, Pergamon Press, London, 1953.

5. Peterson, W. W., T. G. Birdsall, and W. C. Fox, "The Theory of Signal Detectability," *Trans. IRE*, PGAC-4, September, 1954, pp. 171–212.

6. Middleton, D., and D. Van Meter, "Detection and Extraction of Signals in Noise from Point of View of Statistical Decision Theory," *J. Soc. Indust. Appl. Math.*, Part I, Vol. 3, 1955, pp. 192–253; Part II, Vol. 4, 1956, pp. 86–119.

7. Grenander, U., "Stochastic Processes and Statistical Inference," *Ark. Mat.*, Vol. 1, 1950, pp. 195–277.

8. Davenport, W. B., and W. L. Root, *Introduction to the Theory of Random Signals and Noise*, McGraw-Hill Book Co., Inc., New York, 1958.

9. Davis, R. C., "Detectability of Random Signals in the Presence of Noise," *Trans. IRE*, PGIT-3, March, 1954, pp. 52–62.

10. Helstrom, C. W., *Statistical Theory of Signal Detection*, Pergamon Press, London, 1960.

11. Kailath, T., *Optimum Receivers for Randomly Varying Channels*, Proc. Fourth London Symposium on Information Theory, Butterworth Scientific Press, London, 1961.

12. Kailath, T., *Optimum Diversity Combiners*, Research Laboratory of Electronics, Massachusetts Institute of Technology, Quarterly Progress Report No. 58, Cambridge, Mass., July 15, 1960, pp. 197–199.

13. Cramér, H., *Mathematical Methods of Statistics*, Princeton University Press, Princeton, N.J., 1946.

14. Rudnick, P., *Likelihood Detection of Small Signals in Stationary Noise*, Scripps Institute, Report MPL-U-v/58, La Jolla, Calif.

15. Kailath, T., "Correlation Detection of Signals Perturbed by a Random Channel," *Trans. IRE*, PGIT-6, No. 3, June, 1960, pp. 361–366.

16. Kailath, T., *Estimating Filters for Linear Time-Variant Channels*, Research Laboratory of Electronics, Massachusetts Institute of Technology, Quarterly Progress Report No. 58, Cambridge, Mass., July 15, 1960, pp. 185–197.

17. Bellman, R., *An Introduction to Matrix Analysis*, McGraw-Hill Book Co., Inc., New York, 1960.

18. Price, R., *Statistical Theory Applied to Communication Through Multipath Disturbances*, Research Laboratory of Electronics, Massachusetts Institute of Technology, Technical Report 266, Cambridge, Mass., September 3, 1953.

19. Root, W. L., and T. S. Pitcher, "Some Remarks on Statistical Detection," *Trans. IRE*, PGIT-1, No. 3, December, 1955, pp. 33–38.

20. Bode, H. W., and C. E. Shannon, "A Simplified Derivation of Linear Least-Squares Smoothing and Prediction Theory," *Proc. IRE*, Vol. 38, 1950, pp. 417–425.

21. Sherman, S., "Non-Mean-Square Error Criteria," *Trans. IRE*, PGIT-4, No. 3, September, 1958, pp. 125–127.

22. Price, R., "Optimum Detection of Random Signals in Noise, with

Applications to Scatter-Multipath Communication. Part I," *Trans. IRE*, PGIT-2, December, 1956, pp. 125–135.

23. Price, R., and P. E. Green, Jr., "A Communication Technique for Multipath Channels," *Proc. IRE*, Vol. 46, 1958, pp. 555–570.

24. Turin, G. L., "On the Estimation in the Presence of Noise of the Impulse Response of a Random Linear Filter," *Trans. IRE*, PGIT-3, March, 1957, pp. 5–10.

25. Green, P. E., Jr., "Feedback Communication Systems," Chap. 14 *in* E. J. Baghdady (ed.), *Lectures on Communication System Theory*, McGraw-Hill Book Co., Inc., New York, 1961.

26. Faddeeva, V. N., *Computational Methods in Linear Algebra*, Dover Publications, Inc., New York, 1957.

27. Wozencraft, J. M., and B. Reiffen, *Sequential Decoding*, John Wiley & Sons, Inc., New York, 1961.

28. Bellman, R., *Adaptive Decision Processes, A Guided Tour*, Princeton University Press, Princeton, N. J., 1960.

29. Braverman, D., *Machine Learning and Automatic Pattern Recognition*, Stanford Electronics Laboratory, Stanford University, Calif., Technical Report 2003-1, February, 1961.

30. Braverman, D., "Learning Filters for Pattern Recognition," *Trans. IRE*, PGIT-8, July, 1962, pp. 280–285.

31. Wozencraft, J. M., "Sequential Reception of Time-Variant, Dispersive Transmissions," Chap. 12 *in* E. J. Baghdady (ed.), *Lectures on Communication System Theory*, McGraw-Hill Book Co., Inc., New York, 1961.

32. Grenander, U., and G. Szegö, *Toeplitz Forms and Their Applications*, University of California Press, Berkeley, Calif., 1958.

Chapter 7

Optimization Problems in Statistical Communication Theory

DAVID MIDDLETON

1. Introduction

The purpose of this chapter is to outline some of the principal optimization problems of current interest in the field of statistical communication theory, to describe briefly their formulation and techniques
of treatment, and to emphasize the role and *raison d'être* of optimality,
including such important features as optimum structure and optimum
performance. Inasmuch as results achieved to date by these approaches
have been described in detail elsewhere [1], the present discussion is
intended to be mainly expository, taking as its focal point the concept of optimization and referring the reader to standard works for
specific applications [1], [2]. More precisely, our aims are as follows:
(a) to discuss optimization procedures in which the dominant physical
features of the communication environment are incorporated into the
decision process; (b) to point out some of the technical problems involved in their application; and (c) to suggest some continuing problems of importance in this context. Both stochastic and deterministic
signals in noise backgrounds are permitted, and while the general formulation quite naturally includes multilink systems (i.e., those with
more than one message source and sink), attention is directed here for
the most part to the basic single-link cases. These may be compactly
described [1] by the relation

$$\{v\} = T_R^{(N)} T_M^{(N)} T_T^{(N)} \{u\}, \tag{1.1}$$

where $\{u\}$ represents a set or ensemble of possible messages at the source (or transmitting end) that result in a corresponding set $\{v\}$ of received messages, or decisions consequent thereto. The T's are transformations embodying, respectively, the operations of encoding ($T_T^{(N)}$), the effects of the channel or medium of propagation ($T_N^{(N)}$), and the action of the receiver ($T_R^{(N)}$). The superscripts (N) refer to the possible, and usual, injection of noise into the system at the various stages of transmission and reception. Optimization problems of various types accordingly arise when these transformations $T_T^{(N)}$, etc., are constrained or left open to adjustment under one or more criteria of expected performance.

In statistical communication theory we may say that the aims of an adequate theory of system performances and design should include the ability to describe optimum systems. Although such systems are never strictly attainable in practice, they provide limits on attainable performance and guides as to suitable suboptimum structures that can be constructed within the "economy" of the desired application. Thus, we seek optimum structure, that is, optimum modes of processing the information-bearing data at our disposal, and the evaluation of the system's operation. The latter is necessarily statistical, since one deals always with the ensemble of possible signals and interfering noise processes in any meaningful communication situation, where the results are necessarily expressed in terms of various appropriate statistical properties (e.g., error probabilities and moments) of these ensembles. In addition, the comparison of suboptimum systems, which are the ones used in practice, with the corresponding optima provides the required link between the theoretical limits and the actual practice —an essential product of an effective theory.

Optimum structure, optimum performance, and the comparison with suboptimum systems for similar purposes are thus the principal aims of the theory. Among the main applications of optimization methods in communication theory are those that lie in the following areas: (a) *signal detection*, in which the basic problem is to determine the presence or absence of a signal in noise; (b) *signal extraction*, in which some information-bearing feature of the signal, including possibly the signal itself, is to be extracted, that is, measured, from signals corrupted by noise; (c) *cost coding*, in which both operations $T_R^{(N)}$ and $T_T^{(N)}$ are simultaneously adjusted for possible further optimization of the reception process by the further choice of signal alphabets or waveforms; (d) *data processing*, as nondecision examples of processing techniques required for decision-making systems; (e) *coding techniques*, using the methods of information theory; (f) combinations of the analytical procedures required in (a), (b), and (c), together with coding methods. We shall consider here mainly the decision-theory approach, as developed by Middleton and Van Meter [2] for problems of types (a), (b), and (c), to illustrate the present questions of optimization. In Section 2 a concise formulation of the basic single-link communication system is given in decision-theoretic terms, and in Section 3 various principal optimization problems are considered in a general way within this framework. In Sections 4 and 5, optimum signal detection and extraction are examined in more detail, and in Section 6 some of the auxiliary results arising in the optimization procedures are considered briefly. Section 7 concludes the chapter with comments on present approaches and some current problems.

2. Formulation

Let us now construct a concise model of the communication process in the single-link case in which definite decisions in the face of uncertainty are required. The model is naturally based on the methods of statistical decision theory as noted above.†

We begin with abstract signal and data spaces, Ω and Γ, respectively, and consider sampled signal and data processes on a finite interval $(t_0, t_0 + T)$:

$$\mathbf{S} = [S(t_1), \cdots, S(t_n)] = [S_1, \cdots, S_n],$$

$$\mathbf{V} = [V(t_1), \cdots, V(t_n)] = [V_1, \cdots, V_n]$$

$$(t_0 \leq t_1 \leq t_2 \leq \cdots \leq t_n \leq T + t_0), \qquad (2.1)$$

where $\mathbf{S}$ and $\mathbf{V}$ are accordingly n-component vectors in Ω and Γ space.

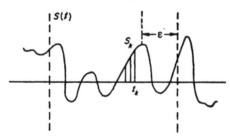

Fig. 1. A signal waveform, with sampled values S_k.

Figures 1 and 2 show typical signal and data waveforms. The times t_λ indicated in Figure 2 refer to instants, usually different from sampling times t_k, at which some estimate of the signal imbedded in the accom-

† See [1], Chap. 18, and [2] for a much fuller discussion.

panying noise is desired. Thus, if $t_\lambda > t_0 + T$ we have a case of *prediction*, while if $t_0 < t_\lambda < t_0 + T$, we speak of *interpolation;* in general, if $t_\lambda < t_0$ or $t_\lambda > t_0 + T$ we have an example of *extrapolation*. The quantity ϵ in Figure 1 is an *epoch*, referring the observer's time scale to that of the signal structure, with ϵ measured from some distinguishing point on the signal waveform. The degree of the observer's knowledge of ϵ strongly influences the type of optimum (and suboptimum) structure for both signal detection and extraction.† If ϵ is known precisely, we have *coherent reception* with coherent sampling; if ϵ is not known and in fact is uncertain over a period comparable to, or larger than, an

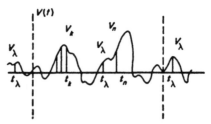

Fig. 2. A data waveform, with sampled values V_n, V_k.

average fluctuation period of the signal, we have *incoherent reception* with a corresponding sample uncertainty. The former is distinguished in the critical situation of threshold or weak-signal reception by a dependence on the square root of the input signal-to-noise (power) ratio, whereas the latter depends on the first power of this ratio—in both instances quite apart from the structure of the signal itself.

The decision situation, where either a detection or estimation regarding the signal is now required, is illustrated in Figure 3. Besides

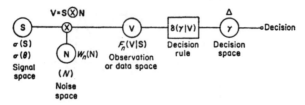

Fig. 3. The decision situation in general terms.

data and signal spaces, we introduce also a noise space (N) and a decision space Δ. The vector $\gamma = [\gamma_1, \cdots, \gamma_m]$ represents a set of m decisions about $\mathbf{S}$, based on $\mathbf{V}$, the received data process. Here $\delta(\gamma \mid \mathbf{V})$ is the *decision rule*, governing the decisions γ and based on the data $\mathbf{V}$;

† See [1], Sec. 19.4-3.

δ is either a probability ($0 \leq \delta \leq 1$), as in detection, where the decisions are of the "yes" or "no" type, or δ is a probability density ($0 \leq \delta \leq \infty$), as in extraction problems, where the decisions are now measured values of an appropriate kind. The quantities $\sigma(S)$ or $\sigma(\theta)$, $F_n(V \mid S)$, and $W_n(N)$ are, respectively, the a priori distribution densities of the following variables: S or the random parameters θ of S when $S = S(\theta)$ is deterministic; the data process V, given S; and the accompanying noise process N, which is combined with S according to the operation $\otimes$ to form V, as indicated in Figure 3. The usual combination of noise with signal is additive, so that $\otimes = +$, but frequently one encounters multiplicative processes, where $V = S \times N$ in some suitable sense. Scatter communications, multipath, and ground and sea clutter are important physical examples.

Having constructed the decision situation in terms of the relevant a priori information, as suggested by Figure 3, we must now introduce an evaluation function whereby we can evaluate performance, not only for particular decisions but also for the ensemble of possible decisions consequent upon the ensemble of received data V and the signal process S. In other words, we seek some statistic of the performance by which to describe the system's behavior. For this purpose we employ the cost or loss function $\mathfrak{F}(S, \gamma)$, which assigns to each signal or signal class (S), and to the decision associated with it, a quantitative value judgment or "cost." Then we can write for the average cost or loss associated with the decision process in our communication model of Section 1,

$$
\begin{aligned}
\mathcal{L}(\sigma, \delta) &= E_{V,S}\big\{\mathfrak{F}(S, \gamma(V))\big\} \\[6pt]
&= \int_{\Omega} dS\, \sigma(S) \int_{\Gamma} F_n(V \mid S)\, dV \int_{\Delta} \delta(\gamma \mid V)\mathfrak{F}(S, \gamma)\, d\gamma \\[6pt]
&= E_S\left\{ \int_{\Gamma} F_n(V \mid S)\, dV \int_{\Delta} d\gamma\, \delta(\gamma \mid V)\mathfrak{F}(S, \gamma) \right\} \\[6pt]
&= E_S\big\{ E_V\{\mathfrak{F}(S, \gamma(V))\} \big\},
\end{aligned}
\tag{2.2}
$$

where the expression $E_V\{\mathfrak{F}(S, \gamma(V))\}$ is the *conditional cost*. When $S = S(\theta)$, (2.2) is modified directly to

$$
\mathcal{L}(\sigma, \delta) = E_{V,\theta}\big\{\mathfrak{F}(S, \gamma(V))\big\} \quad \text{or} \quad E_{V,\theta}\big\{\mathfrak{F}(\theta, \gamma(V))\big\}, \tag{2.3}
$$

which is appropriate in some estimation or signal-extraction cases.

Two cost functions of particular importance in communication applications are the simple cost or loss function,

$$\mathcal{F}_1(\mathbf{S}, \boldsymbol{\gamma}) = C(\mathbf{S}, \boldsymbol{\gamma}), \qquad (2.4)$$

where $\mathcal{F}_1$ (that is, C) is independent of the decision rule δ and associates a preassigned set of fixed costs to the various possible decisions about $\mathbf{S}$, and

$$\mathcal{F}_2(\mathbf{S}, \boldsymbol{\gamma}) = -\log p(\mathbf{S} \mid \boldsymbol{\gamma}), \qquad (2.5)$$

where p is the a posteriori probability of $\mathbf{S}$, given $\boldsymbol{\gamma}$. Unlike $C(\mathbf{S}, \boldsymbol{\gamma})$, $\mathcal{F}_2$ depends (implicitly) on the decision rule δ and therefore cannot be preassigned independently of it. For this reason we may expect the theory of optimum and suboptimum systems based on $\mathcal{F}_2$ to be more complex than that for $\mathcal{F}_1$, which has indeed been found to be the case to date. A motivation for choosing $\mathcal{F}_2$ according to (2.5) above follows from the fact that $-\log p(\mathbf{S} \mid \boldsymbol{\gamma})$ is a measure of uncertainty or equivocation in the information-theory sense.† Accordingly, the average risks for $\mathcal{F}_1$ and $\mathcal{F}_2$, with (2.4) and (2.5) in (2.3), become

$$\mathcal{L}_1 \equiv R(\sigma, \delta) = \int_\Omega \sigma(\mathbf{S}) \, d\mathbf{S} \int_\Gamma F_n(\mathbf{V} \mid \mathbf{S}) \, d\mathbf{V} \int_\Delta \delta(\boldsymbol{\gamma} \mid \mathbf{V}) C(\mathbf{S}, \boldsymbol{\gamma}) \, d\boldsymbol{\gamma} \quad (2.6)$$

$$= \text{average cost,}$$

$$\mathcal{L}_2 \equiv H(\sigma, \delta) = -\int_\Omega \sigma(\mathbf{S}) \, d\mathbf{S} \int_\Gamma F_n(\mathbf{V} \mid \mathbf{S}) \, d\mathbf{V} \int_\Delta \delta(\boldsymbol{\gamma} \mid \mathbf{V})$$

$$\cdot \log p(\mathbf{S} \mid \boldsymbol{\gamma}) \, d\boldsymbol{\gamma} \qquad (2.7)$$

$$= \text{average information loss.}$$

We remark again that the decision rule δ in detection situations is a probability and that the integral representation over decision space Δ is in effect a sum over the discrete and distinct points that in this abstract space represent the decisions $\gamma_1, \cdots, \gamma_m$. In extraction problems, which are characteristically measurement operations, the decision rule δ takes the form of a probability density, for example,

$$\delta(\boldsymbol{\gamma} \mid \mathbf{V}) = \delta(\mathbf{S} - \boldsymbol{\gamma}_\sigma(\mathbf{V})). \qquad (2.8)$$

The right-hand member of (2.8) is the Dirac delta function and $\boldsymbol{\gamma}_\sigma(\mathbf{V}) \equiv \boldsymbol{\gamma}_\sigma(\mathbf{S} \mid \mathbf{V})$ is the *estimator* (here of $\mathbf{S}$) based on the received data $\mathbf{V}$ and subject to the distribution density (d.d.) σ of $\mathbf{S}$. Prediction and extrapolation, as well as simple estimation ($t_\lambda = t_k, \quad k = 1, \cdots, n$), are readily included in the present formalism. We write for (2.6) the somewhat extended form

$$R(\sigma, \delta) = \int_\Omega \sigma(\mathbf{S}, S_\lambda) \, d\mathbf{S} \, dS_\lambda \int_\Gamma F_n(\mathbf{V} \mid \mathbf{S}) C(\mathbf{S}, S_\lambda; \boldsymbol{\gamma}_\sigma) \, d\mathbf{V}, \qquad (2.9)$$

† See [1], Sec. 18.4-1.

where, for example, S_λ is the predicted value of S at t_λ—here for all possible ensemble representations.

Thus, with (2.6) and (2.7), for both detection and extraction we have now a measure of performance for the class of single-link systems embodied in (2.1), above. Our next step in the discussion of optimization problems and procedures is to seek appropriate extrema of $\mathcal{L}_1$ and $\mathcal{L}_2$, with and without constraints, by suitable choices of the decision rules, and, where possible, to find the various distribution densities σ, F_n, etc., governing the average cost or information loss.

3. Optimization and Extrema in General Terms

Let us now consider the central subject of the present chapter: the optimization of communication systems in which definite decisions are required, either in the "yes-no" form characteristic of the signal-detection situation, or as measurements characteristic of signal-extraction operations. Here we outline some of the major classes of questions according to the two principal classes of detection and extraction situations noted above. These questions will be described in somewhat greater detail in Sections 4 and 5. For the basic single-link system (2.1), we may state our problem symbolically as

$$\text{op } \{v\} = \underset{T_R, T_T}{\text{op}} \left\{ T_R^{(N)} T_M^{(N)} T_T^{(N)} \{u\} \right\},$$

where optimization, when possible and meaningful, is to be achieved by selection of the transformations representing reception ($T_R^{(N)}$), transmission (or encoding) ($T_T^{(N)}$), or both. Since in physical applications we do not have control of the properties of the medium through which the communication process is propagated, we cannot expect to adjust $T_M^{(N)}$. From the receiver's viewpoint, often the case in practice, the transmitted processes $y = T_M^{(N)} T_T^{(N)} \{u\}$ are specified, so that the only possibilities of optimizing performance lie in suitable choices of $T_R^{(N)}$. This is, of course, a nontrivial problem because of the inevitable and unavoidable presence of interfering noise, which in effect guarantees that $T_R^{(N)} \neq T_T^{(N)-1}$. From the point of view of decision-theory methods, the question of optimum reception, either as detection or as extraction, is also a natural point of application—in particular, in cases for which the decision itself has primary significance [1], [2]. If, however, the interests of the communication process are more naturally focused on maximum use of channel capacity rather than on the significance of the messages sent and received, that is, on their outcomes, the principal concern in optimization is then with the encoding process, $T_T^{(N)}$, as illustrated in the usual applications of information theory. If both interests are naturally combined, we may seek some sort of simultaneous adjustment of $T_R^{(N)}$ and $T_T^{(N)}$. Even when specific encoding procedures over a span of many individual decisions at the receiver concerning transmitted symbols are not themselves directly incorporated into $T_T^{(N)}$, we may still seek simultaneous adjustments of $T_R^{(N)}$ and $T_T^{(N)}$ for possible further optimization of performance, a procedure we have called *cost coding*.[†]

Among the first and most important optimization procedures is the minimization of the average risk or cost $R(\sigma, \delta)$ [see (2.6)] by suitable choice of decision rule δ. Thus, we have

$$\min_{\delta} R(\sigma, \delta) = R^*(\sigma, \delta^*), \qquad \delta \to \delta^*, \tag{3.1}$$

and R^* is called the *Bayes risk*, while δ^* is known as the *Bayes decision rule*, for which R is a minimum. Accordingly, here

$$\text{op}\,\{v\} = \underset{T_R}{\text{op}}\,\{\,T_R^{(N)} T_M^{(N)} T_T^{(N)}\{u\}\} \tag{3.2}$$

yields R^* through suitable choice of $T_{R-\text{op}}^{(N)} \equiv \delta^*$. A corresponding situation occurs for the average information loss, $\mathcal{L}_2 = H$, although under more restricted circumstances. We can similarly write

$$\min_{\delta} H(\sigma, \delta) = H^*(\sigma, \delta_H^*), \qquad \delta \to \delta_H^*, \tag{3.3}$$

where δ_H^* in general is different from δ^* above, as is H^* from R^*. Equation (3.2) is still representative, with $T_{R-\text{op}}^{(N)} \equiv \delta_H^*$; H^* is called the *Bayes equivocation*, or minimum average information loss.

A third class of extremal systems of considerable importance is provided by the so-called *minimax systems*. These are defined for cost functions of type 1, that is, $\mathcal{F}_1 = C(\mathbf{S}, \boldsymbol{\gamma})$, as the least unfavorable of the worst average costs, the latter being obtained for the "most unfavorable" a priori d.d. σ_0 of the signal process $\mathbf{S}$, that is,

$$R_M^*(\sigma_0, \delta_M^*) = \min_{\delta} \max_{\sigma} R(\sigma, \delta), \tag{3.4}$$

in which $\delta_M^* \equiv T_{R-\text{op}}^{(N)}$ is the minimax decision rule, $\delta \to \delta_M^*$. We may symbolically write (3.2) now as

$$\text{op}\,\{v\} = \underset{T_R}{\text{op}}\,\underset{T_T}{\text{op}}^{-1}\,\{\,T_R^{(N)} T_M^{(N)} T_T^{(N)}\{u\}\} = \min\text{-}\max\,\{v\}, \tag{3.5}$$

[†] See [1], Sec. 23.2.

with

$$\text{op} \equiv \overset{*}{\delta_M}; \qquad \overset{-1}{\text{op}} \equiv \max_{\sigma} \qquad (3.5a)$$
$$\scriptstyle T_R \qquad\qquad\qquad\quad T_T$$

in this case. Minimax systems are, of course, more complicated than the simple Bayes systems δ^* that minimize average risk or cost, since a double extremizing process is now required. Under conditions readily met in physical applications,† it is also true that

$$\min_{\delta} \max_{\sigma} R(\sigma, \delta) = \max_{\sigma} \min_{\delta} R(\sigma, \delta), \qquad (3.6)$$

so that in the language of game theory we have a strictly determined, zero-sum, two-person game. Here the game is between the observer (i.e., the receiver) and nature,‡ with the characteristic saddle-point condition represented by (3.6). In a similar way, although under more restricted conditions, we may expect *minimax equivocation*, that is,

$$\overset{*}{H_M}(\sigma_{0H}, \overset{*}{\delta_{MH}}) = \min_{\delta} \max_{\sigma} H(\sigma, \delta) = \max_{\sigma} \min_{\delta} H(\sigma, \delta), \qquad (3.7)$$

this latter in the strictly determined cases. Hence $\delta \to \overset{*}{\delta_{MH}} \neq \overset{*}{\delta_M}$ for the minimax average risk, and also, in general, $\sigma_{0H} \neq \sigma_0$. Equation (3.5) applies once more, but now with

$$\text{op} \equiv \overset{*}{\delta_{MH}}, \quad \text{and} \quad \overset{-1}{\text{op}} \equiv \max_{\sigma} \qquad (3.7a)$$
$$\scriptstyle T_R \qquad\qquad\qquad\qquad\quad T_T$$

[see (3.5a)], since the cost function here is $\mathfrak{F}_2 = -\log p(\mathbf{S} \mid \boldsymbol{\gamma})$ [see (2.5)] instead of the simpler cost assignment (2.4). The criteria of optimization represented by the Bayes risk and minimax average risks, (3.1), (3.4), (3.6), were first proposed and examined in their general forms by Wald [3] in his original construction of statistical decision theory, along with the principal theorems upon which are based the extremal operations formally presented above.

Until recently, attention has been directed principally to optimization on the basis of the simple cost function, $\mathfrak{F}_1$, and to some examples employing $\mathfrak{F}_2$—in both instances without additional constraints other than those necessarily imposed through the probabilistic nature of the decision rules and a priori probabilities σ, $F_N(\mathbf{V} \mid \mathbf{S})$, etc. Other extremal situations of importance to communication theory applications occur,

---

† See [1], Sec. 18.5-3, for statements of some of the principal theorems and Sec. 18.4-4 for a more detailed discussion.

‡ See [1], Sec. 23.3-1; also see [2], Chap. 6, Sec. 1.

however, in which additional constraints naturally arise. Cost coding, mentioned above, is a typical example. Here we seek a possible further minimization of the average risk $R(\sigma, \delta)$ by suitable choice of waveform S subject, say, to the constraint of fixed signal power. This can be expressed as

$$R_C^* = \underset{S}{\text{ext}} \left\{ R^*(\sigma, \delta^*) \not\subset f(S) \right\} = R_C^*(\sigma, \delta_C^*), \qquad (3.8)$$

where $\not\subset$ denotes the constraint; for fixed power this is

$$f(S) = k \int_0^T \overline{S(t)^2} \, dt.$$

In terms of the transformations $T_T^{(N)}$, etc., this becomes alternatively [see (3.2), (3.5)]

$$\text{op} \{v\} = \underset{T_T}{\text{op}} \, \underset{T_R}{\text{op}} \left\{ T_R^{(N)} T_M^{(N)} T_T^{(N)} \{u\} \right\}, \qquad (3.9)$$

where $\text{op}/T_R \equiv \delta \to \delta^*$. Then op/T_T applied to this, subject to the constraint $f(S)$, yields $\delta^* \to \delta_C^*$, with the "best" receiver for the "best" transmitted waveform. One thus adjusts $T_T^{(N)}$ and $T_R^{(N)}$ to achieve R_C^*, a further minimization of an already Bayes risk. In general, this is not unique; nor is it always possible, as some of the results to date indicate.[†]

Still other, and more elaborate, extremal situations may arise: For instance, we may wish to minimize average risk for a class of sub-optimum systems ($\delta \neq \delta^*$) by proper choice of waveform, subject to fixed signal power, for example,

$$R_C \equiv \underset{S}{\text{min}} \left\{ R(\sigma, \delta) \not\subset f(S) \right\}. \qquad (3.10)$$

Another possibility is the *maximization* of Bayes risk by choice of noise statistics, again subject to the constraint of fixed noise power, for example,

$$R_N^* \equiv \underset{W(N)}{\text{max}} \left\{ R^*(\sigma, \delta^*) \not\subset F(N) \right\}. \qquad (3.11)$$

We may also wish to combine cost coding and maximization of Bayes risk; for example,

$$R_{CN}^* \equiv \underset{S}{\text{min}} \, \underset{W(N)}{\text{max}} \left\{ R^*(\sigma, \delta^*) \not\subset f(S) \not\subset F(N) \right\}, \qquad (3.12)$$

---

† See the examples in [1], Sec. 23.2-1.

a situation characteristic of communication in the presence of interference, where best operation of the communication link is desired on
the one hand in the face of interference on the other. This may be
selected to render transmission (or reception) as ineffective as possible
—a type of minimax situation in which the constraints on both signal
and noise processes now play a key role.

Not much is as yet known about the general solutions to problems
of the above kind. The customary attack is that of the calculus of variations—which is not usually sufficient to guarantee a solution in each
case, although inspection of the special problem frequently permits us
to decide on a maximum or minimum result or to determine whether
the extremum in question is the desired maximum or minimum. In the
next two sections, we shall illustrate some of these remarks with examples of two classes of optimization situations: Bayes signal detection
and Bayes signal extraction.

4. Optimum Detection†

Let us begin by considering the general binary (i.e., two-alternative)
situation of detecting the presence of a signal of type S_2 in noise versus
that of a signal of type S_1, also in noise, which may occur alternatively.
The hypotheses accordingly are $H_2:S_2 \otimes N$ and $H_1:S_1 \otimes N$. We shall
construct the Bayes, that is, minimum average risk, test of H_2 versus
H_1 on the basis of fixed data samples on $(0, T)$. For this purpose, we
have

$\Omega_1 + \Omega_2 = \Omega$, where Ω_1 and Ω_2 are signal spaces appropriate to
$\qquad$ S_1 and S_2, and are disjoint, that is, nonover-lap-
$\qquad$ ping (see Fig. 3); $\hfill$ (4.1a)

$w_1(S)$, $w_2(S) = $ d.d.'s of $S = (S_1, S_2)$, where $w_1(S) = w_1(S_1)$,
$\qquad$ that is, for $S \in \Omega_1$, and $w_2(S) = w_2(S_2)$,
$\qquad$ $S \in \Omega_2$; S, of course, embodies all signals,
$\qquad$ as indicated; $\hfill$ (4.1b)

$$\int_{\Omega_1} w_1(S) \, dS = \int_{\Omega_1} w_1(S_1) \, dS_1 = 1, \text{ etc.;} \qquad (4.1c)$$

p_1, $p_2 = $ a priori probabilities that the data sample V comes
$\qquad$ from an ensemble $V = S_1 \otimes N$, or $V = S_2 \otimes N$,
$\qquad$ respectively. $\hfill$ (4.1d)

---

† This section presents a generalization of some of the results of Chap. 19 of
[1], especially Secs. 19.1–19.3.

Consequently, we have

$$\sigma(\mathbf{S}) = p_1 w_1(\mathbf{S}_1) + p_2 w_2(\mathbf{S}_2) \quad \text{and} \quad \int_\Omega \sigma(\mathbf{S})\, d\mathbf{S} = 1, \qquad (4.2)$$

since $p_1 + p_2 = 1$. For the two possible decisions here, that is, $\gamma_1: H: S_1 \otimes N$, $\gamma_2: H_2: S_2 \otimes N$, we write

$$\delta(\gamma_1 | \mathbf{V}) + \delta(\gamma_2 | \mathbf{V}) = 1, \qquad (4.3)$$

inasmuch as a definite decision is always made, and $0 \le \delta_{1,2} \le 1$, since the δ's are probabilities in this case. Finally, let us represent the cost assignments in the form of a cost matrix $\mathbf{C}(\mathbf{S}, \gamma)$, with rows representing the hypotheses H_1, H_2 and columns the decisions γ_1, γ_2,

$$\mathbf{C}(\mathbf{S}, \gamma) = \begin{bmatrix} C_1^{(1)} & C_2^{(1)} \\ C_1^{(2)} & C_2^{(2)} \end{bmatrix}. \qquad (4.4)$$

The superscripts refer to the hypothesis state and the subscripts to the decisions actually made. Consistent with the meaning of "success" and "failure," or "correct" and "incorrect," with respect to the possible decisions, we require that

$$C_1^{(1)} < C_2^{(1)}, \qquad C_2^{(2)} < C_1^{(2)}; \qquad (4.4a)$$

that is, "failure" costs more than "success." Note that the costs are assigned vis-à-vis the possible hypothesis states and not with respect to any one signal in a signal class (which may contain an infinite number of members).

With the above consideration in mind, we compute next the average risk according to (2.6) by integrating over the two points in the decision space Δ, for γ_1, γ_2. The result is

$$R(\sigma, \delta) = \int_\Gamma \{ [\langle F_n(\mathbf{V} | \mathbf{S}_1)\rangle_1 p_1 C_1^{(1)} + \langle F_n(\mathbf{V} | \mathbf{S}_2)\rangle_2 C_1^{(2)} p_2] \delta(\gamma_1 | \mathbf{V})$$

$$+ [\langle F_n(\mathbf{V} | \mathbf{S}_1)\rangle_1 p_1 C_2^{(1)} + \langle F_n(\mathbf{V} | \mathbf{S}_2)\rangle_2 C_2^{(2)} p_2] \delta(\gamma_2 | \mathbf{V})\}\, d\mathbf{V}, \quad (4.5)$$

where

$$p_1 \langle F_n(\mathbf{V} | \mathbf{S})\rangle_1 = \int_{\Omega_1} \sigma(\mathbf{S}) F_n(\mathbf{V} | \mathbf{S})\, d\mathbf{S}$$

$$= p_1 \int_{S_1} w_1(\mathbf{S}_1) F_n(\mathbf{V} | \mathbf{S}_1)\, d\mathbf{S}_1, \text{ etc.} \qquad (4.6)$$

If the signal processes owe their statistical character to a set of random parameters θ alone, that is, are deterministic, then (4.6) is equivalently expressed as

$$p_1\langle F_n(\mathbf{V}\,|\,\mathbf{S}_1)\rangle_1 = p_1 \int_{\theta_1} w(\theta_1) F_n(\mathbf{V}\,|\,\mathbf{S}_1(\theta_1))\, d\theta_1, \text{ etc.} \qquad (4.6a)$$

At this point, it is convenient to introduce the conditional and total error probabilities associated with the decisions γ_1, γ_2. These are

$$\beta_2^{(1)} \equiv \beta_2^{(1)}(H_2|H_1) = \text{conditional probability of incorrectly}$$
deciding that a signal of class 2 is present when actually a signal of type 1 occurs, $\qquad (4.7a)$

$$\beta_1^{(2)} \equiv \beta_1^{(2)}(H_1|H_2) = \text{the same as (4.7a) except that } S_1 \text{ and } S_2$$
are interchanged. $\qquad (4.7b)$

The corresponding total error probabilities are therefore the following:

$$p_1\beta_2^{(1)} = \text{total probability of incorrectly deciding } H_2 \text{ when}$$
H_1 is the true state, $\qquad (4.8a)$

$$p_2\beta_1^{(2)} = \text{the same as (4.8a), with } H_2 \text{ and } H_1 \text{ interchanged.} \qquad (4.8b)$$

In expanded form, we have

$$\beta_2^{(1)} = \int_\Gamma \langle F_n(\mathbf{V}\,|\,\mathbf{S}_1)\rangle_1 \delta(\gamma_2\,|\,\mathbf{V})\, d\mathbf{V},$$
$$\beta_1^{(2)} = \int_\Gamma \langle F_n(\mathbf{V}\,|\,\mathbf{S}_2)\rangle_2 \delta(\gamma_1\,|\,\mathbf{V})\, d\mathbf{V}, \qquad (4.9)$$

so that, alternatively, the conditional probabilities of correct decisions are

$$\beta_1^{(1)} \equiv \beta_1^{(1)}(H_1|H_1) = 1 - \beta_2^{(1)} = \int_\Gamma \langle F_n(\mathbf{V}\,|\,\mathbf{S}_1)\rangle_1 \delta(\gamma_1\,|\,\mathbf{V})\, d\mathbf{V}, \quad (4.10a)$$

$$\beta_2^{(2)} \equiv \beta_2^{(2)}(H_2|H_2) = 1 - \beta_1^{(2)} = \int_\Gamma \langle F_n(\mathbf{V}\,|\,\mathbf{S}_2)\rangle_2 \delta(\gamma_2\,|\,\mathbf{V})\, d\mathbf{V}. \quad (4.10b)$$

Using the above results in (4.5) we can now rewrite the average risk more compactly in terms of the error probabilities as

$$R(\sigma, \delta) = \{p_1 C_1^{(1)} + p_2 C_2^{(2)}\} + p_1(C_2^{(1)} - C_1^{(1)})\beta_2^{(1)}$$
$$+ p_2(C_1^{(2)} - C_2^{(2)})\beta_1^{(2)}. \qquad (4.11)$$

Optimization is next accomplished, according to (3.2), by suitable choice of decision rule δ, so as to minimize the average risk $R(\sigma, \delta)$. Eliminating $\delta(\gamma_2 | V)$ with the aid of (4.3), we can express (4.5) as

$$R(\sigma, \delta) = \mathfrak{R}_0 + p_1(C_1^{(2)} - C_2^{(2)}) \int_\Gamma \delta(\gamma_1 | V)[\Lambda_{21}(V) - \mathfrak{K}_{12}]$$

$$\cdot \langle F_n(V | S_1) \rangle_1 \, dV, \tag{4.12}$$

where

$$\Lambda_{21}(V) \equiv \frac{p_2}{p_1} \frac{\langle F_n(V | S_2) \rangle_2}{\langle F_n(V | S_1) \rangle_1} \tag{4.13}$$

is a *generalized likelihood ratio*, and

$$\mathfrak{K}_{12} = \frac{C_2^{(1)} - C_1^{(1)}}{C_1^{(2)} - C_2^{(2)}} > 0 \tag{4.14}$$

is a *threshold*, with $\mathfrak{R}_0 = p_1 C_1^{(1)} + p_2 C_2^{(2)}$ the *irreducible risk*. Since $\delta_1 \langle F_n \rangle$, $C_1^{(2)} - C_2^{(2)}$, p_1, etc., are all positive, it is at once clear that we minimize R by choosing $\delta(\gamma_1 | V) \to \delta^*(\gamma_1 | V)$ to be unity when $\Lambda_{21} < \mathfrak{K}_{12}$ and zero when $\Lambda_{21} > \mathfrak{K}_{12}$. In other words, we decide

$\gamma_1 : H_1$ if $\Lambda_{21}(V) < \mathfrak{K}_{12}$; that is, we choose $\delta^*(\gamma_1 | V) = 1$ for any V such that this inequality applies, and take $\delta^*(\gamma_2 | V) = 0$; (4.15)

or

$\gamma_2 : H_2$ if $\Lambda_{21}(V) \geq \mathfrak{K}_{12}$; that is, we choose $\delta^*(\gamma_2 | V) = 1$ for any V satisfying this inequality (and equality), and take $\delta^*(\gamma_1 | V) = 0$.

Note that the $\delta_{1,2}^*$ are nonrandomized decision rules, automatically arrived at here in the minimization process. The error probabilities (4.9) are accordingly

$$\min_\delta \beta_2^{(1)} \to \beta_2^{(1)*} = \int_\Gamma \langle F(V | S_1) \rangle_1 \delta^*(\gamma_2 | V) \, dV, \text{ etc.,} \tag{4.16}$$

with $\beta_2^{(1)*}$ obtained explicitly by evaluating the integral above over that portion of the data space Γ for which $\delta^*(\gamma_2 | V) = 1$, with a similar procedure for obtaining $\beta_1^{(2)*}$.

For actual applications, it is much more convenient to use as the representation of the optimum system $\log \Lambda_{21}$, in place of Λ_{21}. This in no way changes the optimization results because the logarithm is monotonic (and $\Lambda_{21} \geq 0$); (4.15) is simply rewritten as

$$\gamma_1 : H_1 : \quad \text{if} \quad \log \Lambda_{21} < \log \mathcal{K}_{12}, \tag{4.17}$$

or

$$\gamma_2 : H_2 : \quad \text{if} \quad \log \Lambda_{21} \geq \log \mathcal{K}_{12}.$$

The Bayes risk is now

$$R^* = \mathcal{R}_0 + p_2(C_1^{(2)} - C_2^{(2)}) \left(\frac{\mathcal{K}}{\mu_{21}} \beta_2^{(1)*} + \beta_1^{(2)*} \right), \qquad \mu_{21} \equiv \frac{p_2}{p_1}. \tag{4.18}$$

Besides the optimum system structure embodied in $\log \Lambda_{21}$, we need the error probabilities $\beta_2^{(1)*}$, $\beta_1^{(2)*}$ to evaluate the system's performance. Letting

$$x = \log \Lambda_{21}(\mathbf{V}), \tag{4.19}$$

we see that these error probabilities are given by

$$\beta_2^{(1)*} = \int_{\log \mathcal{K}}^{\infty} P_n^{(1)}(x) \, dx, \quad \text{and} \quad \beta_1^{(2)*} = \int_{-\infty}^{\log \mathcal{K}} P_n^{(2)}(x) \, dx, \tag{4.20}$$

where $P_n^{(1),(2)}(x)$ are the d.d.'s of x, respectively under H_1, H_2. For example,

$$P_n^{(1)}(x) = \mathcal{F}^{-1}\{E_{V|H_1}\{e^{i\xi x}\}\}; \qquad P_n^{(2)}(x) = \mathcal{F}^{-1}\{E_{V|H_2}\{e^{i\xi x}\}\}, \tag{4.21}$$

in which $\mathcal{F}^{-1}$ denotes the inverse Fourier transform and the expectations are the two characteristic functions of x, which are determined from

$$F_x^{(1)}(i\xi) = \int_\Gamma d\mathbf{V} \langle F_n(\mathbf{V} \mid \mathbf{S}_1) \rangle_1 e^{i\xi \log \Lambda_{21}(\mathbf{V})}, \quad \text{etc.} \tag{4.22}$$

In terms of these characteristic functions, the error probabilities (4.20) may be alternatively expressed as

$$\beta_2^{(1)*} = \int_{C^{(-)}} \frac{e^{-i\xi \log \mathcal{K}}}{2\pi i \xi} F_x^{(1)}(i\xi) \, d\xi,$$

$$\beta_1^{(2)*} = \int_{C^{(+)}} \frac{e^{-i\xi \log \mathcal{K}}}{-2\pi i \xi} F_x^{(2)}(i\xi) \, d\xi, \tag{4.23}$$

where $C^{(-)}$ and $C^{(+)}$, respectively, are contours extending from $-\infty$ to $+\infty$ along the real axis and indented downward and upward about any singularities on this axis, usually at $\xi = 0$. The optimum detection situation of H_2 versus H_1 is schematically illustrated in Figure 4, while the relationship of the corresponding Bayes error probabilities is sketched in Figure 5. The essential problems at this stage are technical: how to evaluate (4.21)–(4.23). This is by no means a simple task; it requires considerable skill and insight. In the most important and critical case —threshold, or weak signal operation (where the *input* signal is less than the background noise, power-wise)—a fairly comprehensive binary theory has nevertheless been developed.†

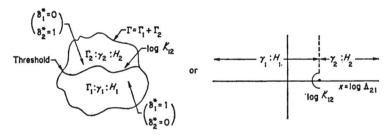

Fig. 4. Optimum detection.

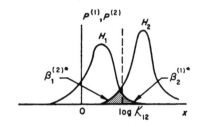

Fig. 5. The Bayes error probabilities.

Finally, in the common case of a signal in noise versus noise alone, we have $H_1:N:H_2:S \otimes N$, and the preceding results can be immediately reduced by inspection to the familiar expressions discussed in earlier work.‡

Various special cases of the general Bayes detection systems described above deserve attention. Among those belonging to the optimum class—that is, minimum average risk class in some sense—are the so-called Neyman–Pearson detectors, wherein the average risk associated with one or the other of the error probabilities $\beta_2^{(1)}$ or $\beta_1^{(2)}$ is held fixed, thereby determining a threshold $\mathcal{K}'$, while the average risk associated with the other ($\beta_1^{(2)}$ or $\beta_2^{(1)}$) is minimized; for example,

$$R^*(\sigma, \delta^*)_{NP} = C_0\{\min_\delta (p_2\beta_1^{(2)}) + \lambda p_1\beta_2^{(1)}\}. \qquad (4.24)$$

The result is readily shown to be a Bayes, that is, likelihood ratio test of the type (4.15) or (4.17), where $\lambda = \mathcal{K}'$, a threshold determined by

$$\beta_2^{(1)} = \int_{\log \mathcal{K}'}^\infty P_n^{(1)}(x)\,dx = \text{constant } (<1), \qquad (4.25a)$$

with

$$\beta_1^{(2)*} = \int_{-\infty}^{\log \mathcal{K}'} P_n^{(2)}(x)\,dx, \qquad (4.25b)$$

for the minimized error probability.

Another optimum class of binary detection systems within the Bayes group is that of the Ideal Observer, where both $\beta_2^{(1)}$ and $\beta_1^{(2)}$ are jointly minimized; for example,

$$R^*(\sigma, \delta^*)_I = C_0 \min_\delta (p_1\beta_2^{(1)} + p_2\beta_1^{(2)}), \qquad (4.26)$$

for which it is found that the decision procedure requires a threshold $\mathcal{K}_I = 1$. [See (4.15), (4.17).] Still another class, also Bayes, is the Minimax detector [see (3.4) *et seq.*], where the a priori probabilities p_1, p_2 are unknown or unspecified. Again, the optimum system is embodied in a generalized likelihood ratio (4.13). The actual evaluation of such systems, however, may be quite involved, particularly if other a priori probabilities or probability densities are open to adjustment—for example, $\sigma(S)$, $\sigma(\theta)$, $F_n(V|S)$. In Section 19.2 of [1] these three types of systems are discussed in more detail and other possible Bayes situations are presented in which other subcriteria for optimization may be employed to deal with the ever-present question of partially or totally unspecified a priori probabilities.

Finally, we mention extensions of the Bayes theory sketched here. An important case is that of variable sample, or sequential procedures, for signal detection [4]. Here sample size is the random variable, and optimization is achieved by minimizing the average cost of experimentation, proportional to sample size. For example, we have

$$R^*_{seq} = \min_\delta R(\sigma, \delta)_{seq} = C_0 \min_\delta f(n, \delta, S), \qquad f = n(\delta, S), \quad (4.27)$$

and in the more general cases f is a nonlinear function of sample size n, or of T, if continuous sampling is employed. Adaptive systems, which adjust themselves on the basis of successive stages of incoming data to be optimal at later stages, provide other more sophisticated examples of optimum systems, of which the general character is still in the process of formulation.

5. Optimum Signal Extraction

As we have noted previously (Sec. 2), the broad class of problems involving the estimation of signal waveform or signal parameters, either in an interpolatory or in a predictive sense, is naturally included within the decision-theory framework. Equally naturally, this class of problems suggests a spectrum of optimization questions wherein minimization of average risk or average equivocation are two important criteria. For our summary discussion here, we shall consider only *simple estimation*, where t_λ coincides with one or more of the sampling instants $t_1, \cdots, t_n$ on a typical interval $(0, T)$. Here, also, suitable classes of cost functions (i.e., convex cost functions†) lead to the desirable nonrandomized decision rules, which are of the form

$$\delta(\gamma \mid \mathbf{V}) = \delta(\gamma - \gamma_\sigma(\mathbf{V})), \qquad (5.1)$$

where $\gamma_\sigma(\mathbf{V}) = T_R^{(N)}(\mathbf{V})$ is the functional operation performed by the receiving system $T_R^{(N)}$ on the received data $\mathbf{V}$ and thus embodies the estimation structure. For particular $\mathbf{V}$, $\gamma_\sigma(\mathbf{V})$ is an *estimate;* for the process $\mathbf{V}$, $\gamma_\sigma(\mathbf{V})$ is called the estimator, based on the a priori signal information embodied statistically (as well as deterministically) in $\sigma(\mathbf{S})$ or $\sigma(\mathbf{\theta})$, where $\mathbf{\theta} = (\theta_1, \cdots, \theta_M)$ is a set of M different parameters. The left-hand member of (5.1) is a probability density for the usual case of continuous values of $\mathbf{V}$ and of either $\mathbf{S}$ or $\mathbf{\theta}$ under estimation; the right-hand member is the Dirac delta function [see (2.8)].

When γ_σ is given, that is, the system is specified, the average risk may be computed according to (2.9) above, which for simple estimation becomes directly

$$R(\sigma, \delta)_S = \int_\Omega \sigma(\mathbf{S}) \, d\mathbf{S} \int_\Gamma F_n(\mathbf{V} \mid \mathbf{S}) C(\mathbf{S}, \gamma_\sigma) \, d\mathbf{V}, \qquad (5.2a)$$

or often in deterministic cases for which $\mathbf{S} = \mathbf{S}(\mathbf{\theta})$,

$$R(\sigma, \delta)_\theta = \int_{\Omega_\theta} \sigma(\mathbf{\theta}) \, d\mathbf{\theta} \int_\Gamma W_n(\mathbf{V} \mid \mathbf{\theta}) C(\mathbf{\theta}, \gamma_\sigma) \, d\mathbf{V}. \qquad (5.2b)$$

The average risk depends, of course, on our choice of cost function, and here we may expect a considerably wider range of possibilities, since

---

† See [1], p. 961.

the number of reasonable and acceptable cost functions is now much expanded, at least potentially, over those available in detection.

As in detection, we may expect a variety of optima, depending on the choice of cost function and on possible constraints. In the unconstrained case—the deterministic case, for example—we seek optima from the solution of

$$\delta R = \int_{\Omega} \sigma(\theta) \, d\theta \, W_n(\mathbf{V} \mid \theta) \left. \frac{\partial C \,(\theta, \, \gamma_\sigma)}{\partial \gamma} \right|_{\gamma = \gamma^*} = 0, \qquad (5.3)$$

whenever C possesses the required derivatives; the γ^* that is the solution of (5.3) can usually be shown to be the desired optimum, either by inspection or by a direct demonstration that $\delta^2 R|_{\gamma^*} > 0$, for minimum average risk. The most common and familiar example occurs for a quadratic cost function,

$$C(\theta, \, \gamma) = C_0 \left| \theta - \gamma_\sigma \right|^2, \qquad (5.4)$$

for which it is easily shown that the Bayes risk is

$$\overset{*}{R}_\theta = C_0 \min_{\gamma_\sigma} \overline{\left| \theta - \gamma_\sigma \right|^2}^{S,N} = C_0 \overline{\left| \theta - \gamma_\sigma^*(\theta \mid \mathbf{V}) \right|^2}^{V}, \qquad (5.5)$$

where the optimum estimator is determined from the set of equations

$$\gamma_\sigma^*(\theta \mid \mathbf{V}) = \int_{\Omega_\theta} \theta \sigma(\theta) W_n(\mathbf{V} \mid \theta) \, d\theta \Big/ \int_{\Omega_\theta} \sigma(\theta) W_n(\mathbf{V} \mid \theta) \, d\theta \qquad (5.6)$$

$$= \int_{\Omega_\theta} \theta w_n(\mathbf{V}, \, \theta) \, d\theta \Big/ \int_{\Omega_\theta} w_n(\mathbf{V}, \, \theta) \, d\theta, \qquad (5.6a)$$

in which $w_n(\mathbf{V}, \theta)$ is the joint d.d. of $\mathbf{V}$ and θ. Note that in general the Bayes estimator γ^* here is a *nonlinear* operator on the received data $\mathbf{V}$; the optimum receiver for estimating θ is a nonlinear system. In fact, for this cost function it is the conditional expectation of θ, given $\mathbf{V}$— a well-known result.†

As an example in which a direct variational procedure is not possible, that is, in which $\partial C/\partial \gamma$ does not exist but nonetheless optimum systems may exist and be obtained, we have the important case of the simple cost function (in extraction),

$$C(\theta, \, \gamma_\sigma) = \sum_{k=1}^{M} [C_E A_k' - (C_E - C_C)\delta(\gamma_k - \theta_k)]. \qquad (5.7)$$

Here the A_k' are positive constants, chosen so that the resulting average risk is positive or zero for each k, C_E is the cost associated with an error,

---

† See [1], Sec. 21.2-2.

and C_C is the cost of a correct estimate of each θ_k. More compactly, (5.7) becomes

$$C(\theta, \gamma_\sigma) = C_0 \sum_{k=1}^{M} [A_k - \delta(\gamma_k - \theta_k)], \qquad (5.7a)$$

with appropriate definitions of C_0 and A_k. Thus, in this simple cost assignment, errors of all sizes exact the same cost C_E, while all correct decisions (i.e., estimates) cost $C_C < C_E$. The average risk, or cost, here is found to be

$$R(\sigma, \delta)_\theta = C_0 \sum_{k=1}^{M} \left[A_k - \int_\Gamma \mathfrak{D}_k(\mathbf{V}; \sigma, \delta) \, d\mathbf{V} \right], \qquad (5.8)$$

where

$$\mathfrak{D}_k(\mathbf{V}; \sigma, \delta) = \int_\Delta \sigma(\gamma_k) W_n(\mathbf{V} \mid \gamma_k) \delta(\gamma_k \mid \mathbf{V}) \, d\gamma_k \qquad (5.8a)$$

and δ is now a probability density. Minimization of average risk leads directly to the condition

$$\delta(\gamma_k \mid \mathbf{V}) = \delta(\gamma_k - \gamma_k^*(\theta \mid \mathbf{V})), \qquad \gamma_k = \theta_k, \qquad k = 1, \cdots, M, \quad (5.9)$$

where the components γ_k^* $(k = 1, \cdots, M)$ of the Bayes estimator, here γ^*, are determined from the relations

$$\sigma(\gamma_k^*) W_n(\mathbf{V} \mid \gamma_k^*) \geq \sigma(\theta_k) W_n(\mathbf{V} \mid \theta_k), \qquad (5.10)$$

for each k and for all θ_k in Ω_{θ_k}. But this is precisely the condition that determines the *unconditional maximum likelihood estimators* (UMLE's) of θ_k, that is, γ_k^*. Thus, equivalently, the γ_k^* are determined by

$$\frac{\partial}{\partial \theta_k} \log \left[\sigma(\theta_k) W_n(\mathbf{V} \mid \theta_k) \right] \big|_{\theta_k = \theta_k = \gamma_k^*} = 0. \qquad (5.10a)$$

For a more detailed discussion, see [1], Sections 21.2-2 and 21.2-1 and equation (21.82).

Although the quadratic cost function (QCF) and simple cost function (SCF) are analytically and historically the most familiar, many other cost functions, more reasonable in special applications, can be constructed. Even if they bear the expected difference form $F(\theta - \gamma)$, however, explicit optimization is not readily achieved for general statistics of $\mathbf{V}$ and θ, and little appears to be known as yet in a systematic way about the Bayes extractors in these cases. Under certain conditions, we can escape from the sometimes too restrictive and inappropriate nature of the QCF and SCF with the help of the following result:

Let cost functions of the type $C(\theta - \gamma_\sigma)$ have the property that for $|(\theta - \gamma_\sigma)_1| < |(\theta - \gamma_\sigma)_2|$, we have

$$C(\theta - \gamma_\sigma) = C(\gamma_\sigma - \theta); \quad C[(\theta - \gamma_\sigma)_1] < C[(\theta - \gamma_\sigma)_2]. \quad (5.11)$$

Then the Bayes estimator $(\gamma_\sigma^*)_{\text{QCF}}$ for the quadratic cost function $(\theta - \gamma_\sigma)^2$ also minimizes the average risk for these other cost functions (5.11), provided the conditional d.d. of θ, given $\mathbf{V}$—that is, $w(\theta|\mathbf{V})$—is unimodal and symmetric about this mode $(\theta = \gamma_\sigma^*)$. Typical cost functions obeying (5.11) and satisfying this condition, and consequently for which $(\gamma_\sigma^*)_{\text{QCF}}$ is the optimum estimator, are

$$C_1(\theta - \gamma_\sigma^*) = |\theta - \gamma_\sigma^*|;$$

$$C_2(\theta - \gamma_\sigma^*) = \begin{cases} 0, & |\theta - \gamma_\sigma^*| < A \quad (\gamma_\sigma^* = (\gamma_\sigma^*)_{\text{QCF}}), \\ 1, & |\theta - \gamma_\sigma^*| > A \quad (>0). \end{cases} \quad (5.12)$$

The resulting Bayes risks for (5.12) are different from $(R^*)_{\text{QCF}}$, of course, although $\gamma_\sigma^* = (\gamma_\sigma^*)_{\text{QCF}}$ here.

Still other cost functions may have desirable properties in applications; when $\delta(\theta)$ is uniform, γ_{QCF}^* may have minimax properties,[†] and if $w_n(\mathbf{V}, \theta)$ possesses certain symmetry properties, then the Bayes estimator often may be directly determined from the symmetry structure. Not only Bayes extraction, but also those minimizing average equivocation—as in detection—are possible. Also, as in detection, we may consider more sophisticated extraction procedures, for example, sequential estimation and adaptive systems employing dynamic-programming techniques. In all these areas, as far as communication systems are concerned, the search for optimality has just begun. This is particularly so in those cases in which it is meaningful to include constraints. For example, in cost coding it may often be profitable to minimize further the average risk, associated now with signal extraction. Equations (3.8), (3.9) apply here also, as do (3.10)–(3.12), for the more elaborate situations in which the communication link—both transmitter and receiver—is in effect playing a game against the medium, represented by either natural or man-made disturbances.

6. Some Consequences of Optimality

As we have indicated above, a direct and important by-product of each optimization procedure—whether in signal detection or extraction—

[†] For further comments, see [1], Secs. 21.2-1, 21.2-3, 4, and Sec. 21.2-5.

is optimum structure, which is embodied in the appropriate decision rule δ. In binary detection, this is expressed in terms of the generalized likelihood ratio or its logarithm [see (4.17)],

$$T_{R-\mathrm{op}}^{(N)}\{V\} = \log \Lambda_n(\mathbf{V}). \tag{6.1}$$

In extraction one has a variety of structures, embodied in the estimators

$$T_{R-\mathrm{op}}^{(N)}\{V\} = \gamma_\sigma^*(\theta \,|\, \mathbf{V}). \tag{6.2}$$

For threshold operation various canonical forms of optimum structure are possible. In detection of a signal in noise—for example, $H_1:N$ alone, $H_2:S + N$, the "on-off" case—these can be expressed as

$$\log \Lambda_n = B_0 + B_1 \tilde{\mathbf{v}} \mathbf{C}^{(1)} \langle \mathbf{s} \rangle + B_2 \tilde{\mathbf{v}} \mathbf{C}^{(2)} \mathbf{v} + O(a_0^3; s^3; v, v^3, v^3, \cdots), \tag{6.3}$$

where the constants B_0, B_1, B_2 depend of course on the signal and noise statistics and are respectively $O(a_0^0, a_0, a_0^2, \cdots)$, $B_1 = O(a_0)$, $B_2 = O(a_0^2)$, with a_0^2 defined as an input signal-to-noise power ratio, $\mathbf{s}$ is a normalized signal wave form, and $\mathbf{v}$ is a normalized data process. In reception processes for which there is sample certainty, for example, in coherent reception (see Sec. 2), $\langle \mathbf{s} \rangle$ does not vanish, and the predominant threshold structure is that of an averaged, generalized *cross-correlation*† of the received data with a known signal waveform,

$$\log \Lambda_n \big|_{\mathrm{coh}} = B_0' + a_0 B_1' \tilde{\mathbf{v}} \mathbf{C}^{(1)} \langle \mathbf{s} \rangle, \qquad a_0^2 < 1, \tag{6.4}$$

dependent on the square root of the input signal-to-noise power ratio. On the other hand, when the observer does not know the epoch ϵ (see Fig. 1), and ϵ is so distributed that $\langle \mathbf{s} \rangle = 0$—a usual condition in this state of *incoherent observation*—we have an example characterized by an averaged, generalized *autocorrelation*‡ of the received data (v) with itself:

$$\log \Lambda_n \big|_{\mathrm{incoh}} = B_0'' + a_0^2 B_2'' \tilde{\mathbf{v}} \mathbf{C}^{(2)} \mathbf{v}. \tag{6.5}$$

Unlike the coherent case, this is now a nonlinear operation on the data. Moreover, the structure depends on the first power of the signal-to-noise power ratio:

---

† See [1], Sec. 19.4.
‡ See [1], Sec. 19.4.

From a design viewpoint, we must next interpret these generalized averaged cross- and autocorrelations in terms of ordered sequences of realizable elements, both linear and nonlinear. This has recently been done [5],† and typical structures consist of a linear time-invariant, realizable Bayes matched filter of the first kind, and similar to the older type of matched filter [6] for maximizing S/N in the coherent cases. With incoherent reception the structure may consist of time-varying linear and realizable Bayes matched filters of the second and third kinds, followed by zero-memory quadratic or multiplicative devices and terminating in an ideal (i.e., uniformly weighting) post-rectification filter [5]. The details are considered in [5]. Similar structures may also occur in signal extraction. In addition to a threshold structure similar to and sometimes identical with that arising in detection, one often has a further mathematical operation on data so processed, represented by some transcendental function, for example, a Bessel or hypergeometric function, of which the processed data are the argument.‡

Finally, it should be pointed out that the above procedures are not confined to such discrete structures, although they are necessarily formulated in a probabilistic sense in discrete terms, that is, with sample values at discrete times $t_1, \cdots, t_n$ on an interval $(0, T)$. One can quite naturally consider continuous sampling on $(0, T)$ in which the various matrix forms above go over into corresponding integral expressions. The generalized likelihood ratios characteristic of optimum binary detection become generalized likelihood ratio *functionals*, and the various quadratic forms in the threshold development of structure [see (6.3)–(6.5)] likewise become linear or quadratic functionals of $V(t)$, the received data on the interval $(0, T)$. In such instances, also, the technical problems of inverting matrices [the $\mathbf{C}^{(1)}$, $\mathbf{C}^{(2)}$ in (6.3) *et seq.* contain one or more inverse matrices] transform into the corresponding problems of solving linear integral equations of the homogeneous and inhomogeneous varieties. In system structure a similar transformation from the discrete sampling filters to the continuous analogue devices occurs without conceptual change, although, of course, the actual realization of the various optimized circuit elements requires different techniques. Section 19.4-2 of [1] considers some of the conditions under which the passage from the discrete to the continuous state of operation may be analytically performed.

---

† See also [1], Sec. 20.1, 2.

‡ See [1], Sec. 21.3-2, in the case of the incoherent estimation of signal amplitude with a quadratic cost function and narrow-band signal processes. Still other examples may be found in [1], Secs. 21.1 and 21.3, and in [7].

7. Concluding Remarks

In the preceding sections we have described formally and rather in detail some of the roles of optimization in statistical communication theory. The principal applications of optimization techniques here lie in the areas of signal detection, signal extraction, and cost coding, with their concomitant by-products of optimum system structure and comparison with suboptimum systems for these and similar purposes. In its simplest forms, optimization is carried out with respect to the criterion of minimizing an average loss of some sort, whether measured in simple units of cost or in the more intricate units of equivocation, and without constraints. More sophisticated models, appropriate to a variety of special communication situations, introduce one or more constraints, with a resulting increase in the complexity of the analysis. Cost coding offers an important class of such problems. The basic technique for finding solutions is the familiar variational one, which of course is not always adequate for this purpose, since it frequently provides minima when maxima are needed, and vice versa. This difficulty is sometimes overcome by an inspection of the physical model, which often reveals the desired extremum, but in general there appear to be no consistently reliable methods for discovering solutions.

Although our attention has been specifically directed to the single-link communication situations, the fundamental ideas and techniques are by no means restricted to such cases. Frequently a more realistic description of a communication environment requires an extension of fixed-sample models to the variable-sample ones (e.g., sequential detection and estimation). Often, too, simple binary decisions are inadequate: A selection between many alternatives may be required, and we are then confronted with a multiple-alternative detection situation. Moreover, the cost functions in common analytic use (the SCF and QCF of Sec. 5, above) may prove unrealistic, and other, more involved cost assignments must be introduced, with a consequent increase in the technical problems of finding optimum structures (decision rules) in these cases. Extensions to adaptive systems, which in various ways adjust themselves to their changing data environment, put a heavy burden on our optimization methods and require of us new approaches and techniques, as yet only hinted at. One central problem here is that of "reduction of dimensionality," whereby a very large number of raw data elements are to be combined (and in part eliminated) in such a way that nearly all of the pertinent information upon which a decision is to be based may be preserved, with the result that a comparatively small number of effective data elements are then needed to obtain an efficient decision. Simple examples of this are shown above in binary detection, where for optimum performance an explicit operation on the received data V is indicated by the likelihood ratio function or functional. But when we attempt to extend this procedure to multiple decisions under an over-all optimality program, as in dynamic programming, the dimensionality question rapidly reaches a prohibitive state. Hence, other as yet undiscovered techniques for reducing the computational demands on the system must be introduced. Accordingly, it is clear that new optimization methods are required if we are to extend the basic notions and models of an adequate statistical theory of communication to many of the pressing problems of the present and future. Here, then, we have described a framework of approach, successful in its basic aims. The extensions lie before us, with optimization inevitably a central aim, so that not only can we refer actual achievements to their theoretical limiting forms, but with these results to guide us, approach them more closely in practice.

References

1. Middleton, D., *Introduction to Statistical Communication Theory*, Part IV, McGraw-Hill Book Company, Inc., New York, 1960.
2. Middleton, D., and D. Van Meter, "Detection and Extraction of Signals in Noise from Point of View of Statistical Decision Theory," *J. Soc. Indust. Appl. Math.*, Part I, Vol. 3, 1955, pp. 192–253; Part II, Vol. 4, 1956, pp. 86–119.
3. Wald, A., *Statistical Decision Functions*, John Wiley & Sons, Inc., New York, 1950.
4. Bussgang, J. J., and D. Middleton, "Optimum Sequential Detection of Signals in Noise," *Trans. IRE*, PGIT-1, No. 3, December, 1955, p. 1.
5. Middleton, D., "On New Classes of Matched Filters and Generalizations of the Matched Filter Concept," *Trans. IRE*, PGIT-6, No. 3, June, 1960, pp. 349–360.
6. Van Vleck, J. H., and D. Middleton, "A Theoretical Comparison of the Visual, Aural, and Meter Reception of Pulsed Signals in the Presence of Noise," *J. Appl. Phys.*, Vol. 17, 1946, pp. 940–971.
7. Middleton, D., "A Note on the Estimation of Signal Waveform," *Trans. IRE*, PGIT-5, No. 2, June, 1959, p. 86.

Chapter 8

Estimators with Minimum Bias[†]

WILLIAM JACKSON HALL

1. Introduction

In statistical decision theory—and, in particular, in estimation theory —the criterion of optimization is *minimum risk*. An estimator is so chosen that the risk when using it is minimized in some sense—for example, the maximum risk or the average (Bayes) risk may be minimized. On the other hand, some other criterion may be imposed prior to minimization of risk. A requirement of unbiasedness frequently plays such a role.

The thesis set forth in the present chapter is that this popular requirement of unbiasedness—more generally, of minimization of the bias—is a completely separate optimization criterion, alternative to minimization of risk, and that optimization by one of these criteria essentially precludes consideration of the other. Because uniqueness of minimum-bias estimators (in particular, unbiased estimators) is so common, secondary consideration of risk is seldom possible, contrary to usual claims.

An advantage of a minimum-bias criterion is that it may enable determination of estimators without complete specification of a probability model, such as in linear estimation theory, much of which is distribution-free—and loss-function-free as well. Here, a minimum-risk criterion may not be feasible. Our attention will be directed, how-

† This research was supported in part by the U.S. Air Force through the Air Force Office of Scientific Research of the Air Research and Development Command under Contract No. AF 49(638)-261. The author appreciates the assistance of Wassily Hoeffding and Walter L. Smith, who made several helpful suggestions. The author is also grateful for use of the computing facilities of the Research Computing Center of the University of North Carolina.

ever, to problems in which a choice between these criteria is possible.

Although a theory for obtaining and characterizing minimum-risk estimators has been available since the classic papers of Wald, a theory for minimizing bias has not been explicitly set forth except when uniform minimization is possible (which *does* cover most applications). It is thus a second purpose of this chapter to present a theory of minimum-bias estimation, completely parallel to the minimum-risk theory and with possible applicability whenever unbiased estimators do not exist.

A third purpose of this exposition is to emphasize how strong and perhaps unreasonable a requirement of minimization of bias may sometimes be. Not only does it usually produce a unique estimator, leaving no opportunity for secondary consideration of risk, but such an estimator may have some unappealing properties unless care is taken to impose additional restrictions. High risk may well accompany low bias, and conversely. These points are made primarily with reference to an example in which minimum-risk and minimum-bias estimators are compared.

Since these two criteria of minimum bias and minimum risk may be at such cross purposes and yet both have strong intuitive appeal, the final conclusion of this chapter is that some new criterion, heeding both bias and risk but minimizing neither, should be pursued. It is suspected that less stringent requirements on bias would satisfactorily meet all practical needs. A minimum-bias theory is presented at some length to provide a background or a basis for some compromise; it may be desirable to know the extent to which bias *can* be reduced even though we decide to use some other optimization criterion.

We shall first review minimum-risk theory, for the benefit of non-statisticians, and reinterpret the role of bias therein (Sec. 2). We then introduce a minimum-bias theory (Sec. 3), paralleling the risk theory, with the role of risk function replaced by a bias function (absolute, squared, or percentage of absolute bias, for example). We thus introduce estimators that minimize the average or maximum of the bias function. Such estimators are obtained by choosing an unbiased estimator of a suitable approximation to the function to be estimated. Relevant aspects of the theory of approximation, in the Chebyshev† and least-squares senses, are therefore reviewed (Sec. 4). Results bearing considerable conceptual similarity to Wald's results on minimax

---

† It is of interest that Chapter 14 of this book also discusses Chebyshev approximation, used by Kiefer and Wolfowitz in the problem of optimal allocation; to the author's knowledge, Chebyshev approximation has not appeared elsewhere in the statistical literature.

and Bayes estimators are thereby derived (Secs. 4 and 5). Finally, an example is treated (Secs. 6–8), and a variety of minimum-bias and minimum-risk estimators are derived and compared.

2. A Review of Minimum-Risk Estimation

We are concerned with data generated by a probability distribution of which the form is completely known except for specification of a parameter θ lying in a space Θ. We denote the data simply by x, a point in a space $\mathfrak{X}$. It is desired to estimate the value assumed by a numerical function γ of θ; that is, we are to choose an estimator δ, a numerical function on $\mathfrak{X}$, and identify the value $\delta(x)$ of δ at the observed value x with the unknown value $\gamma(\theta)$ of the parametric function γ. We shall not consider here randomized estimators, nonparametric estimation, or sequential experimentation. The Neyman–Pearson theory of testing the null hypothesis that θ is in Θ_0 versus θ in $\Theta - \Theta_0$ can be considered as the special case in which $\gamma(\theta)$ is zero in Θ_0 and unity elsewhere, and $\delta(x) = 0$ is equivalent to acceptance of the null hypothesis; our primary concern, however, will be genuine estimation problems.

Let $L(\delta(x), \gamma(\theta))$ represent the monetary loss incurred by estimating $\gamma(\theta)$ by $\delta(x)$. The expression L is usually taken to be a nonnegative function of the *error of estimate* $\delta - \gamma$—for example, squared error, percentage of absolute error, or *simple loss*—that is, zero if the error is small and unity otherwise. Since the loss is a function of x, it has a probability distribution, the mean value of which we denote by $R(\delta, \theta)$. This is called *risk* when we use the estimator δ and when θ is the true parameter value.

What we shall refer to as *minimum-risk theory* consists of those various approaches to estimation theory, developed by A. Wald and his followers (see [1], [2], [3], or [4]), but with foundations in the works of Bayes, Gauss, and Laplace, which evaluate an estimator primarily on the basis of its associated risk, and which consider estimators with small risk (in some sense) to be good estimators. Since the risk is a function of θ, its minimization must take this into account. Moreover, since uniform minimization is possible only in trivial problems, either some other type of minimization is required, or some conditions must first be imposed to reduce the class of estimators under consideration, or both. In any case, attention is frequently restricted to *admissible estimators*—estimators that cannot be uniformly improved upon in terms of a small or smaller risk.

The two types of minimization commonly considered are given

below. Either of these criteria frequently yields unique admissible estimators.

Minimization of average risk.   This criterion leads to *Bayes estimators*—estimators chosen so that a weighted average (over the parameter space) of the risk function is a minimum. This is particularly appropriate if the parameter itself is a random variable with a known a priori probability distribution, in which case the expected risk is thereby minimized. Alternatively, justification may be made in terms of *rational degrees of belief* or *concern* about the parameter (see [5]).

Minimization of maximum risk.   This criterion leads to *minimax estimators*—estimators chosen so that the maximum (over the parameter space) of the risk function is a minimum. This may be considered a conservative approach whereby, in the absence of specific knowledge about the parameter, one guards against the *least favorable* eventuality. Minimax estimators frequently are Bayes estimators relative to a least favorable weight function (prior distribution), as proved by an application of the *fundamental theorem of the theory of games.*

Other possible criteria are these: Subject to the prior distribution belonging to some specified class, minimize the average risk; or, subject perhaps to some global bounds on the risk, minimize the risk in some sense in some particular locality. Neither of these criteria has been developed here to any extent, except for the latter in the case of hypothesis testing.

Four kinds of restrictions, one or more of which are sometimes imposed to reduce the class of estimators under consideration, are given below; subject to such conditions, the risk may be minimized uniformly or otherwise.

a. Restriction to *linear estimators*, the rationale usually being one of simplicity. Its use is generally restricted to problems of estimating parameters in linear models, in which case a normality assumption further justifies a linearity restriction.

b. Restriction to *invariant* (or *symmetric*) *estimators*. For example, the restriction might be that two statisticians using different units of measurement should obtain equivalent estimates, or that estimators should be symmetric functions of independent and identically distributed observations.

c. Restriction to estimators that are functions only of a sufficient statistic—in fact, a *necessary and sufficient statistic* (if existent). From the risk point of view, nothing is lost by this restriction provided L is convex in δ (and this is not required if randomized estimators are allowed). Restriction to *sufficient estimators* is consistent with the Fisherian concept that a sufficient statistic contains all the relevant information.

d. Restriction to *unbiased estimators*—estimators for which the expected value of the error of estimate is everywhere zero. Such a restriction is usually offered in the guise of preliminary reduction of the class of estimators under consideration, as here, after which risk is minimized. As was shown by Lehmann and Scheffé [6], however, in a wide class of problems there is a unique unbiased estimator depending only on a necessary and sufficient statistic. (P. R. Halmos [7] showed that a symmetry requirement may also lead to unique unbiased estimators.) This condition prevails whenever there are no nontrivial unbiased estimators of zero depending only on a necessary and sufficient statistic [8], [9], in which case the statistic is said to be *complete* [6]. Apparently, few practical (nonsequential) problems fall outside this class. Thus, the condition of unbiasedness is a strong one; its imposition really implies that risk is not even considered, not merely that it is put in a position of secondary importance. To claim that an unbiased estimator has minimum risk (or minimum variance) is usually an empty claim except in comparison with *insufficient* estimators. Thus, restriction to unbiased estimators can be thought of as being outside the minimum-risk theory.

It may be noted that it is unusual for minimax or Bayes estimators to be unbiased. In particular, for squared-error loss, Bayes estimators necessarily are biased [3]. Sometimes no unbiased estimator is even admissible; for example, this is true in the estimation of the variance of a normal distribution with unknown mean and squared-error loss (dividing the sum of squared deviations by $n + 1$ rather than $n - 1$ uniformly reduces the risk). Thus the criteria of unbiasedness and minimum risk frequently are incompatible.

3. An Introduction to Minimum-Bias Estimation

An approach to estimation theory paralleling the minimum-risk theory, with the role of the risk function filled by a *bias function*, is here developed. The bias function is some nonnegative function of the *bias*, the expected error of estimate. Thus, the operations of taking expected value and of applying a nonnegative function are interchanged: Instead of minimizing the expected value of a nonnegative function of the error of estimate as in risk theory, we minimize some nonnegative function of the expected value of the error of estimate. A number of concepts and theorems completely analogous to those in risk theory can be stated for the bias theory, but the mechanics of obtaining *minimum-bias estimators* are quite different.

In minimum-bias theory, in contrast to minimum-risk theory, uniform minimization is frequently possible, leading to unbiased estimators. The theory of unbiased estimation is well established [2], [4], [6]. We shall be concerned with a minimum-bias theory applicable to situations in which no unbiased estimator is available. As one example, in nonsequential binomial problems only polynomials of limited degree in the success probability admit unbiased estimators. As another, if the restriction is made that the range of the estimator be within the range of the function to be estimated, then unbiased estimators are less frequently available (for example, see the second paragraph below and also [2], pp. 3–13). If one desires the simplicity of linear estimators, then some bias may be unavoidable. In the hypothesis-testing case, in which the range of γ is only 0 and 1, unbiased estimators are not available. Thus, four situations can be delineated in which bias is frequently unavoidable: (a) The sample space is finite. (b) The range of the estimator is restricted. (c) The functional form of the estimator is specified. (d) The parametric function to be estimated is discontinuous.

When uniform minimization of bias is not possible, we might look for estimators whose maximum bias (absolute, squared, or relative) or average bias is a minimum, or for estimators with locally small bias in some sense. Such minimum-bias estimators will be considered in the sequel. A. Bhattacharyya [10], in discussing binomial estimation problems, considered estimators with minimum average squared bias and estimators with locally small bias in the sense that all derivatives of the bias vanish at a specified parameter point. A. N. Kolmogorov [11] considered a somewhat different approach to minimum-bias estimation; he suggested finding upper and lower estimators, the bias of the former being everywhere nonnegative and of the latter nonpositive. Thus, one obtains two estimators rather than one, but thereby obtains bounds on the bias of any estimator between the two. No theory was offered, however, for obtaining such estimators in any optimal way. S. H. Siraždinov [12] treated the problem of estimating a polynomial of degree $n + 1$ in the binomial parameter with minimum bias in what we shall call the minimax sense; as noted below, such estimators are unbiased estimators of the Chebyshev approximation of the polynomial to be estimated. Siraždinov noted also that if a constant (the maximum error of approximation) is added to and subtracted from the estimator, one obtains upper and lower estimators as defined by Kolmogorov.

Why should one be concerned with bias? The connotations of the words "bias" and "expected error of estimate" certainly make it appear undesirable to the practitioner. In the sense that the distribution of the error of estimate is centered at zero, an unbiased estimator does allow the sample to "speak for itself"; no prior knowledge or opinion of the experimenter is allowed to influence the estimate. (Other definitions of unbiasedness—for example, in terms of medians rather than expectations—would have similar justifications.) In contrast, minimum-risk (Bayes) estimators may be considered as combinations (sometimes linear) of the best a priori guess and the best information based solely on the sample data. Unbiased estimation thus seems to fit more naturally in a theory of statistical inference in which there may be more reason to consider the estimator as a descriptive statistic, than in a theory of statistical decision in which all prior information and the consequences of the decision taken cannot easily be ignored. It is noteworthy in this regard that if the statistician wishes to limit the parameter space to some subset of its natural range and to limit the range of the estimator accordingly, then no unbiased estimator may be available. For example, the success probability in n Bernoulli trials, if limited to any subset of the unit interval, does not admit an unbiased estimator with range similarly restricted.

When making repeated estimations, another justification for requiring small bias is available; namely, that the average error of estimate should be small, with high probability if repetitions are sufficiently numerous. It may be some consolation to know that the overestimates in some trials tend to be compensated for by underestimates in other trials. † If some consumer loses because of the statistician's overestimate on one trial, however, he is not likely to be consoled by knowing that his losses are compensated for by some other consumer's gain on another trial; such consumers might be more concerned with small risk. The justification for requiring small risk is also founded largely on a long-run interpretation of expectation, applied to loss rather than error, and its justification other than in repeated experimentation is not completely satisfactory. The hypothetical example above illustrates the possible inadequacy of a theory based only on the expected error and not on its variability, as is the case with minimum-bias theory. A slight rewording of the illustration would point out the inadequacy of a theory based only on expected loss (risk).

An additional justification for requiring unbiasedness may be that it usually eliminates the need for specifying a loss function, since it frequently leads to unique sufficient estimators; if unbiased estimators

---

† In a series of election polls, for example, a consistent tendency to over-estimate the strength of any one candidate would seem undesirable.

do not exist, however, then specification of some function analogous to loss (a bias function) will be required in the theory developed here.

Whether or not satisfactory justification is available, it is a fact that statisticians frequently spend great efforts in "correcting for bias," and there seems to be only limited acceptance of any minimum-risk estimators with large biases. For example, suppose x denotes the number of successes in n Bernoulli trials. Then the minimax estimator (squared-error loss) of the success probability θ is $(x + \frac{1}{2}\sqrt{n})/(n + \sqrt{n})$, and although for small samples its risk, $(2\sqrt{n} + 2)^{-2}$, is less than that of the unbiased estimator x/n over a wide range of parameter values, it is apparently difficult to persuade the experimenter of its superiority; this may be due to its large bias, $(1 - 2\theta)/(2\sqrt{n} + 2)$, for probabilities near 0 and 1 (or perhaps to the inappropriateness of squared-error loss).

As another example, division of a sum of squared deviations by $n - 1$ gives an unbiased estimator of σ^2, but, for normal variables, division by $n + 1$ uniformly reduces the risk (squared-error loss) at the expense of introducing a bias of $-2\sigma^2/(n + 1)$. Yet how many pages in textbooks have been allocated to justifying division by $n - 1$? (Questions of whether one really wishes to estimate σ or σ^2 and of the choice of loss function also need to be considered, of course.)

Perhaps some compromise between the risk and the bias approach would be more readily accepted by the practitioners—estimators with risk minimized subject to the bias being within bounds, or, conversely, with bias minimized subject to the risk being within bounds. Such an approach is not new in sample-survey theory (for example, see [13]), where it is sometimes suggested that a small bias may be tolerable if reduction in the mean-square error (risk) is achieved. The minimum-bias theory developed herein is offered as a preliminary step toward such developments. Only an example of such compromise approaches is offered.

Suppose it is desired to obtain a linear minimum-risk (squared-error loss) estimator in the minimax sense of the binomial success probability θ, subject to the absolute bias being everywhere bounded by ρ. We shall assume that $\rho < (2\sqrt{n} + 2)^{-1}$, the maximum absolute bias of the unrestricted minimax estimator. Because of symmetry, it is easy to show that we need only to consider estimators of the form $(x + \alpha)/(n + 2\alpha)$ where $\alpha \geq 0$ (which, incidentally, includes all Bayes estimators relative to a symmetric beta prior distribution). For $\alpha = n\rho/(1 - 2\rho)$, the maximum absolute bias is ρ and the maximum risk is $(1 - 2\rho)^2/4n$, which cannot be further reduced. Thus $\delta = x(1 - 2\rho)/n + \rho$ is the desired estimator.

Before considering minimum-bias estimation in further detail, let us review some relevant aspects of the theory of approximation.

4. Some Aspects of the Theory of Approximation

Let Φ represent a specified system of $n+1$ (finite or infinite) linearly independent bounded and continuous functions $\phi_0, \phi_1, \cdots, \phi_n$ of θ where $\theta \in \Theta$, a subset of the real line or of some other metric space. Let $P = P(\Phi)$ denote the class of all linear combinations

$$p = \sum_{i=0}^{n} a_i \phi_i,$$

where the a_i's are real constants. We call such functions p *generalized polynomials of the system* Φ. For example, suppose Θ is a finite interval and $\phi_i(\theta) = \theta^i$; then P is the class of polynomials of degree n on the interval. Let γ be a bounded and continuous function on Θ but not in P, and let λ be a nonnegative bounded and continuous function on Θ.†

DEFINITION 1. A function $p_0 \in P$ is a *best approximation to* γ *in the minimax* (Chebyshev or uniform norm) *sense* if

$$\sup_{\Theta} \lambda \,|\, p_0 - \gamma \,| \ = \inf_{P} \sup_{\Theta} \lambda \,|\, p - \gamma \,|. \tag{4.1}$$

We shall assume in what follows that $\lambda = 1$; if not, transform the problem by multiplying all other functions defined above by λ before proceeding.

Suppose now that Θ is a closed finite or infinite interval and that any generalized polynomial other than the zero polynomial of the system Φ has at most n roots in Θ (n finite), where a root at which the polynomial does not change sign is counted twice. Then Φ is said to be a *Chebyshev system of functions*. An alternative characterization of a Chebyshev system is that the determinant

$$D(\theta) \equiv \begin{vmatrix} \phi_0(\theta) & \cdots & \phi_n(\theta) \\ \phi_0(\theta_1) & \cdots & \phi_n(\theta_1) \\ \cdots\cdots\cdots\cdots \\ \phi_0(\theta_n) & \cdots & \phi_n(\theta_n) \end{vmatrix}$$

vanishes only at n distinct points $\theta_0, \cdots, \theta_n$ in Θ and that D changes sign in passing through successive θ_i's.

S. Bernstein [14] proved the following result as a generalization of Chebyshev's original work on polynomial approximation: *If* Φ *is a Chebyshev system, then there exists a best approximation to* γ *in the minimax sense; moreover, p_0 is unique, and a necessary and sufficient*

---

† Some of the above restrictions can be relaxed in certain parts of the sequel.

condition for $p = p_0$ is that the number of points where $p - \gamma$ attains its extremum, with alternating signs, be at least $n + 2$.

Various extensions of this result and upper bounds on (4.1) are given by Bernstein, by C. de la Vallée Poussin [15], and by others in more recent publications. Many such results appear in [16], [17], [18], and [19], for example, and in current publications by G. G. Lorentz, J. L. Walsh, and T. S. Motzkin. For the polynomial case, the fundamental theorem of Weierstrass should be mentioned; it states that by choosing n sufficiently large, (4.1) can be made arbitrarily small.

If the various functions are differentiable, the function p_0 may be obtained as follows, though this method may be untractable analytically:

Let $b = p - \gamma$ and let $\rho = \sup_\Theta |b|$ when $p = p_0$. Let $\theta_0, \theta_1, \cdots, \theta_{n+1}$ be $n + 2$ successive points in Θ at which ρ is achieved with alternating signs by b, and let $p_0 = \sum a_i \phi_i$. Then with $b' = db/d\theta$, the system

$$b(\theta_i) = \pm \rho(-1)^i, \qquad b'(\theta_i) = 0, \qquad i = 0, 1, \cdots, n + 1, \quad (4.2)$$

gives $2n + 4$ equations in the $2n + 4$ unknowns $\theta_0, \cdots, \theta_{n+1}$, $a_0, \cdots, a_n$, and ρ. [If θ_0 or θ_{n+1} is an endpoint of Θ, then (4.2) need not hold at $i = 0$ or $n + 1$; also, more than $n + 2$ points may be required.]

In particular, if $\phi_i(\theta) = \theta^i$, $(i = 0, 1, \cdots, n)$ and $\gamma(\theta) = \theta^{n+1}$, it can be shown that *Chebyshev polynomials* can readily be used to obtain the best approximation to γ in the minimax sense (see, for example, [20]). If, instead, γ is any function with a series expansion throughout Θ, the expansion of γ in Chebyshev polynomials, truncated after $n + 1$ terms, will yield "almost" the best polynomial approximation to γ. Even if γ has no valid expansion, the tau-method of C. Lanczos [20] may lead to an approximate solution, again using Chebyshev polynomials.

A generalization of Bernstein's theorem for more general parameter spaces has been given by J. Bram [21]. He assumes that Θ is a locally compact space. Then a necessary and sufficient condition that $\sup |b|$ be a minimum is that, for some $r \leq n$, there exist $r + 2$ points θ_0, $\theta_1, \cdots, \theta_{r+1}$ in Θ such that the $(n + 1) \times (r + 2)$ matrix $[\phi_i(\theta_j)]$ has rank $r + 1$ and such that if the subscripts are assigned so that the first $r + 1$ rows are independent, and a_i is the sign of the cofactor of a_i in

$$\begin{vmatrix} a_0 & \cdots & a_{r+1} \\ \phi_0(\theta_0) & \cdots & \phi_0(\theta_{r+1}) \\ \cdots & \cdots & \cdots \\ \phi_r(\theta_0) & \cdots & \phi_r(\theta_{r+1}) \end{vmatrix}$$

then $b(\theta_i) = \pm a_i \rho$ for all i with $a_i \neq 0$.

DEFINITION 2. Let ξ be a probability measure on the Borel sets $\{\omega\}$ of Ω and assume γ and the ϕ_i's to be square-integrable. A function $p_\xi \in P$ is said to be a *best approximation to γ in the least squares sense relative to ξ* if

$$N(\xi) \equiv \int \lambda^2 (p_\xi - \gamma)^2 \, d\xi = \inf_P \int \lambda^2 (p - \gamma)^2 \, d\xi.$$

Again, we shall assume for simplicity that $\lambda \equiv 1$. Analogous developments, using powers other than two, are also possible.

Let $p_0, \cdots, p_n$ constitute an orthonormal set in P w.r.t. the measure ξ (see Szegö [22] or Achieser [16], for example); i.e.,

$$\int p_i p_j \, d\xi = \delta_{ij} \quad \text{(Kronecker } \delta_{ij}).$$

Such a set always exists and can be constructed from Φ. Denoting $c_i = \int p_i \gamma \, d\xi$, then it is well known that $p_\xi = \sum c_i p_i$ is a best approximation to γ in the least-squares sense, and, moreover, that

$$N(\xi) = \int \gamma^2 \, d\xi - \sum c_i^2.$$

As a special case, suppose $\phi_i(\theta) = \theta^i$. Then $p_0, \cdots, p_n$ are the orthonormal polynomials associated with ξ, and p_ξ is the best polynomial approximation to γ (in the sense of least squares).

At this point, we note the analogy in Definitions 1 and 2 with the minimax and Bayes solutions to problems in the theory of games, as treated, for example, by Wald [1] and Blackwell and Girshick [3]. We need only to replace the role of the risk function or the expected payoff in decision or game theory by $\lambda|p - \gamma|$ or its square.

DEFINITION 3. A probability measure ξ_0 on $\{\omega\}$, the Borel sets of Θ, is said to be *least favorable* if $N(\xi_0) = \sup N(\xi)$, where the supremum is over all probability measures on $\{\omega\}$.

Moreover, the fundamental theorem of the theory of games is applicable, so that, with suitable compactness assumptions [1], it readily follows that $p_0 = p_{\xi_0}$, that is, that any best approximation in the minimax sense is also a best approximation in the least-squares sense relative to a least favorable distribution. We need only to note that the operations of squaring and taking sup's (or inf's) can be interchanged when applied to the function $\lambda|p - \gamma|$. Thus there exists a norm $\int \lambda^2 (p - \gamma)^2 d\xi_0$ minimized by the same p_0 that minimizes the norm $\sup \lambda|p - \gamma|$.

Other decision-theory or game-theory results also carry over. For example, a sufficient condition for ξ to be least favorable is that it assign probability 1 to a subset of Θ throughout which

$$\lambda \left| \, p - \gamma \, \right| \; = \sup_{\Theta} \lambda \left| \, p_\xi - \gamma \, \right| .$$

Also, with weaker compactness assumptions, a sequence of least-squares approximations relative, respectively, to a sequence of distributions having certain limit properties will yield a minimax approximation, analogous to Bayes solutions in the wide sense. Precise theorems have not been stated here because of the perfect analogy with those published elsewhere.

The possible relevance of the fundamental theorem here, analogous to its other applications, is that constructive methods for finding best approximations in the least-squares sense are quite generally available whereas approximation in the minimax sense usually is more difficult. There seems, however, to be no constructive, or even intuitive, way of finding least favorable distributions.

As indicated by Bernstein's theorem, ξ_0 will frequently assign probability 1 to a finite point set. In this case a result somewhat similar to the one given above was stated by De la Vallée Poussin [15] and also by J. L. Walsh [23] for the case of polynomial approximation. Consider a set E in Θ consisting of $n + 2$ points and let p_E denote the minimax polynomial approximation on E with maximum absolute error ρ_E. Call E_0 least favorable if ρ_{E_0} is a maximum over all possible E. De la Vallée Poussin proved that $p_0 = \rho_{E_0}$, that is, that the minimax approximation is the minimax approximation on a *least favorable point set*. It is simple to obtain p_{E_0} once E_0 has been found. Only some iterative techniques, however, seem to be available for finding E_0.

Other definitions of best approximation are possible; for example, one might choose p so that $\lambda \left| \, p - \gamma \, \right|$ is minimized in some sense in the neighborhood of θ_0. For example, if γ possesses a valid series expansion in Θ at θ_0, and P is the class of nth-degree polynomials, then one might approximate γ by a truncated expansion at θ_0 lying in P. Alternatively, one may limit consideration to certain classes of polynomials having relevance in the particular problem and look for approximation within this class. Examples of each of these will be mentioned below.

5. Theory of Minimum-Bias Estimation

The problem we consider is that of estimating a numerical function γ of a parameter θ that indexes the family of probability distributions assumed to generate the sample point x. Extensions to more general situations are possible.

We say a numerical parametric function is estimable if there exists an unbiased estimator of it. (Since we are concerned with situations in which no unbiased estimator is available, our subject is the anomalous one of estimating nonestimable functions!) Our approach is to approximate γ by an estimable function and then estimate γ by an unbiased estimator of its approximating function. Bounds on the error of approximation, derived in approximation theory, yield bounds on the bias of estimators. As noted in Section 3, if the maximum error of approximation is added to and subtracted from the estimator, one obtains upper and lower estimators of γ in the Kolmogorov sense.

Since all functions that are estimable are also estimable by functions of sufficient statistics, restriction to estimators depending only on sufficient statistics may be made, if desired, with no resultant increase in bias.

We consider a system Φ of estimable functions generating a class $P(\Phi)$ of functions that clearly are also estimable. Let $\mathfrak{D} = \mathfrak{D}(\Phi)$ be the class of unbiased estimators of functions p in $P(\Phi)$. For $\delta \in \mathfrak{D}$, we denote

$$\mathcal{E}_\theta \delta = p_\delta(\theta) \in P.$$

Then $p_\delta - \gamma$ is the bias b of δ as an estimator of γ.

An estimator δ_0 is said to be a *minimum-bias estimator* of γ in the *minimax sense* if its expectation p_{δ_0} is the best approximation to γ in the minimax sense. For example, with λ identically unity, δ_0 minimizes the maximum absolute bias; for $\lambda = |\gamma|^{-1}$, if finite, δ_0 minimizes the maximum relative or percentage bias. The methods of the previous section are available for finding such estimators or approximations to them.

An estimator δ_ξ is said to be a *minimum-bias estimator of γ in the least-squares sense relative to ξ* if p_ξ is the best approximation to γ in the least-squares sense relative to ξ. Thus, δ_ξ minimizes the expected *quadratic bias* $\lambda^2(p - \gamma)^2$ relative to an a priori distribution over the parameter space. Averaging of other bias functions could be considered analogously.

If $p_0, p_1, \cdots, p_n$ constitute an orthonormal basis for P, then δ_ξ is equal to $\sum c_i \delta_i$, where the c_i's were defined previously and the δ_i's are unbiased estimators of the p_i's. In the case of orthonormal polynomials associated with ξ, δ is an unbiased estimator of the best polynomial approximation to γ in the least-squares sense.

As in risk theory, minimum-bias estimators in the minimax sense are also (under appropriate compactness assumptions) minimum-bias estimators in the least-squares sense relative to a least favorable prior distribution. It is yet to be demonstrated, however, that this result is of any practical significance in minimum-bias theory.

As noted previously, best approximations frequently are unique, and—if confined to functions of a necessary and sufficient statistic—the corresponding estimators of these approximations also often are unique (whenever the said statistic is *complete*). Thus, minimum-bias estimators will frequently be unique, and, as in the case of unbiased estimators, there is no further room for minimization of risk. Again, such estimators need not be admissible in the risk sense.

Estimators with small local bias might also be considered. In binomial estimation problems, Bhattacharyya [10] suggests the unbiased estimation of the truncated Taylor expansion of γ at a point θ_0 as such an estimator. Thus γ and the expected values of the estimator coincide at θ_0, as do a maximal number of their derivatives at θ_0. C. R. Blyth [24] presented such an estimator for the information measure in a multinomial distribution, though he chose θ_0 outside Θ except in the binomial case. He expanded about the point with all probabilities equal to $\frac{1}{2}$, whereas in fact all probabilities should add to unity. He compared this "low bias" estimator with the minimax and maximum-likelihood estimators.

As exemplified in Section 6, fitting so many derivatives at θ_0 may lead to a very poor fit elsewhere; perhaps a better criterion would be to fit one or two derivatives at θ_0 and use any remaining indeterminacy to ensure a satisfactory fit elsewhere.

For reasons of convenience, one might restrict attention to certain types of polynomial approximations to γ and use unbiased estimators of the approximating function. It is interesting to note that if x is binomially distributed with parameters n and θ, then restriction to Bernstein polynomial approximations (G. G. Lorentz [25]) of the function γ leads to maximum-likelihood estimation of γ; that is, the maximum-likelihood estimator of γ is the unbiased estimator of the Bernstein approximation to γ. Results concerning the error of approximation by Bernstein polynomials thus apply to the bias of maximum-likelihood estimators.

Admissibility, in terms of bias rather than of risk, could also be considered and various complete class theorems derived just as with corresponding theorems in risk theory. Some asymptotic results also are possible. For example, if for samples of size n, polynomials of degree n are estimable, then the maximum absolute bias in estimating a continuous function can be made arbitrarily small by choosing n sufficiently large—a rewording of Weierstrass' theorem. It should be noted, however, that bias is reduced with increasing sample size only if the class of estimable functions is increased correspondingly. If the class of estimable functions remains the class of polynomials of degree m, then no increase in sample size can affect the bias.

6. Estimation of a Root of the Binomial Parameter

To exemplify the foregoing theory, we consider the estimation of an integral root r of a binomial parameter θ; thus, $\gamma(\theta) = \theta^{1/r}$, Θ is the unit interval, and r is an integer > 1. Such a problem may arise, for example, in one of the following ways:

(a) Independently, r shots are fired at a target, each having a probability $1 - \gamma$ of hitting it. If one or more shots hit the target, it is destroyed, so all that is observed is whether or not the target is destroyed. Thus the target is destroyed with a probability $(1 - \gamma^r)$. In n repetitions, let x represent the number of targets not destroyed; then x is binomially distributed with parameters n and $\theta = \gamma^r$. It is necessary to estimate $\gamma = \theta^{1/r}$, which is the probability of an individual miss.

(b) Each of a series of independent specimens has probability $1 - \gamma$ of being "positive" (e.g., tests of blood samples for the presence of an antigen). Specimens are pooled into batches and a single test performed for each of n batches of r specimens. The number of negative tests is then binomially distributed with parameters n and $\theta = \gamma^r$. It is required to estimate $\gamma = \theta^{1/r}$, the probability of a "negative" specimen. (Such a testing procedure may be recommended if the tests are expensive and γ is close to unity.)

Extensions to cases with varying batch sizes are also of interest, but they are considerably more complicated and therefore will not be considered here.

Only polynomials in θ of degree n or less are estimable.† We thus seek a polynomial approximation to $\theta^{1/r}$ and use the unique (since the binomial family is complete) unbiased estimator of it. Reference should be made to Bhattacharyya [10], who considered binomial estimation problems in general (also Lehmann [2]).

Any function $\delta(x)$, defined for $x = 0, 1, \cdots, n$, can be expressed in the form

$$\delta(x) = a_0 + a_1 \frac{x}{n} + a_2 \frac{x_{(2)}}{n_{(2)}} + \cdots + a_n \frac{x_{(n)}}{n_{(n)}}, \qquad (6.1)$$

where $x_{(i)}$ denotes $x(x - 1) \cdots (x - i + 1)$, and similarly for $n_{(i)}$. This form is convenient for finding the expectation of the function, since $\mathcal{E} x_{(i)}/n_{(i)} = \theta^i$ $(i = 1, 2, \cdots, n)$. By standard finite-difference methods [10], the coefficients in (6.1) may be obtained as

$$a_i = \binom{n}{i} \Delta^i \delta(0),$$

where

---

† The value γ can be estimated without bias by inverse sampling; see [26].

$$\Delta^i \delta(0) = \sum_{j=0}^{i} (-1)^{i-j} \binom{i}{j} \delta(j).$$

More conveniently, in matrix form with

$$\delta = (\delta(0), \delta(1), \cdots, \delta(n))',$$

the coefficient vector $\mathbf{a} = (a_0, a_1, \cdots, a_n)'$ may be found from

$$\boldsymbol{\alpha} = A\delta, \tag{6.2}$$

where $\boldsymbol{\alpha}$ has coordinates $a_i/\binom{n}{i}$ and A is a square matrix of order $n + 1$ with elements $a_{ij} = (-1)^{i-j}\binom{i}{j}$ for $i \geq j$ and 0 for $i < j$. Inversely $A^{-1} = (a^{ij})$, where $a^{ii} = \binom{i}{j}$ for $i \geq j$ and 0 for $i < j$, so that the values of $\delta(x)$ corresponding to a specified coefficient vector $\mathbf{a}$ may be found from

$$\delta = A^{-1}\boldsymbol{\alpha}, \tag{6.3}$$

a formula equivalent to (6.1). This relation defines the unbiased estimator of the polynomial

$$p(\theta) = \sum_{i=0}^{n} a_i \theta^i.$$

Thus, any estimator $\delta(x)$ is a polynomial in x of degree at most n. Unless some other convenient functional form is available, as is the case with the maximum-likelihood estimator and certain Bayes estimators, the expression for the estimator will be ponderous if n is not small. Thus it is of interest to consider polynomial estimators of small degree $m(\leq n)$ in x; their expectations are then polynomials of degree m in θ.

In the following section, we illustrate the techniques of obtaining various kinds of minimum-bias estimators of $\gamma = \theta^{1/r}$; the maximum-likelihood estimator and various minimum-risk estimators are also considered for comparison. Wherever appropriate and possible, we consider linear, quadratic, and nth-degree estimation ($m = 1, 2, n$) for general n and r—in particular, for $r = 2$ or 5, and $n = 5$. The usual approach is to derive $p(\theta)$, the expectation of the estimator (which frequently is independent of n); the corresponding δ can then be obtained from (6.3). The coefficients in p and the values of δ appear in Tables 1 ($r = 2, n = 5$) and 2 ($r = n = 5$).

We thus take Φ as the powers of θ from 0 through $m(\leq n)$ and $P(\Phi)$ as polynomials δ in x of degree $\leq m$. We restrict attention to absolute bias $|p - \gamma|$ and quadratic bias $(p - \gamma)^2$ as bias functions, always taking $\lambda = 1$.

The variance of estimators δ can be found directly by using the values

TABLE 1. ESTIMATION OF $\sqrt{\theta}$ FROM A SAMPLE† OF 5 ($r=2$, $n=5$)

Criterion	Type of estimation	Value of $\delta(x)$ at $x=$						Value of a_i for $p = \mathcal{E}\delta = \Sigma a_i \theta^i$				
		0	1	2	3	4	5	a_0	a_1	a_2	a_3	a_4
		Linear Estimators										
Minimum bias	minimax bias	0.125	0.325	0.525	0.725	0.925	1.125	0.1250	0.6667			
	r-method approximate	0.333	0.467	0.600	0.733	0.867	1	0.3333	0.8000			
	least squares (uniform ξ)	0.267	0.427	0.587	0.747	0.907	1.067	0.2667	0.5000			
	local at $\theta = 1$	0.5	0.6	0.7	0.8	0.9	1	0.5000	0.5000			
	local at $\theta = \frac{1}{2}$	0.354	0.495	0.636	0.778	0.919	1.061	0.3536	0.7071			
Minimum risk	minimax risk	0.206	0.372	0.538	0.704	0.870	1.035	0.2060	0.8294			
	Bayes (uniform ξ)	0.381	0.495	0.610	0.724	0.838	0.952	0.3810	0.5714			
		Quadratic Estimators										
Minimum bias	minimax bias	0.068	0.454	0.733	0.906	0.972	0.932	0.0676	1.9303	-1.0656		
	r-method approximate	0.158	0.411	0.621	0.789	0.916	1	0.1579	1.2632	-0.4211		
	least squares (uniform ξ)	0.171	0.446	0.663	0.823	0.926	0.971	0.1714	1.3714	-0.5714		
	local at $\theta = 1$	0.375	0.525	0.663	0.788	0.900	1	0.3750	0.7500	-0.1250		
	local at $\theta = \frac{1}{2}$	0.265	0.477	0.654	0.795	0.902	0.972	0.2652	1.0607	-0.3536		
Minimum risk	Bayes (uniform ξ)	0.347	0.502	0.637	0.751	0.845	0.918	0.3469	0.7755	-0.2041		
		nth-Degree Estimators										
Maximum likelihood		0	0.447	0.632	0.775	0.894	1	0	2.2361	-2.6197	2.1887	-0.9904	0.1853
Minimum bias	r-method approximate	0.064	0.701	0.488	0.854	0.891	1	0.0637	3.1850	-8.4934	14.2690	-11.6481	3.6239
	least squares (uniform ξ)	0.084	0.671	0.476	0.906	0.856	1.007	0.0839	2.9371	-7.8322	14.0979	-12.5874	4.3077
	local at $\theta = 1$	0.246	0.492	0.656	0.788	0.900	1	0.2461	1.2305	-0.8203	0.4922	-0.1758	0.0273
	lower at $\theta = 1$	0	0.492	0.656	0.788	0.900	1	0	2.4609	-3.2813	2.9531	-1.4063	0.2734
	local at $\theta = \frac{1}{2}$	0.174	0.522	0.638	0.800	0.890	1.008	0.1740	1.7401	-2.3202	2.7842	-1.9887	0.6187
Minimum risk	Bayes (uniform ξ)	0.341	0.511	0.639	0.746	0.839	0.923	0.3410	0.8525	-0.4262	0.2131	-0.0666	0.0093
	Bayes (beta ξ, $\alpha = \beta = 2$)	0.449	0.562	0.655	0.737	0.811	0.878	0.4493	0.5616	-0.1872	0.0702	-0.0176	0.0020

† For the minimum-bias estimators, only the values of $\delta(x)$ and the MSE's (mean-square errors) depend on n (RMSE is root-mean-square error).

TABLE 2. ESTIMATION OF $\theta^{1/5}$ FROM A SAMPLE† OF 5 ($r=5$, $n=5$)

Criterion	Type of estimation	Value of $\delta(x)$ at $z =$						Value of a_i; where $p = \mathcal{E}\delta = \Sigma a_i \theta^i$					
		0	1	2	3	4	5	a_0	a_1	a_2	a_3	a_4	a_5
Linear Estimators													
Minimum bias	minimax bias	0.267	0.467	0.667	0.867	1.067	1.267	0.2675	1				
	r-method approximate	0.667	0.733	0.800	0.867	0.933	1	0.6667	0.3333				
	least squares (uniform ξ)	0.606	0.697	0.788	0.879	0.970	1.061	0.6061	0.4545				
	local at $\theta = 1$	0.800	0.840	0.880	0.920	0.960	1	0.8000	0.2000				
	local at $\theta = \frac{1}{3}$	0.580	0.725	0.870	1.015	1.160	1.305	0.5798	0.7248				
Minimum risk	minimax risk	0.293	0.493	0.693	0.893	1.093	1.293	0.2929	1				
	Bayes (uniform ξ)	0.671	0.736	0.801	0.866	0.931	0.996	0.6710	0.3247				
Quadratic Estimators													
Minimum bias	minimax bias	0.207	0.781	1.126	1.243	1.132	0.793	0.2071	2.8686	−2.2828			
	r-method approximate	0.474	0.663	0.811	0.916	0.979	1	0.4737	0.9474	−0.4211			
	least squares (uniform ξ)	0.511	0.716	0.864	0.955	0.989	0.966	0.5114	1.0227	−0.5682			
	local at $\theta = 1$	0.720	0.792	0.856	0.912	0.960	1	0.7200	0.3600	−0.0800			
	local at $\theta = \frac{1}{3}$	0.522	0.783	0.899	0.870	0.696	0.377	0.5218	1.3046	−1.4496			
Minimum risk	Bayes (uniform ξ)	0.637	0.743	0.828	0.893	0.938	0.962	0.6372	0.5276	−0.2029			
nth-Degree Estimators													
Maximum likelihood		0	0.725	0.833	0.903	0.956	1	0	3.6239	−6.1701	5.7957	−2.7950	0.5455
Minimum bias	r-method approximate	0.331	1.159	0.515	1.048	0.947	1	0.3311	4.1384	−14.7144	26.4860	−22.3040	7.0629
	least squares (uniform ξ)	0.386	1.061	0.536	1.125	0.880	1.013	0.3857	3.3753	−12.0009	23.1466	−21.3174	7.4256
	local at $\theta = 1$	0.613	0.766	0.851	0.912	0.960	1	0.6129	0.7661	−0.6810	0.4378	−0.1613	0.0255
	lower at $\theta = 1$	0	0.766	0.851	0.912	0.960	1	0	3.8304	−6.8096	6.5664	−3.2256	0.6384
	local at $\theta = \frac{1}{3}$	0.444	0.999	0.321	2.374	−3.486	15.321	0	2.7762	−12.3386	39.6599	−73.0578	57.8374
Minimum risk	Bayes (uniform ξ)	0.629	0.755	0.831	0.886	0.931	0.968	0.6294	0.6294	−0.5035	0.3021	−0.1057	0.0161
	Bayes (beta ξ, $\alpha = \beta = 2$)	0.716	0.788	0.841	0.883	0.920	0.949	0.7164	0.3582	−0.1910	0.0860	−0.0241	0.0030

† For the minimum-bias estimators, only the values of $\delta(x)$ and the MSE's (mean-square errors) depend on n (RMSE is root-mean-square error).

of δ and tables of the binomial probability distribution, or alternatively by using a formula given by Bhattacharyya:

$$\sum_{i=1}^{n}\left(\frac{d^i p}{d\theta^i}\right)^2 \frac{\theta^i (i-\theta)^i}{i!\, n_{(i)}},$$

where the ith derivative of p is $\sum_{j=1}^{n} j_{(i)} a_j \theta^{j-i}$. The mean-square error (risk for squared-error loss) is obtained by adding the square of the bias to the variance.

Summary comparisons of the biases and risks of the various estimators can be made from Tables 3 and 4, which show the zeros of the bias $b(\theta)$, the signs of the values assumed by b, the maximum absolute bias ρ and the θ-value(s) at which the maximum occurs, the square root of the average squared bias, the maximum value of the root-mean-square error $\sqrt{R}(\theta)$ and the θ-value(s) at which the maximum occurs, and the square root of the average mean-square error. Figures 1, 2, and

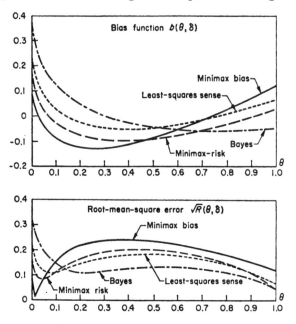

Fig. 1. Bias and risk (RMS error) functions for minimum-bias and minimum-risk linear estimators of $\sqrt{\theta}$ ($n=5$).

3 present the bias function and the root-mean-square (RMS) error function for some of the estimators.†

---

† The tables and figures were prepared with the assistance of students learning to program the Univac 1105.

TABLE 3. ESTIMATION OF $\sqrt{\theta}$ FROM A SAMPLE† OF 5 ($r=2$, $n=5$)

Criterion	Type of estimation	Bias = 0 at θ =	Sign of bias	ρ = max|bias| occurs at θ =	ρ = max|bias| bias =	Square root of ∫(bias)² dθ	Max RMS error at θ =	Max RMS error RMSE	Square root of ∫ (MSE) dθ
Linear Estimators									
Minimum bias	minimax bias	0.021, 0.729	±	0, 0.25, 1	±0.125‡	0.085	0.391	0.244	0.202
	τ-method approximate	0.25, 1	±	0	±0.333	0.061	0	0.333	0.136
	least squares (uniform ξ)	0.149, 0.747	+	0	±0.267	0.047‡	0	0.267	0.153
	local at θ = 1	0.5	+	0	±0.500	0.129	0	0.500	0.158
	local at θ = ⅓		+	0	±0.354	0.068	0	0.354	0.146
Minimum risk	minimax risk	0.070, 0.887	±	0	±0.206	0.066	0, 0.446	0.206‡	0.165
	Bayes (uniform ξ)	0.314	±	0	±0.381	0.081	0	0.381	0.132‡
Quadratic Estimators									
Minimum bias	minimax bias	0.006, 0.292, 0.886	±	0, 0.081, 0.604, 1	±0.068‡	0.047	0.236	0.279	0.196
	τ-method approximate	0.047, 0.718, 1	±	0	±0.158	0.031	0.305	0.215	0.162
	least squares (uniform ξ)	0.070, 0.432, 0.866	+	0	±0.171	0.020‡	0.283	0.215	0.158
	local at θ = 1	1	+	0	±0.375	0.076	0	0.375	0.138
	local at θ = ⅓	0.5	±	0	±0.265	0.039	0	0.265	0.139
Minimum risk	Bayes (uniform ξ)	0.358	±	0	+0.347	0.074	0	0.347	0.130‡
nth-Degree Estimators									
Maximum likelihood		0, 1	−	0.069	−0.120	0.057	0.165	0.284	0.196
Minimum bias	τ-method approximate	0.008, 0.122, 0.379, 0.669, 0.947, 1	±	0	+0.064	0.007	0.129	0.305	0.207
	least squares	0.018, 0.127, 0.330, 0.579, 0.810, 0.962	±	0	+0.084	0.005‡	0.136	0.284	0.205
	local at θ = 1	1	+		+0.246	0.034	0.154	0.246	0.142
	lower at θ = 1	0, 1	−	0.055	−0.109	0.045		0.292	0.195
	local at θ = ⅓	0.5	±		+0.174	0.017	0.197	0.205	0.154
Minimum risk	Bayes (uniform ξ)	0.367	±	0	+0.341	0.074	0	0.341	0.130‡
	Bayes (beta ξ, α = β = 2)	0.447	±	0	+0.449	0.116	0	0.449	0.140

† For the minimum-bias estimators, only the values of δ(x) and the MSE's (mean-square errors) depend on n (RMSE is root-mean-square error).

‡ Minimum value among all estimators of the same degree.

TABLE 4. ESTIMATION OF $\theta^{1/5}$ FROM A SAMPLE† OF 5 ($r=5$, $n=5$)

Criterion	Type of estimation	Bias = 0 at θ =	Sign of bias	ρ = max|bias| occurs at θ =	ρ = max|bias| bias =	Square root of ∫(bias)² dθ	Max RMS error at θ =	Max RMS error RMSE	Square root of ∫(MSE) dθ
Linear Estimators									
Minimum bias	minimax bias	0.001, 0.650	±	0, 0.134, 1	±0.267‡	0.178	0.188	0.314	0.255
	τ-method approximate	0.224, 1	+	0	+0.667	0.062	0	0.667	0.087
	least squares (uniform ξ)	0.130, 0.736	+	0	+0.606	0.051‡	0	0.606	0.098
	local at θ = 1	1	+	0	+0.800	0.112	0	0.800	0.117
	local at θ = ½	0.2	+	0	+0.580	0.143	0	0.580	0.195
Minimum risk	minimax risk	0.002, 0.614	±	0, 1	+0.293	0.170	0, 0.195, 1	0.293‡	0.255
	Bayes (uniform ξ)	0.231	+	0	+0.671	0.063	0	0.671	0.087‡
Quadratic Estimators									
Minimum bias	minimax bias	0.001, 0.229, 0.872	±	0, 0.039, 0.558, 1	±0.207‡	0.144	0.151	0.374	0.290
	τ-method approximate	0.033, 0.707, 1	+	0	+0.474	0.045	0	0.474	0.114
	least squares (uniform ξ)	0.061, 0.420, 0.862	+	0	+0.511	0.029‡	0	0.511	0.105
	local at θ = 1	1	+	1	+0.720	0.079	0	0.720	0.095
	local at θ = ½	0.2	+	1	−0.623	0.235	1	0.623	0.278
Minimum risk	Bayes (uniform ξ)	0.242	+	0	+0.637	0.055	0	0.637	0.083‡
nth-Degree Estimators									
Maximum likelihood		0, 1	−	0.030	−0.393	0.155	0.072	0.498	0.271
	τ-method approximate	0.006, 0.115, 0.375, 0.664, 0.947, 1	+	0	+0.331	0.016	0.135	0.389	0.265
	least squares	0.016, 0.123, 0.325, 0.576, 0.808, 0.961	+	0	+0.386	0.011‡	0	0.386	0.239
Minimum bias	local at θ = 1	1	+	0	+0.613	0.046	0	0.613	0.086
	lower at θ = 1	0, 1	−	0.028	−0.387	0.146	0.072	0.502	0.271
	local at θ = ½	0.2	+	1	+14.3	3.77	1	14.3	0.617
Minimum risk	Bayes (uniform ξ)	0.260	±	0	+0.629	0.054	0	0.629	0.083‡
	Bayes (beta ξ, α = β = 2)	0.409	+	0	+0.716	0.081	0	0.716	0.092

† For the minimum-bias estimators, only the values of δ(x) and the MSE's (mean-square errors) depend on n (RMSE is root-mean-square error).

‡ Minimum value among all estimators of the same degree.

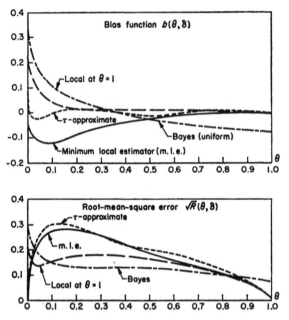

Fig. 2. Bias and risk (RMS error) functions for various nth-degree estimators of $\sqrt{\theta}$ ($n = 5$).

Upper (lower) estimators can be obtained from any one of the estimators given by subtracting from it the minimum (maximum) bias, or $-(+)\rho$. For example, the minimax-bias upper linear estimator of $\sqrt{\theta}$ is $x/n + \frac{1}{4}$; and the lower linear estimator is x/n (since $\rho = \frac{1}{8}$); each has maximum absolute bias of $\frac{1}{4}$. Any other upper or lower linear estimator would have a greater maximum absolute bias.

7. Various Estimators of a Root of p

Maximum-Likelihood Estimators

The maximum-likelihood estimator $\delta(x)$ of $\gamma = \theta^{1/r}$ is $(x/n)^{1/r}$, which, for $r = 2$, $n = 5$, yields, by (6.2),

$$
\begin{bmatrix} a_0 \\ a_1/5 \\ a_2/10 \\ a_3/10 \\ a_4/5 \\ a_5 \end{bmatrix} = \begin{bmatrix} 1 & 0 & 0 & 0 & 0 & 0 \\ -1 & 1 & 0 & 0 & 0 & 0 \\ 1 & -2 & 1 & 0 & 0 & 0 \\ -1 & 3 & -3 & 1 & 0 & 0 \\ 1 & -4 & 6 & -4 & 1 & 0 \\ -1 & 5 & -10 & 10 & -5 & 0 \end{bmatrix} \begin{bmatrix} 0 \\ \sqrt{.2} \\ \sqrt{.4} \\ \sqrt{.6} \\ \sqrt{.8} \\ 1 \end{bmatrix}. \tag{7.1}
$$

Therefore $p(\theta) = \mathcal{E}\delta(x) = \sum a_i\theta^i = 2.24\theta - 2.62\theta^2 + 2.19\theta^3 - 0.99\theta^4 + 0.19\theta^5$. For $r = 5$, $n = 5$, we obtain, substituting $\frac{1}{5}$ powers for square roots in (7.1), $p(\theta) = 3.62\theta - 6.17\theta^2 + 5.80\theta^3 - 2.80\theta^4 + 0.5\bar{5}\theta^5$. It should be borne in mind that these polynomials are Bernstein approximations to γ.

In all instances, the bias is zero at the extremes and negative elsewhere (see Figs. 2 and 3 and the tables). The absolute bias, of course, would decrease with increasing n, as would the RMS error.

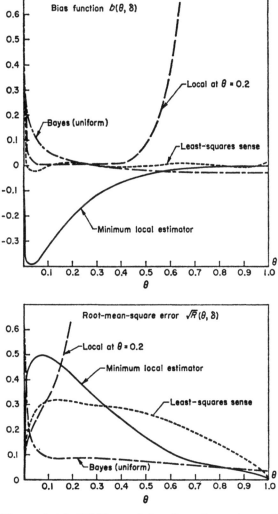

Fig. 3. Bias and risk (RMS error) functions for various nth-degree estimators of $\theta^{1/5}$ ($n=5$).

Minimax-Bias Estimators

The minimum-bias estimator in the minimax sense, $\delta_0(x)$, is the unbiased estimator of the minimax polynomial approximation to $\theta^{1/r}$. As indicated earlier, $2m + 2$ nonlinear equations in $2m + 2$ unknowns ($\theta_0 = 0$ and $\theta_{m+1} = 1$ in addition) can be written down, the solution of which provides the minimax mth-degree approximation p_0, as well as the maximum error of approximation and the $m + 2$ points at which this error obtains. These equations are amenable to solution, however, only in the simplest cases; other cases would apparently require iterative techniques using a computer.

The minimax linear approximation to $\theta^{1/r}$ is simple enough, since there are but four equations; one finds

$$p_0 = \tfrac{1}{2}(r - 1)r^{-r(r-1)} + \theta,$$

and the minimax-bias linear estimator is obtained by replacing θ by x/n. The maximum absolute bias, or maximum error of approximation ρ is given by the constant term in p_0, and it is attained at $\theta = 0$, $r^{-r/(r-1)}$, and 1.

Minimax quadratic approximation to $\theta^{\cdot/r}$ leads to the following equations, after a considerable amount of algebraic manipulation:

$$p_0 = a_0 + a_1\theta + a_2\theta^2,$$

where

$$a_2 = - (r - 1)r^{-1}v^{2r-1}, \qquad a_1 = (2r - 1)r^{-1}v^{r-1},$$
$$a_0 = \rho = (1 - a_2 - a_1)/2, \qquad \theta = v^{-r}, \qquad \theta_1 = \theta_2 t^r,$$

and t and v are solutions (readily obtainable by iteration) of

$$2(r - 1)t^{2r} - (2r - 1)t^r + t = 0,$$
$$(r - 1)(v^{2r} - t^{2r}) - (2r - 1)(v^r - t^r) + r(v - t) = 0,$$

satisfying $0 < t < 1 < v$. The solutions for $r = 2$ and $r = 5$ appear in the tables (see Fig. 1).

These approximations, of course, do not depend on $n(> 1)$, and thus the biases of their unbiased estimators do not depend on n, as would be the case if nth-degree approximations were used. It should be noted that these quadratic approximations are not monotone in θ near $\theta = 1$; therefore, we cannot expect δ_0 necessarily to be monotone in x. Moreover, for small n, the range of δ_0 is not necessarily confined to the unit interval. Thus, these estimators are of questionable use.

Consideration of these estimators as least-squares estimators relative to a least favorable distribution or minimax-bias estimators on a least favorable point set appears in a later section.

Approximate Minimax-Bias Estimators

Since γ does not have a valid series expansion throughout Θ, minimax approximation by expansion in Chebyshev polynomials is not possible; nevertheless, the "τ-method" of Lanczos may be employed, still using Chebyshev polynomials.

We note that the function $\gamma = \theta^{1/r}$ satisfies the differential equation $r\theta\gamma' - \gamma = 0$, where the prime denotes differentiation. Letting p_n denote an nth-degree polynomial approximation to γ and τ a real number, we choose p_n and τ so that

$$r\theta p_n' - p_n = \tau T_n^*,$$

where T_n^* is the nth Chebyshev polynomial shifted to $(0, 1)$ (see [20]). Since we cannot require that $r\theta\gamma' - \gamma$ equal zero after substitution of p_n for γ, we set it equal to an nth-degree approximation to zero. Equating coefficients, we have $n + 1$ equations in $n + 2$ unknowns (τ and the coefficients in p_n). An additional equation is obtained by imposing the boundary condition $p_n(1) = \gamma(1) = 1$. It follows also that the maximum error of approximation is $|\tau|$.

When we introduce the canonical polynomials $Q_m(\theta)$, ($m = 0$, $1, \cdots, n$), defined by

$$r\theta Q_m' - Q_m = \theta^m, \tag{7.2}$$

it follows readily that $Q_m = (mr - 1)^{-1}\theta^m$ and

$$p_n = \tau \sum_{m=0}^{n} c_n^m Q_m, \qquad \tau^{-1} = \sum_{m=0}^{n} c_n^m (mr - 1)^{-1},$$

where

$$T_n^* = \sum_{m=0}^{n} c_n^m \theta^m.$$

The Q_m's can be found successively from (7.2).

For $\gamma = \sqrt{\theta}$, $r = 2$, we find from Table 3 of the National Bureau of Standards *Tables of Chebyshev Polynomials* that

$$p_1 = (1 + 2\theta)/3, \qquad p_2 = (3 + 24\theta - 8\theta^2)/19,$$

and for $\gamma = \theta^{1/5}$, $r = 5$, that

$$p_1 = (2 + \theta)/3, \qquad p_2 = (9 + 18\theta - 8\theta^2)/19;$$

the coefficients in p_5 appear in the tables.

The unbiased estimators of these approximations have zero bias at $\theta = 1$ and maximum absolute bias at $\theta = 0$ (given by the constant terms in the p_m's). Again, the estimators are not necessarily monotone in x, nor necessarily confined to the unit interval for small n. The corresponding bias and RMS error appear in Figure 2 for $r = 2$; for $r = 5$, these functions are similar to those for the least-squares sense estimator shown in Figure 3.

The maximum absolute biases for the quadratic cases are substantially larger than those of the exact minimax quadratic approximations (see tables); higher-order approximations are feasible, however, by the τ-method but not by the exact method. For $m = n = 5$, it is found that, although the maximum absolute bias is not low, the bias is exceedingly close to zero (< 0.025 for $r = 5$) for most θ-values ($\theta > 0.1$). An erratic estimator ($\delta \geq 1$ at $x = 1, 3,$ and 5) is required to achieve this low bias, however.

Minimum-Bias Estimators in the Least-Squares Sense

We shall derive the mth-degree approximation of $\gamma = \theta^{1/r}$ in the least-squares sense relative to the uniform distribution ξ on $(0, 1)$. Subsequently, we shall consider a case in which ξ is a three-point distribution. Such approximation relative to other distributions ξ could be derived analogously. Unbiased estimators δ_ξ of these approximations are readily obtained from (6.3).

The orthogonal polynomials $p_m(\theta)$ relative to the uniform distribution ξ (Legendre polynomials on the unit interval) are readily constructed successively from the relations

$$\int p_m(\theta)\theta^i \, d\theta = 0, \qquad i = 0, 1, \cdots, m - 1,$$

where p_m is an mth-degree polynomial, and then normalized by requiring that $\int p_m^2 \, d\theta = 1$ and that the coefficient of θ^m in p_m be positive. These polynomials may be expressed as

$$p_m(\theta) = \sum_{i=0}^{m} c_{m,i}\theta^i$$

where

$$c_{m,i} = \sqrt{2m+1}\,(-1)^{m+i}\binom{m+i}{m}\binom{m}{m-i}.$$

Then we have

$$c_m \int \theta^{1/r}p_m(\theta) \, d\theta = \sum c_{m,i} \int \theta^{(ri+1)/r} \, d\theta = r\sum c_{m,i}/(ri+1+1).$$

The least-squares mth-degree approximation to $\theta^{1/r}$ is accordingly

$$p_{\xi,m}^{(\theta)} = \sum_{i=0}^{m} c_i p_i(\theta);$$

the average squared error of approximation (squared bias of δ_ξ) is

$$N(\xi) = \frac{r}{(r+2)} - \sum c_i^2.$$

Using an $(m+1)$th-order approximation decreases $N(\xi)$ by c_{m+1}^2.

Explicitly, the least-squares linear and quadratic approximations are

$$p_{\xi,1}(\theta) = c_0 p_0(\theta) + c_i p_i(\theta) = \frac{2r(r-1+3\theta)}{(r+1)(2r+1)},$$

$$p_{\xi,2}(\theta) = \frac{3r[(r-1)(2r-1) + 8(2r-1)\theta - 10(r-1)\theta^2]}{(r+1)(2r+1)(3r+1)}.$$

See tables and figures for $r = 2, 5$ and $m = 1, 2, 5$; the approximations are similar to those by the τ-method.

Recall that a minimax approximation is a least-squares approximation relative to a least favorable distribution ξ_0; moreover, ξ_0 assigns probability 1 to the set of points at which the maximum error of approximation is attained. For the minimax-bias linear estimator given previously, we shall solve for the corresponding least favorable distribution ξ_0.

The minimax linear approximation attains its maximum error ρ at three points, denoted $(\theta_0, \theta_1, \theta_2)$. Thus, ξ_0 assigns probability 1 to this set of three points; denote the corresponding probabilities, the components of ξ_0, by $(\xi_0^0, \xi_0^1, \xi_0^2)$.

The orthogonal polynomials $p_0(\theta)$ and $p_1(\theta)$ relative to any distribution ξ are found to be

$$p_0(\theta) = 1, \qquad p_1(\theta) = \frac{\theta - \mathcal{E}\theta}{\sigma_\theta}$$

where $\mathcal{E}$ denotes integration relative to ξ and σ_θ^2 is the variance of θ relative to ξ. Thus

$$c_0 = \mathcal{E}\theta^{1/r}, \qquad c_1 = (\mathcal{E}\theta^{(r+1)/r} - \mathcal{E}\theta^{1/r}\mathcal{E}\theta)\sigma_\theta,$$

and the least-squares linear approximation is $p_{\xi_0} = c_0 p_0 + c_1 p_1$.

Evaluating p_{ξ_0} for a three-point distribution on $(0, r^{-r/(r-1)}, 1)$, that is, for the values of the θ_i's given previously, and equating it with p_0 (the minimax linear approximation given previously), we can solve for the components of ξ_0. We thus find

$$\overset{0}{\xi_0} = \frac{1 - r^{-r/(r-1)}}{2}, \qquad \overset{1}{\xi_0} = \frac{1}{2}, \qquad \overset{2}{\xi_0} = \frac{r^{-r/(r-1)}}{2} .$$

For $r = 2$, we have $\xi_0 = (\frac{3}{8}, \frac{1}{2}, \frac{1}{8})$, and for $r = 5$, we have $\xi_0 = (0.4331, 0.5, 0.0669)$.

In addition, p_0 is the minimax linear approximation to γ on the point set E consisting of the three points θ_0, θ_1, θ_2; thus, E is a least favorable point set. To show that p_0 is minimax (linear) on E, we follow the method given by De la Vallée Poussin [15]:

$$A_0 = \theta_2 - \theta_1 = 1 - r^{-r/(r-1)}, \qquad A_1 = \theta_2 - \theta_0 = 1,$$

$$A_2 = \theta_1 - \theta_0 = r^{-r/(r-1)}; \qquad \gamma(\theta_i) \equiv \gamma_i = 0, r^{-1/(r-1)}, 1,$$

respectively;

$$\rho = \frac{A_0\gamma_0 - A_1\gamma_1 + A_2\gamma_2}{A_0 + A_1 + A_2} = \frac{1}{2}(r - 1)r^{-r/(r-1)}.$$

Then, setting $\gamma_i = (-1)^{i+1}\rho + a_0 + a_1\theta_i$ and solving, we find $a_0 = \rho$, $a_1 = 1$; that is, the minimax linear approximation on E is $a_0 + a_1\theta = \rho + \theta = p_0$, obtained earlier.

Minimum Local-Bias Estimators at $\theta = 1$

The minimum local-bias estimator of order m at $\theta = 1$ is the unbiased estimator of the truncated Taylor series approximation to γ at $\theta = 1$, $p_m^*(\theta) = \sum_{i=0}^m a_i\theta^i$; that is, the a_i's are chosen so that $p_m^*(1)$ and $\gamma(1)$ coincide, as to their first m derivatives, at $\theta = 1$.

Using superscripts to denote derivatives and $a_i^* = i!a_i$, we find

$$p_m^{*(i)}(1) = \sum_{j=i}^m a_j^*/(j - i)!$$

and

$$\gamma^{(i)}(1) \equiv c_i = (-1)^{i-1}r^{-i}(r - 1)(2r - 1) \cdots [(i - 1)r - 1].$$

Define the $(m + 1)$th-order matrix $D = (d_{ij})$ by $d_{ij} = 1/(j - 1)!$ if $i \leq j$ and by $\delta_{ij} = 0$ otherwise; then $D^{-1} = (d^{ij})$, where

$$d^{ij} = (-1)^{j-1}/(j - 1)! \quad \text{if } i \leq j, \quad \text{and} \quad d^{ij} = 0 \text{ otherwise.}$$

The vector of a_i^*'s must satisfy $\mathbf{c} = D\mathbf{a}^*$ or $\mathbf{a}^* = D^{-1}\mathbf{c}$, from which the coefficients a_i in p_m^* may be obtained.

The unbiased estimators of these approximations turn out to be upper estimators of γ. *Lower estimators* can be obtained by fitting only the first $m - 1$ derivatives, rather than m, at $\theta = 1$, and imposing the additional restriction that $p_m(0) = \gamma(0) = 0$. Analogous equations can be established to solve for the coefficients a_{*i}, or a_i; the inversion of the analogue of the D matrix is not so trivial. The estimator obtained for $m = n = 5$ is almost identical with the minimum local-bias estimator (see tables and figures).

Minimum Local-Bias Estimators at $\theta = 1/r$

Let $p_m^0(\theta)$ be the Taylor expansion of γ at $\theta = 1/r$, truncated to terms through the mth power. Using notation analogous to that in the previous section, we find

$$c_0 = (1/r)^{1/r}, \qquad c_i = (-1)^{i-1}r^{-1/r}(r-1)(2r-1) \cdots [(i-1)r - 1],$$

$$d_{ij} = \begin{cases} r^{j-i}/(i-j)! & \text{if } i \leq j, \\ 0 & \text{otherwise,} \end{cases}$$

$$d^{ij} = \begin{cases} (-r)^{j-i}/(i-j)! & \text{if } i \leq j, \\ 0 & \text{otherwise,} \end{cases}$$

$$\mathbf{a}_* = D^{-1}\mathbf{c} \quad \text{and} \quad a_i = a_{*i}/i!.$$

The solutions for $r = 2$ and 5, $n = m = 5$, appear in the tables. For the case $r = n = 5$, to obtain such a good fit (small bias) near $\theta = \frac{1}{5}$, the δ-values must range from -3.5 to 15.

Bayes Estimators

We shall first derive Bayes linear and quadratic estimators relative to a uniform distribution on the unit interval with squared-error loss. We obtain these estimators by finding the average-risk function for an estimator and choosing the coefficients in the estimator to minimize the average risk. We then derive unrestricted (i.e., of nth-degree) Bayes estimators relative to a beta distribution by standard techniques and consider in particular the special case of a uniform distribution.

For the linear case, we have

$$\delta(x) = a_0 + a_1 x/n \quad \text{and} \quad R(\delta, \theta) = \mathcal{E}(\delta - \theta^{1/r})^2;$$

the average risk relative to a uniform distribution is $\int R(\delta, \theta) \, d\theta$, from which minimizing values of a_0 and a_1 are found by differentiation:

$$a_0 = \frac{2r[(r-1)n + 2r + 1]}{(n+2)(r+1)(2r+1)}, \qquad a_1 = \frac{6rn}{(n+2)(r+1)(2r+1)}.$$

Thus, the Bayes linear estimator for $r = 2$, $n = 5$ is the unbiased estimator of $4(2 + 3\theta)/21$ and for $r = 5$, $n = 5$, of $5(31 + 15\theta)/231$.

Analogous results for the case of quadratic estimation were obtained in a similar fashion (see tables).

The maximum bias of these estimators (see tables) is greater than for some of the linear and quadratic minimum-bias estimators, but the average mean-square error is reduced. Note that the expectation of these linear and quadratic estimators depends on n, in contrast to the analogous minimum-bias estimators. The quadratic estimators have the unreasonable property of not being monotone in x.

We now derive Bayes estimators relative to a beta distribution. We let ξ denote a beta distribution on the unit interval with density given by

$$B^{-1}(\alpha, \beta)\theta^{\alpha-1}(1 - \theta)^{\beta-1},$$

where $B(\alpha, \beta)$ is the beta function and α and β are positive numbers. For $\alpha = \beta = 1$, ξ is a uniform distribution.

The posterior distribution of θ (the conditional distribution of θ, given x) when x is binomial is readily found to be a beta distribution with α and β replaced by $x + \alpha$ and $n - x + \beta$, respectively (see [2], for example). The Bayes estimator of γ with squared-error loss is the expected value of γ relative to the posterior distribution of θ; this estimator is here found to be

$$\delta_\xi(x) = \frac{\Gamma\left(x + \alpha + \dfrac{1}{r}\right)\Gamma(n + \alpha + \beta)}{\Gamma(n + \alpha)\Gamma\left(n + \alpha + \beta + \dfrac{1}{r}\right)},$$

which, if β is integral, reduces to

$$\delta_\xi(x) = \prod_{i=1}^{n-x+\beta} \frac{n + \alpha + \beta - i}{n + \alpha + \beta + \dfrac{1}{r} - i}.$$

Otherwise, tables of the log gamma function or, if $r = 2$, the National Bureau of Standards *Tables of n! and* $\Gamma(n + \frac{1}{2})$ (see [27]) can be used to calculate values of δ_ξ. Alternatively, Stirling's formula could be used to derive adequate approximations to δ_ξ. The range of δ_ξ (and therefore of its expectation) is confined to the unit interval, and it is monotone in δ_ξ. For the cases $r = 2$ and 5, $n = 5$, see the tables and figures. It is noteworthy that use of a fifth-degree Bayes estimator gives very little reduction in average risk compared with a quadratic, and in fact a linear, Bayes estimator (Tables 2 and 4). The average risk is only slightly smaller than that for the minimum local-bias estimator at $\theta = 1$, but the maximum and average squared bias are increased (considerably so for $r = 2$).

Minimax-Risk Estimators

General expression for minimax-risk estimators could not be obtained. Linear estimators ($m = 1$, $n = 5$, $r = 2$ and 5), for which the maximum risk is a minimum, were found by trial and error and then iteration (Newton's method) by means of a Univac 1105; such primitive techniques are not feasible for approximations of higher degree.

As seen in the tables and Figure 1, some reduction in maximum risk was achieved as compared with other linear estimators, with a corresponding increase in maximum bias. Whether this reduction can be considered worth the extensive effort required is a matter of opinion!

8. Conclusion

Maximum-likelihood estimators are readily available, for arbitrary r and n, but the bias and risk may be large unless n is large.

Among minimum-risk estimators, Bayes estimators relative to the beta family are readily available, but their biases are large, especially near $\theta = 0$ and 1, compared with other methods of estimation, including maximum likelihood. Minimax-risk estimators are not generally available.

Among minimum-bias estimators, a variety are fairly readily available (minimax not so readily). All are expressed, however, as polynomials in x and are thus ponderous unless the degree is a priori limited. Moreover, efforts to reduce bias may result in (a) δ-values greater than unity (and perhaps negative values also), and (b) lack of monotonicity of δ, as well as increased risk, suggesting that too much emphasis has been placed on the bias. In particular, the use of unbiased estimators of Taylor expansions seems especially perilous without additional restrictions, since negligible bias is achieved locally at the possible expense of all other reasonable properties.

References

1. Wald, A., *Statistical Decision Functions*, John Wiley & Sons, Inc., New York, 1950.
2. Lehmann, E. L., *Notes on the Theory of Estimation* (mimeographed), available at Associated Students Stores, University of California, Berkeley, 1950.

3. Blackwell, David, and M. A. Girshick, *Theory of Games and Statistical Decisions*, John Wiley & Sons, Inc., New York, 1954.
4. Fraser, D. A. S., *Nonparametric Methods in Statistics*, John Wiley & Sons, Inc., New York, 1957.
5. Savage, L. J., *The Foundations of Statistics*, John Wiley & Sons, Inc., New York, 1954.
6. Lehmann, E. L., and Henry Scheffé, "Completeness, Similar Regions, and Unbiased Estimation—Part I," *Sankhyā*, Vol. 10, 1950, pp. 305–340.
7. Halmos, P. R., "The Theory of Unbiased Estimators," *Ann. Math. Statist.*, Vol. 17, 1946, pp. 34–43.
8. Bahadur, R. R., "On Unbiased Estimates of Uniformly Minimum Variance," *Sankhyā*, Vol. 18, 1957, pp. 211–224.
9. Dynkin, E. G., "Necessary and Sufficient Statistics for a Family of Probability Distributions," *Uspehi Mat. Nauk*, Vol. 6, 1951, pp. 68–90 (Russian); English translation by Statistical Laboratory, Cambridge, England.
10. Bhattacharyya, A., "Notes on the Use of Unbiased and Biased Statistics in the Binomial Population," *Calcutta Statist. Assoc. Bull.*, Vol. 5, 1954, pp. 149–164.
11. Kolmogorov, A. N., "Unbiased Estimates," *Izv. Akad. Nauk SSSR, Ser. Mat.*, Vol. 14, 1950, pp. 303–326; American Mathematical Society Translation No. 98, 1953.
12. Siraždinov, S. H., "Concerning Estimations with Minimum Bias for a Binomial Distribution," *Theory of Probability and Its Applications, Akad. Nauk SSSR* (Russian), Vol. 1, 1956, pp. 174–176; English summary.
13. Hansen, Morris H., William N. Hurwitz, and William G. Madow, *Sample Survey Methods and Theory*, Vol. I, John Wiley & Sons, Inc., New York, 1953.
14. Bernstein, S., "Leçons sur les propriétés extremales," *Borel Monograph Series*, Gauthier-Villars, Paris, 1926.
15. La Vallée Poussin, C. de, "Leçons sur l'approximation des fonctions," *Borel Monograph Series*, Gauthier-Villars, Paris, 1919.
16. Achieser, N. I., *Theory of Approximation* (translated by C. J. Hyman), Frederick Ungar Publishing Company, New York, 1956.
17. Favard, J., "Sur l'approximation des fonctions d'une variable réelle," *Analyse Harmonique*, Centre National de la Recherche Scientifique, Paris, 1947.
18. Clement, P. R., "The Chebyshev Approximation Method," *Quart. Appl. Math.*, Vol. 11, 1953, pp. 167–183.
19. Jackson, Dunham, *The Theory of Approximation*, American Mathematical Society, New York, 1930.
20. Lanczos, C., Introduction to *Tables of Chebyshev Polynomials* $S_n(x)$ *and* $C_n(x)$, National Bureau of Standards Applied Mathematics Series, No. 9, Government Printing Office, Washington, D.C., 1952.
21. Bram, Joseph, "Chebychev Approximation in Locally Compact Spaces," *Proc. Amer. Math. Soc.*, Vol. 9, 1958, pp. 133–136.
22. Szegö, G., *Orthogonal Polynomials*, American Mathematical Society, New York, 1939.

23. Walsh, J. L., "Best-approximation Polynomials of Given Degree," *Proc. Symposium Appl. Math.*, Vol. 6, McGraw-Hill Book Co., Inc., New York, 1956, pp. 213–218.

24. Blyth, Colin R., "Note on Estimating Information," *Ann. Math. Statist.*, Vol. 30, 1959, pp. 71–79.

25. Lorentz, G. G., *Bernstein Polynomials*, University of Toronto Press, Toronto, 1953.

26. DeGroot, Morris H., "Unbiased Sequential Estimation for Binomial Populations," *Ann. Math. Statist.*, Vol. 30, 1959, pp. 80–101.

27. National Bureau of Standards, *Tables of n! and $\Gamma(n+1/2)$ for the First Thousand Values of* n, Applied Mathematics Series, No. 16, Government Printing Office, Washington, D.C., 1951.

Chapter 9

On Optimal Replacement Rules When Changes of State Are Markovian[†]

C. DERMAN

1. Introduction

A common industrial and military activity is the periodic inspection of some system, or one of its components, as part of a procedure for keeping it operative. After each inspection, a decision must be made as to whether or not to alter the system at that time. If the inspection procedure and the ways of modifying the system are fixed, an important problem is that of determining, according to some cost criterion, the optimal rule for making the appropriate decision. This chapter is an outgrowth of a problem considered by Derman and Sacks [1] and is concerned with a problem such that the only possible way to alter the system is to replace it.

Suppose a unit (a system, a component of a system, a piece of operating equipment, etc.) is inspected at equally spaced points in time and that after each inspection it is classified into one of $L + 1$ states, $0, 1, \cdots, L$. A unit is in state $0(L)$ if and only if it is new (inoperative). Let the times of inspection be $t = 0, 1, \cdots$, and let X_t denote the observed state of the unit in use at time t. We assume that $\{X_t\}$ is a Markov chain with stationary transition probabilities,

$$q_{ij} = P(X_{t+1} = j \mid X_t = i),$$

for all i, j, and t. The actual values of the q_{ij}'s are functions of the nature of the unit and the replacement rule in force. For example, if replacement of an inoperative unit is compulsory, then $q_{L0} = 1$; otherwise

---

† Research sponsored by the Office of Naval Research.

$q_{j0} = 1$, if and only if j is a state at which the unit is also replaced.

We suppose that the costs are as follows: A cost of amount $c(c > 0)$ is incurred if the unit is replaced before it becomes inoperative; a cost of amount $c + A$ $(A > 0)$ is incurred if the unit is replaced after becoming inoperative; otherwise, no cost is incurred. The criterion for comparing replacement rules is the *average cost per unit time* averaged over a large (infinite) interval of time.

The problem is, then, to determine in the sense of the above cost criterion, the optimal replacement rule, that is, an optimal partitioning of the state space into two categories: states at which the unit is replaced and states at which it is not replaced.†

Since there are at most a finite number $(2^{L-1} - 1)$ of possible partitionings, an optimal one exists. For small values of L, it is a matter of enumerating and computing in order to select the optimal partitioning. When L is even moderately large, however, solution by enumeration becomes impracticable. Thus it is of interest to know conditions under which a certain relatively small (small enough for enumeration) subclass of rules contains the optimal one.

Let p_{ij} denote the transition probabilities associated with the rule: *Replace only when the unit is inoperative.* Since these transition probabilities $\{p_{ij}\}$ usually are not precisely known, it is important that the conditions for reducing the problem to manageable size be relatively indifferent to the precise values of the p_{ij}'s. The principal result of this chapter is in this direction; namely, conditions are given on the transition probabilities $\{p_{ij}\}$ guaranteeing that the optimal replacement rule is of the simple form: *Replace the item if and only if the observed state is one of the states i, $i + 1$, $\cdots$, L for some i.* Henceforth, such rules will be referred to as *control-limit rules* and the above state i, the *control limit.*

When the p_{ij}'s are not precisely known, empirical methods are necessary in order to arrive at the optimal replacement rule. A method is suggested in Section 4.

2. Statement of Problem

Let $\{p_{ij}\}$ denote the transition probabilities of a Markov chain with states $0, 1, \cdots, L$. The transition probabilities satisfy, in addition to the usual conditions, the further conditions

---

† We restrict our attention to nonrandomized stationary rules. We have shown in [2] that an optimal rule over all possible rules is a member of this restricted class of rules.

$$p_{j0} = 0 \quad \text{for} \quad j < L,$$
$$p_{jL}^{(t)} > 0 \quad \text{for some} \quad t \geq 1 \quad \text{for each} \quad j < L,$$

and

$$p_{L0} = 1,$$

where $p_{jL}^{(t)}$ denotes the t-step transition probability from j to L.

We can conceive of modifying the chain by setting $p_{j0} = 1$ for one or more (but not all) of the j's for $0 < j < L$. Such a modification corresponds, in the replacement context, to replacing the item if it is observed to be in state j. Thus there are $2^{L-1} - 1$ such possible modifications, each corresponding to a possible replacement rule. Let $\mathcal{C}$ denote the class of such rules. For each rule R in $\mathcal{C}$, let q_{ij} (we suppress the letter R for typographical convenience) denote the resulting set of transition probabilities and consider the cost function

$$
\begin{aligned}
g(j) &= 0 &&\text{if} \quad q_{j0} = 0 \\
g(j) &= c &&\text{if} \quad q_{j0} = 1
\end{aligned}
\Bigg\}, \quad j < L;
$$
$$g(j) = c + A, \qquad\qquad\qquad j = L.$$

That is, $g(j)$ denotes the cost incurred at any given time t when the Markov chain is in state j.

It is well known from Markov-chain theory that

$$\phi_R = \lim_{T \to \infty} \frac{1}{T} \sum_{t=1}^{T} g(X_t) = \sum_{j=0}^{L} \pi_j g(j), \qquad (2.1)$$

with probability one, where the quantities π_j (steady-state probabilities) satisfy the equations

$$\pi_j = \sum_{i=0}^{L} \pi_i q_{ij}, \qquad j = 0, \cdots, L,$$

$$\sum_{j=0}^{L} \pi_j = 1,$$

and the inequalities

$$0 \leq \pi_j \leq 1, \qquad j = 0, \cdots, L.$$

The limit ϕ_R is the average cost per unit time, the criterion of interest, using the rule R.

We can evaluate ϕ_R in another way. Let N_k denote the kth recurrence time to state 0 (i.e., the length of the kth replacement cycle) and C_k the cost (either c or $c + A$) associated with the kth replacement cycle.

Then $\{N_k\}$ and $\{C_k\}$ $(k=1, 2, \cdots)$ are sequences of independent and identically distributed random variables. It can be shown by a straightforward application of the law of large numbers that

$$\phi_R = \frac{EC}{EN} . \tag{2.2}$$

Expression (2.1) or (2.2) for ϕ_R can be rewritten, depending on R, as a function of the original transition probabilities $\{p_{ij}\}$, and can be evaluated, at least theoretically, for each R in $\mathfrak{C}$. Neither of the representations (2.1) nor (2.2) seems, however, to be informative enough to allow us to arrive at reasonable conditions on the p_{ij}'s in order to imply that an optimal rule will have a simple structure.†
   If the p_{ij}'s are of the form

$$p_{ij} = 0 \quad \text{for} \quad |j - i| > 1,$$

that is, if transitions are possible only to adjacent states, then it is obvious that an optimal rule will be a control-limit rule such that

$$q_{j0} = 0, \qquad j \le L - 2,$$

and

$$q_{L-1,0} = 1 \quad \text{or} \quad 0,$$

according to the values of c, A, and the p_{ij}'s. For more general chains, however, where states can be skipped in the transitions, the situation is not so transparent. To arrive at workable conditions for reducing the problem, we use the method of functional equations [4].

3. Functional Equation Approach

Suppose that $P(X_0 = i) = 1$; that is, suppose that at time $t = 0$ the unit is in state i with probability 1. As an intermediate step in our argument, consider the function

$$\phi_R(i, \alpha) = E \sum_{t=0}^{\infty} \alpha^t g(X_t), \qquad 0 < \alpha < 1,$$

for any R in $\mathfrak{C}$. Later we shall use, in the way suggested by Arrow, Karlin, and Scarf (see [5], p. 35), the fact that

$$\lim_{\alpha \to 1} (1 - \alpha)\phi_R(i, \alpha) = \phi_R.$$

---

† Linear programming and dynamic programming methods are available [2], [3] for computing an optimal rule.

The factor α can be considered, as in inventory theory, to be a discount factor; $\phi_R(i, \alpha)$ is then a meaningful cost criterion.

Suppose that R_α^* in $\mathcal{C}$ is such that for all i we have

$$\phi(i, \alpha) = \phi_{R_\alpha^*}(i, \alpha) = \min_{R \in \mathcal{C}} \phi_R(i, \alpha);$$

that is, that R_α^* is the optimal replacement rule when $\phi_R(i, \alpha)$ is the cost criterion. Then by standard arguments it can be shown that $\phi(i, \alpha)$ must satisfy the functional equations,

$$\phi(i, \alpha) = \min \left\{ \alpha \sum_{j=0}^{L} p_{ij}\phi(j, \alpha), c + \alpha \sum_{j=0}^{L} p_{0j}\phi(j, \alpha) \right\} \quad \text{for } i \neq L, \quad (3.1)$$

$$\phi(L, \alpha) = c + A + \alpha \sum_{j=0}^{L} p_{0j}\phi(j, \alpha),$$

where the relationship between (3.1) and R_α^* is apparent. Also, the recursively defined functions (method of successive approximations),

$$\phi(i, \alpha, 0) = 0 \qquad \text{if } i \neq L,$$
$$\phi(i, \alpha, 0) = c + A \quad \text{if } i = L,$$

and

$$\phi(i, \alpha, N) = \min \left\{ \alpha \sum_{j=0}^{L} p_{ij}\phi(j, \alpha, N-1), c + \alpha \sum_{j=0}^{L} p_{0j}\phi(j, \alpha, N-1) \right\}$$
$$\text{if } i \neq L,$$

$$\phi(i, \alpha, N) = c + A + \alpha \sum_{j=0}^{L} p_{0j}\phi(j, \alpha, N-1) \quad \text{if } i = L,$$

for $N \geq 1$, can be shown (see [4]) to converge to $\phi(i, \alpha)$; that is,

$$\lim_{N \to \infty} \phi(i, \alpha, N) = \phi(i, \alpha), \qquad i = 0, 1, \cdots, L.$$

We shall impose, to some advantage, the following monotonicity-preserving condition on the transition probabilities $\{p_{ij}\}$.

Condition A. For every nondecreasing function $h(j), j = 0, 1, \cdots, L$, the function

$$k(i) = \sum_{j=0}^{L} p_{ij}h(j), \qquad i = 0, 1, \cdots, L-1,$$

is also nondecreasing.

We now state and prove the following result:

THEOREM 1. *If Condition A holds, then there exists a control-limit rule R^* such that*

$$\phi_{R^*} = \min_{R \in \mathcal{C}} \phi_R.$$

PROOF: The proof proceeds in three steps. First we prove that, for each $N \geq 1$, there is an i_N such that

$$\phi(i, \alpha, N) = \alpha \sum_{j=0}^{L} p_{ij}\phi(j, \alpha, N - 1), \qquad\qquad i < i_N,$$

$$= c + \alpha \sum_{j=0}^{L} p_{0j}\phi(j, \alpha, N - 1), \qquad i_N \leq i < L, \quad (3.2)$$

$$= c + A + \alpha \sum_{j=0}^{L} p_{0j}\phi(j, \alpha, N - 1), \qquad i = L.$$

By definition $\phi(i, \alpha, 0)$ is a nondecreasing function. From the definition of $\phi(i, \alpha, 1)$ we can easily deduce, using Condition A, that $\phi(i, \alpha, 1)$ is of the form (3.2) and that $\phi(i, \alpha, 1)$ is also nondecreasing. The argument then proceeds by induction, establishing (3.2).

Secondly, since

$$\phi(i, \alpha) = \lim_{N \to \infty} \phi(i, \alpha, N),$$

it also follows that $\phi(i, \alpha)$ is nondecreasing. Hence, using this fact, Condition A, the functional equation (3.1), and its interpretation in terms of R_α^*, we can establish that R_α^* is a control-limit rule.

Finally, let i_α denote the control limit of R_α^*. Let $\{\alpha_v\}$, with

$$\lim_{v \to \infty} \alpha_v = 1,$$

be a sequence such that $i_{\alpha_v} = i^*$ for all v. Since there is at most a finite number of possible states, such a sequence and i^* exist. Now let R be any rule that is not a control-limit rule. Let R^* denote the control-limit rule with i^* as its control limit (i.e., $R^* = R_{\alpha_v}^*$ for all v). Then

$$\phi_R(i, \alpha_v) \geq \phi(i, \alpha_v), \qquad v = 1, 2, \cdots,$$

and hence

$$\phi_R = \lim_{v \to \infty} (1 - \alpha_v)\phi_R(i, \alpha_v) \geq \lim_{v \to \infty} (1 - \alpha_v)\phi(i, \alpha_v) = \phi_{R^*},$$

which proves the theorem.

Condition A as stated is not verifiable. The following condition is more satisfactory from this point of view.

Condition B: For each $k = 0, 1, \cdots, L$, the function

$$r_k(i) = \sum_{j=k}^{L} p_{ij}, \qquad i = 0, 1, \cdots, L - 1,$$

is nondecreasing.

We have the following lemma:

LEMMA. *Conditions A and B are equivalent.*

PROOF: Assume Condition A. Then in particular the function

$$h_k(j) = \begin{cases} 0, & j < k, \\ 1, & j \geq k, \end{cases}$$

is nondecreasing. But then we have

$$k(i) = \sum_{j=0}^{L} p_{ij} h_k(j) = \sum_{j=k}^{L} p_{ij} = r_k(i),$$

and hence Condition B holds.

Assume Condition B. Any nondecreasing function $h(j)$ can be expressed in the form

$$h(j) = \sum_{i=0}^{L} c_i h_i(j),$$

where $c_i \geq 0$ for $i = 0, \cdots, L$, and

$$h_i(j) = \begin{cases} 0, & j < i, \\ 1, & j \geq i. \end{cases}$$

Then

$$K(i) = \sum_{j=0}^{L} p_{ij} h(j) = \sum_{j=0}^{L} p_{ij} \sum_{k=0}^{L} c_k h_k(j) = \sum_{k=0}^{L} c_k \sum_{j=0}^{L} p_{ij} h_k(j)$$

$$= \sum_{k=0}^{L} c_k \sum_{j=k}^{L} p_{ij}.$$

Since $c_k \geq 0$ and, by Condition B, $\sum_{j=k}^{L} p_{ij}$ is nondecreasing for each k, it follows that $K(i)$ is also nondecreasing. This proves the lemma.

The equivalence of the two conditions allows us to restate Theorem 1:

THEOREM 2. *If Condition B holds, then the conclusion of Theorem 1 holds.*

As an application of Theorem 2, consider the following example in which the transitions from state to state are generated by cumulative sums of identically distributed lattice variables. More precisely, let $\{a_v\}, v = \cdots -1, 0, 1, \cdots$, be a sequence of nonnegative numbers such that

$$\sum_{v=-\infty}^{\infty} a_v = 1.$$

We define

$$p_{01} = p_{L0} = 1,$$

and otherwise

$$p_{i0} = 0,$$

$$p_{ij} = a_{j-i}, \qquad 1 < j < L,$$

$$p_{i1} = \sum_{v=-\infty}^{1} a_{v-i},$$

$$p_{iL} = \sum_{v=L}^{\infty} a_{v-1}.$$

Then

$$r_0(i) = 1, \qquad\qquad i = 0, \cdots, L - 1,$$
$$r_1(i) = 1, \qquad\qquad i = 0, \cdots, L - 1,$$
$$r_k(0) = 0, \qquad\qquad k \geq 2,$$
$$r_k(i) = \sum_{v=k-i}^{\infty} a_v, \qquad k \geq 2, \qquad i = 2, \cdots, L - 1;$$

therefore, Condition B is satisfied and the conclusion of Theorem 2 holds.

4. Empirical Method

Frequently, when the p_{ij}'s cannot be assumed to be known, it may still be reasonable to assume that Condition A holds. If so, then we know by Theorem 1 that a control-limit rule is optimal. One approach to obtaining the optimal rule is to estimate the p_{ij}'s from observations taken on units in operation and then, using (2.1), compute the optimal rule from these estimates. Because of (2.2), however, it is not necessary to estimate the p_{ij}'s; we need only estimate $\phi_R = EC/EN$ for each R in the class $\mathcal{C}'$ of control-limit rules and select the rule that appears to have the smallest ϕ_R.

We conceive of the process of observation and replacement going on indefinitely and, in fact, the cost criterion is calculated on this basis; this suggests the existence of a rule $\tilde{R}$ (not in $\mathcal{C}$) that uses the past history of observations in such a way that it converges rapidly enough to be equivalent (in accordance with the cost criterion) to R^*. (See [1] for a similar result.) Many such rules are possible. We mention one as an example of this approach.

Let R_i $(i = 2, \cdots, L)$ denote the control-limit rule that has i for its control limit. We now define $\tilde{R}$. On the kth replacement cycle, $k = 1, \cdots, L - 1$, use R_{k+1} as the replacement rule. Thereafter $(k \geq L)$, choose R_i randomly such that

$$P(R_i \text{ is used during the } k\text{th cycle}) = 1 - \frac{1}{k}$$

if†

$$\hat{\phi}_{R_i,k-1} = \min(\hat{\phi}_{R_2,k-1}, \cdots, \hat{\phi}_{R_L,k-1}),$$

and otherwise

$$P(R_i \text{ is used during the } k\text{th cycle}) = \frac{1}{(L-1)k},$$

where

$$\hat{\phi}_{R_i,k} = \frac{\sum\limits_{v=1}^{k_i} C_v}{\sum\limits_{v=1}^{k_i} N_v};$$

the expression

$$\sum_{v=1}^{k_i} C_v \left(\sum_{v=1}^{k_i} N_v \right)$$

is to be interpreted as being the summation of costs (lengths) of those of the first k cycles during which R_i is used.

It is easily seen that each of the $L - 1$ control-limit rules will be

_____
† If the minimum is achieved by more than one R_i, choose one arbitrarily.

used infinitely often as $k \to \infty$ with probability 1; further, by the strong law of large numbers, we have

$$\lim_{k \to \infty} \hat{\phi}_{R_i k} = \hat{\phi}_{R_i}$$

with probability 1. Hence, with probability 1,

$$\lim_{k \to \infty} \min (\hat{\phi}_{R_2 k}, \cdots, \hat{\phi}_{R_L k}) = \min (\phi_{R_2}, \cdots, \phi_{R_L}) = \phi_{R^*}.$$

It now follows (essentially, as shown in [5]) that $\phi_{\bar{R}} = \phi_{R^*}$ with probability 1.

References

1. Derman, C., and J. Sacks, "Replacement of Periodically Inspected Equipment (An Optimal Optional Stopping Rule)," *Naval Res. Logist. Quart.*, Vol. 7, 1960, pp. 597–607.
2. Derman, C., "On Sequential Decisions and Markov Chains," *Management Sci.*, Vol. 9, No. 1, 1962, pp. 16–24.
3. Howard, R. A., *Dynamic Programming and Markov Processes*, Technology Press, Massachusetts Institute of Technology, Cambridge, Mass., and John Wiley & Sons, Inc., London, 1960.
4. Bellman, R., *Dynamic Programming*, Princeton University Press, Princeton, N.J., 1957.
5. Arrow, K. J., S. Karlin, and H. Scarf, *Studies in the Mathematical Theory of Inventory and Production*, Stanford University Press, Stanford, California, 1958.
6. Karlin, S., "The Structure of Dynamic Programming Models," *Naval Res. Logist. Quart.*, Vol. 2, 1955, pp. 285–294.

PART THREE:

PROGRAMMING, COMBINATORICS, AND DESIGN

Chapter 10

Simplex Method and Theory

A. W. TUCKER

1. Introduction

The simplex method (1947) of G. B. Dantzig [1] is much more than the basic computational tool of linear programming. It is a combinatorial algorithm that provides constructive means of establishing fundamental theorems of linear programming [2]—as well as like theorems in cognate areas, such as von Neumann's minimax theorem for matrix games [3] and Farkas' theorem for linear inequalities. Its characteristic pivot transformations are related in an essential way to Gauss–Jordan elimination [4] and to a combinatorial equivalence of matrices [5].

This chapter discusses the simplex method in a format designed to exhibit over-all structure rather than specific operational details. The various terminal possibilities are represented schematically and geometrically. Also, it is shown that transposition-duality theorems [6], such as the classical ones of Gordan, Farkas, Stiemke, and Motzkin, can be regarded as corollaries of the duality theorem for a "homogeneous linear program."

The schemata and block-pivot transformations used in this chapter seem to be important methodological devices. They follow closely along lines developed by the author in a previous paper concerned with solutions of matrix games by linear programming [7].

2. Dual Linear Systems

This section and the next develop underlying concepts and format for use in later sections.

The schema

$$
\begin{array}{c|cccc}
 & -y_1 & -y_2 & \cdots & -y_n \\
\hline
\xi_1 & a_{11} & a_{12} & \cdots & a_{1n} & = x_1 \\
\xi_2 & a_{21} & a_{22} & \cdots & a_{2n} & = x_2 \\
\vdots & \vdots & \vdots & & \vdots & \vdots \\
\xi_m & a_{m1} & a_{m2} & \cdots & a_{mn} & = x_m \\
\hline
 & = \eta_1 & = \eta_2 & \cdots & = \eta_n
\end{array}
\tag{2.1}
$$

is a convenient device for the joint presentation of two systems of linear equations: a *column* system

$$
\left.
\begin{aligned}
\xi_1 a_{11} + \xi_2 a_{21} + \cdots + \xi_m a_{m1} &= \eta_1 \\
\xi_1 a_{12} + \xi_2 a_{22} + \cdots + \xi_m a_{m2} &= \eta_2 \\
&\;\;\vdots \\
\xi_1 a_{1n} + \xi_2 a_{2n} + \cdots + \xi_m a_{mn} &= \eta_n
\end{aligned}
\right\} \quad \Xi\, A = \mathrm{H}, \tag{2.2}
$$

and a *row* system

$$
\left.
\begin{aligned}
-a_{11}y_1 - a_{12}y_2 - \cdots - a_{1n}y_n &= x_1 \\
-a_{21}y_1 - a_{22}y_2 - \cdots - a_{2n}y_n &= x_2 \\
&\;\;\vdots \\
-a_{m1}y_1 - a_{m2}y_2 - \cdots - a_{mn}y_n &= x_m
\end{aligned}
\right\} \quad -A\,Y = X. \tag{2.3}
$$

These two systems are *dual* in the sense that

$$
[\Xi,\ \mathrm{H}]\begin{bmatrix} X \\ Y \end{bmatrix} = \Xi X + \mathrm{H}Y = \Xi(-A\,Y) + (\Xi A)\,Y = 0 \tag{2.4}
$$

for any Ξ, H satisfying the column system (2.2) and any X, Y satisfying the row system (2.3).

The column system (2.2) consists of n linear equations in $m + n$ variables; these n equations are linearly independent because each η occurs with nonzero coefficient in just one equation. The row system (2.3) consists of m linear equations in $n + m$ variables; these m equations are linearly independent because each x occurs with nonzero coefficient in just one equation. If the Greek variables Ξ, H are regarded as (row) coordinates in a space of $m + n$ dimensions and the Latin variables X,

Y as (column) coordinates in the same space, then the solution sets of (2.2) and (2.3) are linear subspaces of complementary dimensions m and n, respectively, in the space of $m + n$ dimensions. Because of (2.4), these are complementary *orthogonal* linear subspaces. Thus the "duality" of linear systems has the geometric interpretation of "orthogonal complementarity."

3. Block-Pivot Transformation

Let A_{11} be a *nonsingular* square submatrix of A, and A_{12}, A_{21}, A_{22} the remaining submatrices of A. Then the schema (2.1) can be rewritten as

$$
\begin{array}{cc}
 & \begin{array}{cc} -Y_1 & -Y_2 \end{array} \\
\begin{array}{c} \Xi_1 \\[1em] \Xi_2 \end{array} &
\begin{array}{|c|c|}
\hline
A_{11} & A_{12} \\
\hline
A_{21} & A_{22} \\
\hline
\end{array}
\begin{array}{c} = X_1 \\[1em] = X_2 \end{array} \\
 & \begin{array}{cc} = H_1 & = H_2 \end{array}
\end{array}
\qquad (3.1)
$$

Since A_{11}^{-1} exists, the subsystems

$$\Xi_1 A_{11} + \Xi_2 A_{21} = H_1 \quad \text{and} \quad -A_{11}Y_1 - A_{12}Y_2 = X_1$$

can be solved for Ξ_1 and Y_1 to obtain

$$\Xi_1 = H_1 A_{11}^{-1} - \Xi_2 A_{21} A_{11}^{-1} \quad \text{and} \quad Y_1 = -A_{11}^{-1}X_1 - A_{11}^{-1}A_{12}Y_2.$$

Substitution for Ξ_1 and Y_1 in the subsystems

$$\Xi_1 A_{12} + \Xi_2 A_{22} = H_2 \quad \text{and} \quad -A_{21}Y_1 - A_{22}Y_2 = X_2$$

yields

$$H_1 A_{11}^{-1}A_{12} + \Xi_2 (A_{22} - A_{21}A_{11}^{-1}A_{12}) = H_2$$

and

$$A_{21}A_{11}^{-1}X_1 - (A_{22} - A_{21}A_{11}^{-1}A_{12})Y_2 = X_2.$$

These results are exhibited by the column and row systems of the schema

$$
\begin{array}{c|c|c|}
 & -X_1 & -Y_2 & \\
\hline
\mathrm{H}_1 & A_{11}^{-1} & A_{11}^{-1}A_{12} & = Y_1 \\
\hline
\Xi_2 & -A_{21}A_{11}^{-1} & A_{22} - A_{21}A_{11}^{-1}A_{12} & = X_2 \\
\hline
 & = \Xi_1 & = \mathrm{H}_2 &
\end{array}
\tag{3.2}
$$

The schema (3.2) is *equivalent* to the schema (3.1) in the sense that the column equation systems of (3.1) and (3.2) have the same solutions Ξ, H and the row equation systems of (3.1) and (3.2) have the same solutions X, Y.

Let r be the order of the nonsingular square submatrix A_{11}, the choice of which determines uniquely the transformation from the schema (3.1) to the equivalent schema (3.2). Then the transformation from (3.1) to (3.2) is called a *block-pivot transformation of order r*, the nonsingular square submatrix A_{11} of order r being called the *block pivot*. It can readily be verified that the inverse of the block-pivot transformation from (3.1) to (3.2) is a block-pivot transformation from (3.2) to (3.1), the block pivot being A_{11}^{-1}.

Any nonzero entry of the matrix A determines a block pivot A_{11} of order one; the corresponding pivot transformation of order one is called an *elementary pivot transformation*. Elementary pivoting, utilized so effectively in the simplex method, has its roots in the classical process of Gauss-Jordan (complete) elimination.

Note that the block-pivot transformation of order r from (3.1) to (3.2) exchanges r of the individual marginal labels at the left with r labels at the bottom and r parallel labels at the right with r parallel labels at the top, signs being reversed in the latter exchange. Such a block-pivot transformation can always be decomposed into a succession of elementary pivot transformations, exchanging just one label on a margin at a time; conversely, any finite succession of elementary pivot transformations is summarized by a single block-pivot transformation (as explained in [5] and illustrated in [7]).

The m by n matrices in (3.1) and (3.2), or any row and/or column permutations thereof, are *combinatorially equivalent* in a sense discussed by the author in [5]. In fact, the relationship between (3.1) and (3.2) can be taken as defining combinatorial equivalence.

4. Dual Linear Programs

Here the format developed in the two previous sections will be used, with some change of symbols, to discuss dual linear programs.

The schema

$$
\begin{array}{c|cccc|cc}
 & -x_1 & -x_2 & \cdots & -x_N & 1 & \\
\hline
\lambda_1 & a_{11} & a_{12} & \cdots & a_{1N} & b_1 & = 0 \\
\lambda_2 & a_{21} & a_{22} & \cdots & a_{2N} & b_2 & = 0 \\
\vdots & \vdots & \vdots & & \vdots & \vdots & \vdots \\
\lambda_M & a_{M1} & a_{M2} & \cdots & a_{MN} & b_M & = 0 \\
\hline
1 & c_1 & c_2 & \cdots & c_N & d & = w \\
\hline
 & = \xi_1 & = \xi_2 & \cdots & = \xi_N & = \omega &
\end{array}
\tag{4.1}
$$

exhibits row and column equation systems

$$
\left. \begin{array}{c} -AX + B = 0 \\ -CX + d = w \end{array} \right\} \quad \text{and} \quad \left\{ \begin{array}{c} \Lambda A + C = \Xi \\ \Lambda B + d = \omega \end{array} \right. ,
$$

which pertain to the following pair of linear programs:

Primal program: To maximize $w = d - CX$

constrained by $AX = B, \qquad X \geqq 0.$ (4.2)

Dual program: To minimize $\omega = d + \Lambda B$

constrained by $\Lambda A + C = \Xi \geqq 0.$ (4.3)

(In this chapter only, vector inequalities are used. The inequality $X \geq 0$ means that each of the components $x_1, x_2, \cdots, x_N$ of X is nonnegative and at least one of them is positive. The inequality $X \geqq 0$ means merely that each component is nonnegative. For conformity, we shall also use the symbol $\geqq$ rather than $\geq$ for scalar inequalities.) The "parameters"

$$(\lambda_1, \lambda_2, \cdots, \lambda_M) = \Lambda$$

in the dual program are unrestricted in sign.

Let A_{11} be a *nonsingular* square submatrix of the matrix A above.

Then the schema (4.1) can be recast as

$$
\begin{array}{c|cc|c}
 & -X_1 & -X_2 & 1 \\
\hline
\Lambda_1 & A_{11} & A_{12} & B_1 & = 0 \\
\Lambda_2 & A_{21} & A_{22} & B_2 & = 0 \\
\hline
1 & C_1 & C_2 & d & = w \\
\hline
 & = \Xi_1 & = \Xi_2 & = \omega
\end{array}
\qquad (4.4)
$$

(Of course, the Λ_2-headed row in (4.4) will be vacuous if the submatrix A_{11} omits no row of A, and the X_2-headed column in (4.4) will be vacuous if A_{11} omits no column of A.) Define

$$
\begin{aligned}
\overline{A}_{11} &= A_{11}^{-1}, & \overline{A}_{12} &= A_{11}^{-1}A_{12}, & \overline{B}_1 &= A_{11}^{-1}B_1, \\
\overline{A}_{21} &= -A_{21}A_{11}^{-1}, & \overline{A}_{22} &= A_{22} - A_{21}A_{11}^{-1}A_{12}, & \overline{B}_2 &= B_2 - A_{21}A_{11}^{-1}B_1, \\
\overline{C}_1 &= -C_1A_{11}^{-1}, & \overline{C}_2 &= C_2 - C_1A_{11}^{-1}A_{12}, & \overline{d} &= d - C_1A_{11}^{-1}B_1.
\end{aligned}
$$

Then the schema

$$
\begin{array}{c|cc|c}
 & 0 & -X_2 & 1 \\
\hline
\Xi_1 & \overline{A}_{11} & \overline{A}_{12} & \overline{B}_1 & = X_1 \\
\Lambda_2 & \overline{A}_{21} & \overline{A}_{22} & \overline{B}_2 & = 0 \\
\hline
1 & \overline{C}_1 & \overline{C}_2 & \overline{d} & = w \\
\hline
 & = \Lambda_1 & = \Xi_2 & = \omega
\end{array}
\qquad (4.5)
$$

results from the schema (4.4) by the block-pivot transformation having A_{11} as block pivot.

The new schema (4.5) is equivalent to the old schema (4.4). That is, the row equation system of one schema is satisfied by any X, w satisfying the row equation system of the other schema, and the column equation system of one schema is satisfied by any Λ, Ξ, ω satisfying the column equation system of the other schema. Hence the primal program (4.2) calls now for maximizing w subject to the row equation system of (4.5) and the inequalities

$$ X_1 \geq 0, \qquad X_2 \geq 0, $$

and the dual program (4.3) calls now for minimizing ω subject to the column equation system of (4.5) and the inequalities

$$ \Xi_1 \geq 0, \qquad \Xi_2 \geq 0. $$

If the schema (4.5) is such that

$$\bar{A}_{22} = 0, \qquad \bar{B}_2 = 0 \qquad\qquad (4.6)$$

(or are vacuous) and

$$\bar{B}_1 \geqq 0, \qquad \bar{C}_2 \geqq 0, \qquad\qquad (4.7)$$

then optimal (basic) solutions of the primal and dual programs, (4.2) and (4.3), can be read directly from (4.5) by setting variables at top and left margins equal to zero. These optimal solutions are

$$X_1 = \bar{B}_1(\geqq 0), \qquad X_2 = 0; \qquad w = \bar{d}$$

and

$$\Lambda_1 = \bar{C}_1, \qquad \Lambda_2 = 0; \qquad \Xi_1 = 0, \qquad \Xi_2 = \bar{C}_2(\geqq 0); \qquad \omega = \bar{d}.$$

That $\bar{d}$ is the maximal w follows from $\bar{C}_2 \geqq 0$, because

$$w = \bar{d} - \bar{C}_2 X_2 \leqq \bar{d} \quad \text{for all} \quad X_2 \geqq 0;$$

and that $\bar{d}$ is the minimal ω follows from $\bar{B}_1 \geqq 0$, because

$$\omega = \bar{d} + \Xi_1 \bar{B}_1 \geqq \bar{d} \quad \text{for all} \quad \Xi_1 \geqq 0.$$

The Dantzig simplex method, starting from an initial "presentation" of the pair of linear programs (4.2) and (4.3), employs a finite succession of elementary pivot transformations to achieve, if possible, a terminal "re-presentation" corresponding to a schema (4.5) for which (4.6) and (4.7) hold.

5. Canonical Representation

A *canonical* representation ("re-presentation") of the pair of linear programs (4.2) and (4.3) is provided by any schema

	0	$-X_2$	1	
Ξ_1	$\bar{A}_{11}$	$\bar{A}_{12}$	$\bar{B}_1$	$= X_1$
Λ_2	$\bar{A}_{21}$	0	0	$= 0$
1	$\bar{C}_1$	$\bar{C}_2$	$\bar{d}$	$= w$
	$= \Lambda_1$	$= \Xi_2$	$= \omega$	

$$(5.1)$$

for which (4.6) holds. To have

$$\overline{A}_{22} = A_{22} - A_{21}A_{11}^{-1}A_{12} = 0,$$

it is necessary and sufficient that the order of the block pivot A_{11} be equal to the rank m of the matrix A, since then

$$[A_{21}, A_{22}] = A_{21}A_{11}^{-1}[A_{11}, A_{12}].$$

If $\overline{A}_{22} = 0$, then $\overline{B}_2 = 0$ also, unless $AX = B$ is an inconsistent system of linear equations.

A partly reduced canonical schema

$$
\begin{array}{cccc}
 & 0 & -X_2 & 1 \\
\hline
\Xi_1 & \overline{A}_{11} & \overline{A}_{12} & \overline{B}_1 & = X_1 \\
\hline
1 & \overline{C}_1 & \overline{C}_2 & \overline{d} & = w \\
\hline
 & = \Lambda_1 & = \Xi_2 & = \omega
\end{array}
\tag{5.2}
$$

results from (5.1) through deletion of the Λ_2-headed row in (5.1). The schema (5.2) contains the same information as (5.1) with redundant parameters Λ_2 set equal to zero.

A fully reduced canonical schema

$$
\begin{array}{ccc}
 & -X_2 & 1 \\
\hline
\Xi_1 & \overline{A}_{12} & \overline{B}_1 & = X_1 \\
\hline
1 & \overline{C}_2 & \overline{d} & = w \\
\hline
 & = \Xi_2 & = \omega
\end{array}
\tag{5.3}
$$

results through further deletion of the 0-headed column of (5.2). The schema (5.3) contains all the parameter-free information in (5.1) or (5.2). This corresponds to the "canonical form" in which dual linear programs were originally studied.

If $\overline{B}_1 \geqq 0$ in the above schemata, the canonical representation is *primal feasible* with $X_1 = \overline{B}_1(\geqq 0)$, $X_2 = 0$ yielding a feasible (basic) solution of the primal program (4.2). If $\overline{C}_2 \geqq 0$ in the above schemata, the canonical representation is *dual* feasible with $\Lambda_1 = \overline{C}_1$, $\Lambda_2 = 0$ and $\Xi_1 = 0$, $\Xi_2 = \overline{C}_2(\geqq 0)$ yielding a feasible (basic) solution of the dual program (4.3). If both $\bar{B}_1 \geqq 0$ and $\bar{C}_2 \geqq 0$, the canonical representation is *optimal* with the above-stated feasible (basic) solutions as optimal (basic) solutions of the primal and dual programs.

6. Geometric Interpretation

Let the matrix A in schema (4.1) have rank m and let the number of columns of A be $N = m + n$. Let $[A, B]$ also have rank m, so that $AX = B$ is a consistent system of linear equations. Let S be a space of $N = m + n$ dimensions with a specified coordinate system, so that there is a one-to-one correspondence between points (or vectors) of S and ordered coordinate N-tuples, written as $\xi_1, \xi_2, \cdots, \xi_N$ for row usage and as $x_1, x_2, \cdots, x_N$ for column usage. Then the solution sets

$$P = \{\Xi \mid \Xi = \Lambda A + C, \text{ all } \Lambda\} \quad \text{and} \quad Q = \{X \mid AX = B\}$$

are linear manifolds of complementary dimensions m and n in the space S. Let $\Xi = \Lambda A + C$ and $\Xi' = \Lambda'A + C$ be any two points of P, and X and X' any two points of Q. Then the equation

$$(\Xi' - \Xi)(X' - X) = (\Lambda'A - \Lambda A)(X' - X)$$
$$= (\Lambda' - \Lambda)(AX' - AX) = 0$$

shows that P and Q are complementary *orthogonal* linear manifolds in S.
Let

$$R = \{\Xi \mid \Xi \geqq 0\} = \{X \mid X \geqq 0\}$$

be the *nonnegative orthant* in S. Then the feasible-solution sets

$$\{\Xi \mid \Xi = \Lambda A + C, \Xi \geqq 0\} \quad \text{and} \quad \{X \mid AX = B, X \geqq 0\}$$

of the dual and primal programs (4.3) and (4.2) are polyhedral convex sets $P \cap R$ and $Q \cap R$, respectively.

In a canonical schema (5.3), the complementary orthogonal linear manifolds P and Q are represented by equation systems in the "slope-intercept" form,

$$P: \quad \Xi_2 = \Xi_1 \bar{A}_{12} + \bar{C}_2 \tag{6.1}$$

and

$$Q: \quad \Xi_1 = \Xi_2(-\bar{A}_{12}^T) + \bar{B}_1^T, \tag{6.2}$$

the latter being obtained by transposing

$$X_1 = -\bar{A}_{12}X_2 + \bar{B}_1$$

and substituting Ξ_1 and Ξ_2 for X_1^T and X_2^T. In (6.1) the m by n matrix $\overline{A}_{12}$ is the "Ξ_2:Ξ_1-slope" of P (with Ξ_2 as "rise" and Ξ_1 as "run") and $\overline{C}_2$ is the "Ξ_2-intercept" of P. In (6.2) the negative-transpose matrix $-\overline{A}_{12}^T$ is the "Ξ_1:Ξ_2-slope" of Q (with Ξ_1 as "rise" and Ξ_2 as "run") and $\overline{B}_1^T$ is the "Ξ_1-intercept" of Q. This canonical "slope-intercept" representation of P and Q, introduced by the author in [8], generalizes the relation between the equations $y = mx + b$ and $x = -my + a$ of orthogonal straight lines in plane analytic geometry.

Let $\bar{p}$ and $\bar{q}$ be the intercept points (vectors),

$$\Xi_1 = 0, \quad \Xi_2 = \overline{C}_2 \quad \text{and} \quad \Xi_1 = \overline{B}_1^T, \quad \Xi_2 = 0$$

determined by (6.1) and (6.2). Note that the inner (scalar) product of $\bar{p}$ and $\bar{q}$ satisfies the equation

$$\bar{p} \cdot \bar{q} = [0, \overline{C}_2] \begin{bmatrix} \overline{B}_1 \\ 0 \end{bmatrix} = 0.$$

As canonically represented in schema (5.3), the dual program is to minimize

$$\omega = \bar{d} + \Xi_1 \overline{B}_1 = \bar{d} + [\Xi_1, \Xi_2] \begin{bmatrix} \overline{B}_1 \\ 0 \end{bmatrix} = \bar{d} + p \cdot \bar{q}$$

for p in $P \cap R$, and the primal program is to maximize

$$w = \bar{d} - \overline{C}_2 X_2 = \bar{d} - [0, \overline{C}_2] \begin{bmatrix} X_1 \\ X_2 \end{bmatrix} = \bar{d} - \bar{p} \cdot q$$

for q in $Q \cap R$. If $\bar{p}$ belongs to $P \cap R$ and $\bar{q}$ belongs to $Q \cap R$, then $p \cdot \bar{q} \geqq 0$ for every p in $P \cap R$, and $\bar{p} \cdot q \geqq 0$ for every q in $Q \cap R$ (since any two vectors in R have a nonnegative inner product). Hence, since $\bar{p} \cdot \bar{q} = 0$, it is clear that

$$\omega = \bar{d} + p \cdot \bar{q} \geqq \bar{d} + \bar{p} \cdot \bar{q} = \bar{d} \qquad \text{for every } p \text{ in } P \cap R,$$

and that

$$w = \bar{d} - \bar{p} \cdot q \leqq \bar{d} - \bar{p} \cdot \bar{q} = \bar{d} \qquad \text{for every } q \text{ in } Q \cap R.$$

That is, the desired minimum and maximum are attained at $p = \bar{p}$ and $q = \bar{q}$ if these points both belong to R. (The intercept points $\bar{p}$ or $\bar{q}$ belonging to R are the extreme points of the polyhedral convex set $P \cap R$ or $Q \cap R$.)

In summary, this geometric interpretation of a pair of linear programs (4.2) and (4.3) involves complementary orthogonal linear manifolds P and Q in a space S with nonnegative orthant R. If $P \cap R$ is nonvacuous, the dual program is feasible; if $Q \cap R$ is nonvacuous, the primal program is feasible. A canonical representation of these programs involves a joint "slope-intercept" representation of P and Q. The resulting intercept points $\bar{p}$ and $\bar{q}$ yield optimal solutions if they both belong to R.

7. Simplex Method; Terminal Possibilities

Let $AX = B$ have a solution $X \geq 0$; that is, suppose $Q \cap R$ is nonvacuous and the primal program (4.2) is feasible. Then a proof of the validity of the simplex method, such as the one given in [9], demonstrates the existence of a finite succession of elementary pivot transformations that terminates in a canonical representation for which the matrix

$$
\begin{array}{c|c}
\bar{A}_{12} & \bar{B}_1 \\
\hline
\bar{C}_2 & \bar{d}
\end{array}
\tag{7.1}
$$

of the schema (5.3) has *either* the schematic form

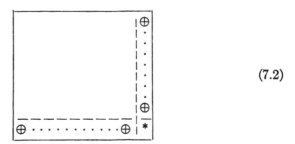

$$\tag{7.2}$$

or the schematic form

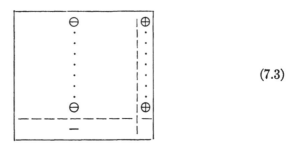

$$\tag{7.3}$$

where each $\oplus$ denotes a positive or zero entry, each $\ominus$ a negative or zero entry, and $-$ a negative entry. The $\oplus$ row and $\oplus$ column in (7.2) determine optimal extreme points of $P \cap R$ and $Q \cap R$, the corner entry * being the common minimum and maximum value. In (7.3) the $\oplus$ column determines an extreme point $\bar{q}$ of $Q \cap R$ and the $\ominus$ column determines the direction of an extreme ray of $Q \cap R$ issuing from $\bar{q}$, along which the objective function w satisfies $w \to + \infty$ because of the corresponding minus entry at the bottom. At the same time the $(\ominus, -)$ column in (7.3) shows that $P \cap R$ is vacuous and the dual program is infeasible.

If $AX = B$ is a consistent system having no solution $X \geqq 0$, so that Q exists but $Q \cap R$ is vacuous and the primal program (4.2) is infeasible, then it can be shown that there exists a finite succession of elementary pivot transformations terminating in a canonical representation for which the matrix (7.1) of the schema (5.3) has *either* the form

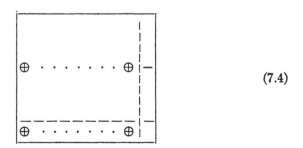

$$(7.4)$$

or the form

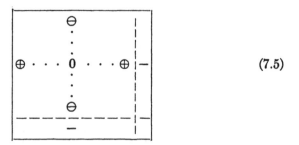

$$(7.5)$$

In (7.4) the $\oplus$ row at the bottom determines an extreme point $\bar{p}$ of $P \cap R$ and the other $\oplus$ row determines the direction of an extreme ray of $P \cap R$ issuing from $\bar{p}$, along which the objective function ω satisfies $\omega \to - \infty$ because of the corresponding minus entry at the right. In (7.5) the nonpositive column with negative entry at bottom shows that $P \cap R$ is vacuous and the dual program is infeasible. The $(\oplus, -)$

row in (7.4) and (7.5) confirms that $Q \cap R$ is vacuous and the primal program is infeasible.

In summary, the terminal possibilities for the simplex method are, in the format of this chapter:

Form (7.2)—primal feasible $(Q \cap R \neq \phi)$,

dual feasible $(P \cap R \neq \phi)$;

Form (7.3)—primal feasible $(Q \cap R \neq \phi)$,

dual infeasible $(P \cap R = \phi)$;

Form (7.4)—primal infeasible $(Q \cap R = \phi)$,

dual feasible $(P \cap R \neq \phi)$;

Form (7.5)—primal infeasible $(Q \cap R = \phi)$,

dual infeasible $(P \cap R = \phi)$.

From any initial presentation (4.1) of the pair of linear programs (4.2) and (4.3), provided $AX = B$ is a consistent system of linear equations (so that Q exists), it is possible through a finite succession of elementary pivot transformations to reach a terminal canonical representation for which the matrix (7.1) of the schema (5.3) has one of the above four forms (7.2), (7.3), (7.4), (7.5).

8. Homogeneous Linear Programs and Transposition-Duality Theorems

In the pair of linear programs (4.2) and (4.3), take $B = 0$ and $d = 0$ to get a *homogeneous linear program,*

$$\text{Minimize } CX \text{ constrained by } AX = 0, \quad (X \geq 0), \qquad (8.1)$$

and its dual program,

$$\text{Solve } UA + C \geq 0. \qquad (8.2)$$

(Here it seems convenient to minimize $CX = -w$ rather than to maximize $w = -CX$, to replace the parametric Λ by U, and to omit Ξ.) The programs (8.1) and (8.2) are jointly exhibited by the schema

$$
\begin{array}{c|c|c}
 & X & (\geq 0) \\
\hline
U & A & = 0 \\
\hline
1 & C & = \min \\
\hline
 & \geq 0 &
\end{array}
\qquad (8.3)
$$

The homogeneous linear program (8.1) is clearly feasible, since $X = 0$ satisfies $AX = 0$. There are just two possibilities (corresponding to the two cases set forth in the first paragraph of Sec. 7): *either CX has a zero minimum and (8.2) is feasible or CX is unbounded below for feasible X and (8.2) is infeasible.* These two possibilities establish a "theorem of alternatives" for a homogeneous linear program (8.1) and its dual (8.2):

THEOREM 1. *Either $UA + C \geq 0$ for some U or $CX < 0$ for some $X \geq 0$ such that $AX = 0$ (but not both).*

This theorem can be regarded as a fundamental existence theorem for an arbitrary system $UA + C \geq 0$ of nonhomogeneous linear inequalities:

THEOREM 2. *The inequality $UA + C \geq 0$ holds for some U if and only if there is no $X \geq 0$ for which $AX = 0$ and $CX < 0$.*

Take $C < 0$. Then $UA + C \geq 0$ implies $UA \geq -C > 0$. Also, $CX < 0$ for $X \geq 0$ if and only if $X \neq 0$. Hence Theorem 1 yields the following classical theorem of Gordan (and later Stiemke), which seems to have been the earliest known transposition-duality theorem (see [6]):

THEOREM 3. *The equality $AX = 0$ holds for some $X \geq 0$ (i.e., $X \geq 0$ and $\neq 0$) if and only if $UA > 0$ for no U.*

Now form the schema

$$
\begin{array}{c|cc|c}
 & x_0 & X' \ (\geq 0) & \\
\hline
U & -B & A & = 0 \\
\hline
1 & -1 & 0 & = \min \\
\hline
 & \geq 0 & \geq 0 &
\end{array}
\qquad (8.4)
$$

where A is a matrix and $-B$ an additional column. Clearly the inequality $-UB - 1 \geq 0$ implies $UB \leq -1 < 0$, and the equality $-Bx_0 + AX' = 0$ for $x_0 > 0$, $X' \geq 0$ implies $AX = B$ for $X = (X'/x_0) \geq 0$. Hence the alternatives of Theorem 1, applied to (8.4), establish the following classical theorem of Farkas concerning "convex-linear dependence":

THEOREM 4. *If $UB \geq 0$ for all U such that $UA \geq 0$, then $B = AX$ for some $X \geq 0$ (and conversely).*

Next form the schema

$$
\begin{array}{c|cccc|l}
 & X_1 & X_2 & X_3^+ & X_2^- & (\geq 0) \\
\hline
U & A_1 & A_2 & A_3 & -A_3 & = 0 \\
\hline
1 & -1 & 0 & 0 & 0 & = \min \\
\hline
 & \geq 0 & \geq 0 & \geq 0 & \geq 0
\end{array}
\tag{8.5}
$$

where -1 denotes a row of -1's. Observe that $UA_1 \geq 1$ implies $UA_1 > 0$ and that $UA_3 \geq 0$, $-UA_3 \geq 0$ imply $UA_3 = 0$. Let $X_3 = X_3^+ - X_3^-$. Then Theorem 1, applied to (8.5), establishes the general transposition theorem of T. S. Motzkin:

THEOREM 5. *Either $UA_1 > 0$, $UA_2 \geq 0$, $UA_3 = 0$ for some U or $A_1X_1 + A_2X_2 + A_3X_3 = 0$ for some $X_1 \geq 0$, $X_2 \geq 0$, X_3 unrestricted.*

9. Theorems for Skew and Dual Linear Systems

Let K be a skew-symmetric (square) matrix, that is, let $K^T = -K$, and let I be the identity matrix of equal order. Form the homogeneous linear program and its dual:

$$
\begin{array}{c|ccc|l}
 & X & Y & Z & (\geq 0) \\
\hline
U & K + I & K & I & = 0 \\
\hline
1 & -1 & 0 & 0 & = \min \\
\hline
 & \geq 0 & \geq 0 & \geq 0
\end{array}
\tag{9.1}
$$

where -1 denotes a row of -1's. Premultiply

$$(K + I)X + KY + IZ = 0$$

by $(X+Y)^T$ to get

$$(X + Y)^T K(X + Y) + (X + Y)^T I(X + Z) = 0.$$

Then since

$$(X + Y)^T K (X + Y) \equiv 0,$$

it follows that

$$X^T X + X^T Z + Y^T X + Y^T Z = 0.$$

However, this holds for $X \geq 0$, $Y \geq 0$, $Z \geq 0$ if and only if each term is zero; and $X^T X = 0$ if and only if $X = 0$. Hence the homogeneous linear program specified by the rows of (9.1) has a zero minimum, and the dual program specified by the columns of (9.1) is feasible. That is, there exists some U^* satisfying the column inequalities of (9.1):

$$U(K + I) \geq 1 > 0, \qquad UK \geq 0, \qquad UI \geq 0.$$

This establishes the following "skew-symmetric matrix theorem" (see [6], Theorem 5):

THEOREM 6. *The system* $UK \geq 0$ *of homogeneous linear inequalities, where* $K^T = -K$, *possesses a solution* $U^* \geq 0$ *such that* $U^* + U^* K > 0$.

Apply Theorem 6 to the matrix

$$K = \begin{bmatrix} 0 & A \\ -A^T & 0 \end{bmatrix}.$$

Then the inequality

$$[\Xi, Y^T] \begin{bmatrix} 0 & A \\ -A^T & 0 \end{bmatrix} \geq 0$$

possesses a solution $\Xi^* \geq 0$, $Y^* \geq 0$ such that

$$[\Xi^*, Y^{*T}] + [\Xi^*, Y^{*T}] \begin{bmatrix} 0 & A \\ -A^T & 0 \end{bmatrix} > 0.$$

This establishes the following theorem (see [6], Theorem 3) concerning the dual linear systems of schema (2.1) in Section 2:

THEOREM 7. *The column and row equation systems of the schema*

$$\begin{array}{c} -Y \\ \Xi \begin{array}{|c|} \hline A \\ \hline \end{array} = X \\ = H \end{array}$$

possess solutions

$$\Xi^* \geq 0, \qquad H^* \geq 0 \quad and \quad X^* \geq 0, \qquad Y^* \geq 0$$

such that

$$\Xi^* + X^{*T} > 0 \quad and \quad H^* + Y^{*T} > 0.$$

Apply Theorem 7 to

$$
\begin{array}{c|ccc|c}
 & -Y_1 & -Y_2^+ & -Y_2^- & \\
\hline
\Xi_1 & A_{11} & A_{12} & -A_{12} & = X_1 \\
\Xi_2^+ & A_{21} & A_{22} & -A_{22} & = X_2^+ \\
\Xi_2^- & -A_{21} & -A_{22} & A_{22} & = X_2^- \\
\hline
 & = H_1 & = H_2^+ & = H_2^- &
\end{array}
\qquad (9.2)
$$

where A_{11} is an arbitrary submatrix of a matrix A and A_{12}, A_{21}, A_{22} are the remaining submatrices. Then there exist nonnegative solutions (starred) of the column and row equation systems of (9.2) such that

$$[\Xi_1^*, \Xi_2^{+*}, \Xi_2^{-*}] + [X_1^{*T}, X_2^{+*T}, X_2^{-*T}] > 0$$

and

$$[H_1^*, H_2^{+*}, H_2^{-*}] + [Y_1^{*T}, Y_2^{+*T}, Y_2^{-*T}] > 0.$$

Since the sum of the last two columns of (9.2) is zero, and also the sum of the last two rows, it follows that

$$H_2^{+*} + H_2^{-*} = 0 \quad and \quad X_2^{+*} + X_2^{-*} = 0.$$

Hence H_2^{+*}, H_2^{-*} and X_2^{+*}, X_2^{-*}, being nonnegative, are all zero. Now set

$$\Xi_2 = \Xi_2^+ - \Xi_2^-, \qquad H_2 = H_2^+ - H_2^-$$

and

$$X_2 = X_2^+ - X_2^-, \qquad Y_2 = Y_2^+ - Y_2^-$$

to obtain the following general transposition-duality theorem for dual linear systems (see [6], Theorem 6):

THEOREM 8. *The column and row equation systems of the schema*

$$
\begin{array}{c}
\quad -Y_1 \quad -Y_2 \\
\begin{array}{c}
\Xi_1 \\
\Xi_2
\end{array}
\left.
\begin{array}{cc}
A_{11} & A_{12} \\
A_{21} & A_{22}
\end{array}
\right|
\begin{array}{c}
= X_1 \\
= X_2
\end{array} \\
\quad = H_1 \quad = H_2
\end{array}
$$

possess solutions

$$
\Xi_1^* \geqq 0, \qquad \Xi_2^* \begin{array}{c} > \\ = \\ < \end{array} 0, \qquad H_1^* \geqq 0, \qquad H_2^* = 0
$$

and

$$
X_1^* \geqq 0, \qquad X_2^* = 0, \qquad Y_1^* \geqq 0, \qquad Y_2^* \begin{array}{c} > \\ = \\ < \end{array} 0
$$

such that

$$
\Xi_1^* + X_1^{*T} > 0 \quad and \quad H_1^* + Y_1^{*T} > 0.
$$

References

1. Dantzig, G. B., "Maximization of a Linear Function of Variables Subject to Linear Inequalities," *in* T. C. Koopmans (ed.), *Activity Analysis of Production and Allocation*, The RAND Corporation, R-193, June, 1951; Cowles Commission Monograph No. 13, John Wiley & Sons, Inc., New York, 1951, pp. 339–347.
2. Dantzig, G. B., A. Orden, and P. Wolfe, "The Generalized Simplex Method for Minimizing a Linear Form under Linear Inequality Restraints," The RAND Corporation, P-392, April 5, 1954; *Pacific J. Math.*, Vol. 5, No. 2, June, 1955, pp. 183–195.
3. Dantzig, G. B., "Constructive Proof of the Min-max Theorem," The RAND Corporation, RM-1267-1, December 18, 1953; *Pacific J. Math.*, Vol. 6, No. 1, 1956, pp. 25–33.
4. Stiefel, E., "Note on Jordan Elimination, Linear Programming and Tchebycheff Approximation," *Numer. Math.*, Vol. 2, No. 1, January, 1960, pp. 1–17.
5. Tucker, A. W., "A Combinatorial Equivalence of Matrices," *in* R. Bellman and Marshall Hall, Jr. (eds.), *Combinatorial Analysis*, Proceedings of Symposia in Applied Mathematics, Vol. X, American Mathematical Society, Providence, R.I., 1960, pp. 129–140.
6. Tucker, A. W., "Dual Systems of Homogeneous Linear Relations," *in* H. W. Kuhn and A. W. Tucker (eds.), *Linear Inequalities and Related*

Systems, Annals of Mathematics Study No. 38, Princeton University Press, Princeton, N.J., 1956, pp. 3–18.

7. Tucker, A. W., "Solving a Matrix Game by Linear Programming," *IBM Journal*, Vol. 4, No. 5, November, 1960, pp. 507–517.

8. Tucker, A. W., "Abstract Structure of the Simplex Method," *The RAND Symposium on Linear Programming*, The RAND Corporation, R-351, June 21, 1959, pp. 33–34.

9. Dantzig, G. B. "Inductive Proof of the Simplex Method," The RAND Corporation, P-1851, December 28, 1959; *IBM Journal*, Vol. 4, No. 5, November, 1960, pp. 505–506.

Chapter 11

The Present Status of Nonlinear Programming†

P. WOLFE

1. Introduction

This chapter is devoted to a survey of certain computational procedures for the solution of the "convex nondiscrete" mathematical programming problem. By a mathematical programming problem we shall mean the problem of minimizing a function $f(x)$ of n variables, $(x_1, \cdots, x_n) = x$, subject to the constraints $g_i(x) \leq 0$ $(i = 1, \cdots, n)$. We shall impose the restriction that the function f and all the functions g_i be convex. The point of this restriction, to which we shall later return, is that it seems to define the largest class of functions for which efficient general computational methods can be devised. From now on, when we refer to a nonlinear programming problem, we shall mean one in which the functions involved are so restricted.

In Figure 1 we have illustrated three principal types of computational problems. The functions defining the *constraints* of the problem—the g_i—may be linear or not; and the so-called *objective function*, f, may also be linear or not. If both f and the g_i are linear, we have the most well-known case—that of linear programming. This problem is labeled A in Figure 1. In the next case, which is as close to linear programming as possible, the constraints are linear, while f is not. This case we have labeled B. It is a more difficult type of problem to solve than A, of course; yet, we shall see that some of the techniques used in completely linear problems may be carried over to problems of this type. This does

---

† An early version of this chapter has appeared as "Computational Techniques for Non-linear Programs," privately printed for members of the Princeton University Conference on Linear Programming, March 13–15, 1957.

not seem to be the case with the class of problems C, those in which the constraints are not linear. We have not distinguished here between linearity or nonlinearity of the objective function, because if the constraints of a problem are not linear, linearity of the objective function is not of help. Nevertheless there are methods, although less efficient than those for problems A and B, for dealing with problems of this class.

Figure 2 indicates several types of computational methods that can be used for these problems. One basic distinction is that between *primal* and *Lagrangian* methods.

A *primal* computing method uses only the variables x and directly related quantities, such as the gradient of f, in the course of a compu-

Objective function	Constraint	
	Linear	Nonlinear
Linear	A	
		C
Nonlinear	B	

Type of step	Method	
	Primal	Lagrangian
Walk	A	B (quadramatic)
Hop	B	
Creep	C	C

Fig. 1. (*left*). Types of computational problems.   **Fig. 2.** (*right*). Types of computational methods.

tation. A *Lagrangian* method uses, in addition to these quantities, the generalized Lagrange multipliers to be discussed below.

These various methods may also be distinguished according to the nature of the steps used in proceeding to a solution. The steps may be large ones, with only a finite and possibly small number of them needed to arrive at an exact solution of the problem; such a method we call a "walk." Another method can be said to "hop"; although it takes fairly large steps most of the way, one does not know how many steps will be required to arrive sufficiently close to a solution. Finally there is the type we call "creep," which is characteristic of most gradient methods and involves taking a large number of very small steps. Figure 2 shows the type of problem—A, B, or C—to which the indicated computational style seems best suited.

2. Linear Programming

We shall begin our discussion with a primal walking method. As indicated by Figure 2, the class of linear programming problems is nearly the largest that can be tackled by this method. (Actually, the exact class seems to be the one for which the objective function has the property that, for any k, the set of x such that $f(x) = k$ is a hyperplane. This class includes not only linear functions but also, as pointed out by John Isbell, quotients of linear functions.)

In Figure 3 we have given a geometric visualization of a linear programming problem. There are seven constraints effective in this particular problem: Three constraints, $x_1 \geq 0$, $x_2 \geq 0$, and $x_3 \geq 0$, which we have not written out explicitly in terms of functions of x (programming problems are conventionally taken to deal with non-negative variables of this kind, although it is a simple matter to transform a problem having unrestricted variables to one having non-negative variables, or vice versa), and the remaining four constraints,

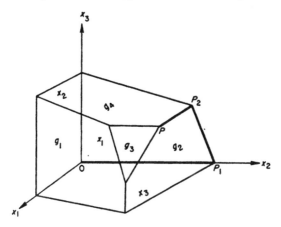

Fig. 3. Linear programming "walking."

illustrated geometrically by the four skew faces of the polyhedron. The whole polyhedron represents the constraint set—the set of all points for which $x \geq 0$ and all $g_i(x) \leq 0$ $(i = 1, \cdots, 4)$. Since these constraints are linear, the faces of the polyhedron are planes. The planes are as follows: The plane i, associated with the constraint $g_i(x) \leq 0$, is the set of points x for which $g_i(x) = 0$. In the diagram the faces have been identified by showing which of the seven functions is equal to zero on each face. The *vertices* of the polyhedron are identifiable by listing the faces that meet in them; for example, the vertex P_2 can be identified as lying on the planes $x_1 = 0$, $g_2 = 0$, and $g_4 = 0$. Such a point of the constraint set, for which as many of the x's and g's vanish as possible, is called a *basic feasible point*. Actually, it is customary to refer to such a point by specifying the complementary functions as a minimal set of *non*vanishing variables.

The linear programming problem, and Dantzig's simplex method for solving it, can be visualized in this way: Let $f(x) = \sum_j c_j x_j$. Then the gradient of f, $\nabla f = c$, is constant. Since the gradient points in the direction of maximum *increase* of f, a solution to the problem of minimizing f will be a point located as far in the constraint set as possible in the direction opposite to the vector c. Take any vertex of the constraint set. The direction numbers of all the edges leading out of the point may be calculated, so we can determine which edges make an obtuse angle with c. Following such an edge, we arrive at another extreme point yielding a lower value of f. The process is repeated until we reach a point at which all edges make acute angles with c; that point is the solution of the problem.

The foregoing process is sketched in Figure 3. Beginning at 0, we find the path $0P_1P_2P_3$ around the constraint set, terminating in a solution point for the problem. The simplex method gives the means of performing the numerical processes corresponding to this description [1].

3. Nonlinear Programming with Linear Constraints

For the problem-type B—the problem of minimizing a nonlinear function subject to linear constraints—a picture can be drawn very much like that of Figure 3. The linear constraints for this problem are the same as for the linear programming problem; see Figure 4. Since the objective function is not linear, however, the gradient vector of the objective function is no longer constant, and at each point a *local*

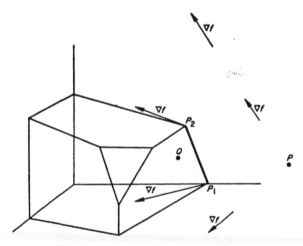

Fig. 4. Minimum-distance problem.

objective vector must be drawn. The fact that there is no single direction of fastest decrease of $f(x)$ makes it impossible to use a simple walking method for the problem; indeed, the solution no longer needs to be a vertex of the constraint set. Figure 4 illustrates the problem of finding the closest point in the constraint polyhedron to an outside point P. Suppose that we attempted to use the gradient $\nabla f(x)$ at each vertex x to establish a direction of motion; we would eventually just circulate among some set of vertices—say P_1 and P_2. To make progress, we must be able to enter a proper face of the constraint set, where the solution point Q lies.

A difficulty of a more general type is that, at the current stage of development, any efficient computational method for attacking a large-scale problem must work almost entirely in terms of *local* information: It must be possible to decide whether to stop the computation, or to continue with it, on the basis of knowledge concerning only the immediate vicinity of the point we have reached, because knowledge of conditions everywhere in the constraint set will generally demand more information than can be stored. Hence a condition such as the following must be imposed on the function f: If a point x gives a minimum of f in some region—no matter how small—surrounding x (i.e., x is a *local minimum*), then x is a solution of the entire problem (i.e., x is a *global minimum*). The most convenient assumption about f that will ensure this is that f must be convex, that is, f must satisfy the inequality

$$f(\alpha x + (1 - \alpha)y) \leq \alpha f(x) + (1 - \alpha)f(y)$$

for any x, y, and $0 \leq \alpha \leq 1$. That this condition is sufficient follows from the fact that if x is not the minimum sought, all points on the line segment $\alpha x + (1 - \alpha)y$ joining it to some lower point y lying in the constraint set give lower values. Figure 5 illustrates the convexity

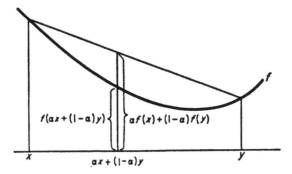

Fig. 5. Convexity.

of a function along a line segment; we require similar behavior along
every line segment in the constraint set. (Any linear function is convex.
The sum of squares of linear functions is also convex, so that the least-
distance problem of Fig. 4 can be solved with local information.)

One method for minimizing a convex function under linear constraints is one that hops. It operates as follows [2]: We will generate
a sequence z^0, z^1, $\cdots$ of points of the constraint set that will converge
to a solution and, for use in the calculation, an auxiliary sequence x^0,

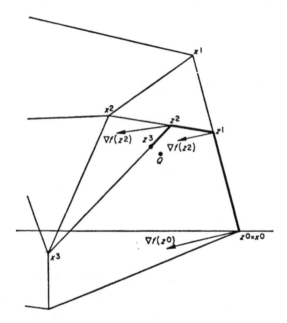

Fig. 6. "Hopping" to a minimum.

x^1, $\cdots$ of *extreme* points. Initially, let x^0 be any extreme point of the
constraint set and let $z^0 = x^0$. Now suppose that n steps have been
taken, and a point z^n and extreme point x^n are at hand (see Fig. 6).
Perform the following operations:

(a) Calculate $\nabla f(z^n)$.

(b) Using $\nabla f(z^n)$ as objective vector and x^n as initial extreme point,
take one step of the simplex method in the minimization of $\nabla f(z^n) \cdot x$,
to the extreme point x^{n+1}.

(c) Choose z^{n+1} so as to minimize f on the segment joining x^{n+1} to z^n.

(d) Repeat with z^{n+1} and x^{n+1}.

The justification of this process lies in the following result:

THEOREM. *There is a constant K such that, if M is the minimum of f as constrained, then*

$$f(z^n) - M \leq \frac{K}{n}.$$

Note that step (c) above amounts to solving a one-dimensional minimization problem, which is easy to do. For modern machine computation, this hopping method has the considerable advantage that the major part of the computational work—that of performing the simplex change of basis from one extreme point to another—is one that is well understood and very likely already coded for the machine one wants to use. The amount of additional routine that has to be written for this method is small.

Another, and very successful, type of hopping method for problems with linear constraints is that of the "projected gradient" [3]. Figure 7 illustrates such a procedure, beginning at the point x^0 and generating

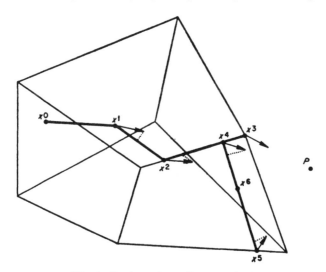

Fig. 7. Projected-gradient method.

the sequence of points x^1, x^2, $\cdots$. Starting with the point x^k, either one or two successors of x^k are determined by the following steps:

(a) Calculate $\nabla f(x^k)$.

(b) Find the *projection* of $\nabla f(x^k)$ onto the *face* of the constraint set on which the point x^k lies. Here "face" is used to denote the intersection of any collection of bounding hyperplane, so that the face for x^0 is the constraint set itself, the projection of $\nabla f(x^k)$ is $\nabla f(x^k)$ itself, and the face for x^2 is the line through x^2 and x^3.

(c) Extend a ray from x^0 in the direction of the projection of $\nabla f(x^k)$. Define x^{k+1} to be the farthest point of the constraint set along this ray.

(d) If $f(x^{k+1}) < f(x^k)$, then the cycle is complete. Otherwise, choose x^{k+2} so as to minimize the function f on the segment $\overline{x^k x^{k+1}}$; this completes the cycle.

As with the gradient-corrected simplex method, it is assumed here that the one-dimensional minimization problem that may have to be solved in step (d) is not a difficult one; this is indeed the case. In Figure 7, the points x^4 and x^6 have been obtained as the result of minimizing on the segments $\overline{x^2 x^3}$ and $\overline{x^4 x^6}$; at these minima, ∇f is, of course, perpendicular to the segment in question.

Convergence of the procedure to a solution of the nonlinear problem is not difficult to establish. Unlike the previous gradient methods, this procedure does not completely reduce to the simplex method for a linear problem, but it does so reduce if the points x^k are vertices of the constraint set.

4. Nonlinear Programming with Nonlinear Constraints: Primal

Figure 8 adds the final complication we want to introduce into programming problems: Besides a nonlinear objective function, we now have nonlinear constraints $g_i(x) \leq 0$. We have said that in general the g_i must be taken to be convex functions. Actually, it is only in such a case that the constraint set is convex, that is, a set that contains the entire line segment joining any two of its points. The nonplanar faces of this constraint set will bulge outward. The necessity for this requirement is implied in our earlier discussion of the convex objective function: To show that any local minimum was global, we made use of the fact that the segment joining two points was in the set. That argument now applies to this more general case, so that the local methods we discuss will solve the global problem.

Most methods for this type of problem are of the creeping kind (the nonlinear boundaries of the constraint set prevent us from taking any bold steps) and use the following general scheme: At any point x of the constraint set, calculate ∇f. Start moving x in the direction $-\nabla f(x)$, modifying this as necessary as x changes. We shall set up a computational scheme that does this, attempting at the same time to satisfy the constraints of the problem. The gradient method has a continuous flavor that may perhaps best be illustrated by setting up a differential equation. This equation will not be solved explicitly, but it can be used

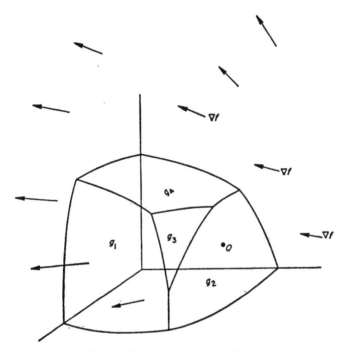

Fig. 8. Very nonlinear problem.

either to indicate analog methods of solution for the problem or to
yield difference equations for digital computation.

By simply proceeding in the direction of fastest decrease, we get

$$\frac{dx}{dt} = -\nabla f(x), \qquad \text{i.e.,} \qquad \frac{dx}{dt^j} = -\frac{\partial f(x)}{\partial x_j}.$$

If we think of the motion of x as taking place in time, the velocity of
the point x is the negative of the gradient. This equation, of course,
ignores the constraints. Under these differential equations, x would
soon leave the constraint set. To inhibit this, we shall try to send x
back into the constraint set whenever it touches the boundary, that is,
whenever one of the functions $g_i(x)$ becomes positive. In that case, the
direction in which to send x is given by the inward normal to the
boundary, $-\nabla g_i(x)$. Our prescription, then, is as follows: If the point
is at $g_i(x) = 0$, send it *in* by adding a vector proportional to $-\nabla g_i(x)$
at that point. We have done this for each g_i by means of the second
term on the right-hand side of the equation below, in which we define
$\delta_i(x)$ to vanish if x satisfies the constraint $g_i(x) \leq 0$, $\delta_i(x) = 1$ other-

wise, and K is a constant of proportionality:

$$\frac{dx}{dt} = -\nabla f(x) - \sum_i K\delta_i(x)\nabla g_i(x),$$

or

$$\frac{dx_j}{dt} = -\frac{\partial f(x)}{\partial x_j} - K\sum_i \delta_i(x)\frac{\partial g_i(x)}{\partial x_j}.$$

One further condition must be noted, however. Since we have the constraints $x_j \geq 0$, we must add the following requirement to the differential equations: If the indicated rate of change of x_j is negative, but x_j is already zero, we must not change x_j. Thus we have

$$\frac{dx_j}{dt} = \begin{cases} \text{same as above, if the above expression or } x_j \text{ is positive,} \\ 0 \text{ otherwise.} \end{cases}$$

It can be shown that any trajectory yielded by these differential equations converges to a solution of the minimization problem if K is so large that at any boundary point x the sum of the inward-pointing vectors, $-\nabla g_i(x)$, is greater than $\nabla f(x)$. Accordingly, any system we can set up that obeys these differential equations will lead to a solution of this type of problem.

The digital means of handling these equations is very simple. It consists of replacing dx_j/dt above by $\Delta x_j/\Delta t$, and choosing t to be a fixed, sufficiently small number. Then one begins with arbitrary x_j's and uses the equation to calculate the amounts Δx_j by which the x_j's must be increased in each time period. After the solution has been found as well as possible using a given value of Δt, it will then be necessary to use a smaller value to obtain more accurate results.

These equations also prompt one to attempt an analog method for solving this problem. It is particularly easy to see how this is done in the linear programming problem. If

$$f(x) = \sum_j c_j x_j \quad \text{and} \quad g_i(x) = \sum_j a_{ij}x_j - b_i \leq 0,$$

then the system of differential equations becomes

$$\frac{dx_j}{dt} = \begin{cases} -c_j - K\sum_i \delta_i(x)a_{ij} \text{ if this expression or } x_j \text{ is positive,} \\ 0 \text{ otherwise.} \end{cases}$$

These equations can be set up in this form on conventional electronic differential-analyzer equipment. This has been done by Pyne [4] for

small-scale problems and works with relatively good accuracy and surprisingly high speed. One can view the trajectories of several x_j on
oscilloscopes and see the solution of a linear programming problem
traced out from an arbitrary initial point in a matter of seconds. In
addition to giving a satisfying graphic account of a solution of the
problem, the analog method has the notable feature that the parameters
occurring in the problem can be varied with a great deal of ease. One
can explore large areas of parameter values quite quickly by this
means; thus it provides a good method for rough sensitivity analysis
in linear problems. It is also possible to wire nonlinearities into the
problem in accordance with the general differential equations, but this
is not easy to do with conventional equipment.

5. Nonlinear Programming with Nonlinear Constraints; Lagrange Multipliers

The remaining methods to be described are those that use the "generalized Lagrange multipliers" of Kuhn and Tucker [5], as well as the
variables x_j, in the computational process. These multipliers u_i are
introduced, as in the classical case, through the Lagrange function

$$L(x, u) = f(x) + \sum_i u_i g_i(x).$$

Also as in the classical case, a necessary condition that x solve the
given extremum problem is that x and some $u = (u_1, \cdots, u_m)$ solve
an extremum problem involving the Lagrangian. In programming,
however, the new problem has a novel formulation:
If x solves the programming problem, then there exists u so that
(x, u) solves the problem

$$\min_{x \geq 0} \ \max_{u \geq 0} L(x, u).$$

This says that simultaneously x must minimize L and u maximize it,
or that (x, u) is a *saddle-point* of L. (In the classical problem, which
involves only setting derivatives equal to zero, it is irrelevant whether
the extrema of the Lagrangian are maxima or minima.)
Under the convexity assumptions of our programming problems,
the necessary condition given above proves to be sufficient. Hence a
method that will enable one to find the saddle-point of a function
having nonnegative variables can be used to solve such problems. A
method along the lines of the primal method can be devised; namely,
differential equations can be set up that will cause x to move so as to

decrease L, and cause u to move so as to increase L, as follows:

$$\frac{dx_j}{dt} = \begin{cases} -\dfrac{\partial L}{\partial x_j} & \text{if this or } x_j \text{ is positive,} \\ 0 & \text{otherwise;} \end{cases}$$

$$\frac{du_i}{dt} = \begin{cases} \dfrac{\partial L}{\partial u_i} & \text{if this or } u_i \text{ is positive,} \\ 0 & \text{otherwise.} \end{cases}$$

As before, we interdict the decreasing of a zero variable.

The foregoing approach is associated primarily with the work of Arrow and Hurwicz, which has appeared in a series of papers. Uzawa [6] has shown that if the objective function f is strictly convex, then a solution of the above equations starting from any initial (x, u) exists, and that the x_j's obtained converge to the solution of the program. Kose [7] has also obtained graphical solutions of these equations on an electronic differential analyzer.

An interesting feature of these equations is that, suitably interpreted, they yield a model for the attainment of efficient production in a competitive economy. Let the constraints be linear again:

$$g_i(x) = \sum_j a_{ij}x_j - b_i \leq 0.$$

The differential equations above then become

$$\frac{dx_j}{dt} = -\frac{\partial f(x)}{\partial x_j} - \sum_i u_i a_{ij}, \qquad \frac{du_i}{dt} = \sum_j a_{ij}x_j - b_i,$$

with "zero" conditions. As usual, x_j is viewed as the level of some production activity. Then $\partial f(x)/\partial x_j$ is a marginal cost, since we are minimizing. Each i denotes a resource needed in production, and b_i is the average amount of resource i available in the market. The coefficient a_{ij} is the amount of resource i that is consumed by carrying on activity j at unit level. Finally, u_i is the market price the producer must pay per unit of the resource i he uses.

The terms on the right-hand side of the first equation can be interpreted in this way: $-\partial f(x)/\partial x_j$ is the profit accruing to the producer for increasing the jth activity level one unit, and the summation is the payment he must make for the additional resources thus consumed. The first equation then says: If a net profit can be made by increasing x_j, then do so; if increasing x_j would make a net loss, then decrease it unless it is already zero. The second equation says simply: If the total

amount of resource i consumed in all the activities exceeds the average
supply, then its price will rise; and if the amount is less than the supply,
then its price will fall unless the price is already zero.

The theorem on the convergence of the solution of the differential
equations to a solution of the programming problem thus says, in
economic terms, that the behavior of the market prices will force the
producer into the optimum production program. It is interesting to
note that the usefulness of this kind of interpretation has led to the
adoption of the term "price" for "generalized Lagrange multiplier" in
the programming literature.

6. Lagrange Multipliers in General

It should be pointed out that these multipliers are present, although
concealed, in the primal methods with which we have dealt. For example, a valuable feature of the simplex method for linear programming
is that in the last step of the simplex calculation one obtains as a by-
product the solution of the "dual problem," that is, the Lagrange multipliers for the extremum.

The multipliers can also be found in the primal creeping process.
After a sufficient length of time, the x_j that solve the equation

$$\frac{dx_j}{dt} = - \frac{\partial f(x)}{\partial x_j} - \sum_i K\delta_i(x) \frac{\partial g_i(x)}{\partial x_j},$$

with nonnegativity condition, will be essentially stationary, and their
average value during an extended time period will be zero. The point
x will, in fact, be tracing out small loops, being kicked back and forth
by the discontinuous terms $\delta_i(x)g_i(x)$. The only quantities on the right-
hand side that vary much will be $K\delta_i(x)$. Denoting their time-average
values by u_i, which is then proportional to the amount of time the constraint $g_i(x) \leq 0$ is called into action, we have

$$0 = - \frac{\partial f(x)}{\partial x_j} - \sum_i u_i \frac{\partial g_i(x)}{\partial x_j}, \tag{6.1}$$

(unless the right-hand side of this equation is negative and $x_j = 0$).
It is easy to see that these u_i are indeed the multipliers, because this is
precisely the condition under which the Lagrangian $L(x, u)$ cannot
be decreased by changing x_j.

The discussion above constitutes the outline of a proof of the Kuhn–
Tucker saddle-point theorem. Through another line of ideas we can
obtain another type of proof, and also suggestions for another computational scheme. In the above equation (6.1), the right-hand side may

be negative, but never positive; in vector notation, if we let

$$v = \nabla f(x) + \sum_i u_i \nabla g_i(x),$$

we need

$$v \geq 0.$$

Now the condition in parentheses following equation (6.1) has a simple paraphrase:

$$\text{If } v_j > 0, \quad \text{then } x_j = 0.$$

Since the variables are all nonnegative, this can be written as

$$vx = \sum_j v_j x_j = 0.$$

We shall now make a (surprisingly slight) modification of our programming problem. Replace the constraints $g_i(x) \leq 0$ by equalities,

$$g_i(x) = 0.$$

The original constraints could be rewritten in this form without loss of generality, if one new variable were added for each constraint, thus:

$$g_i(x) + y_i = 0, \quad y_i \geq 0.$$

Then the saddle-point problem becomes simply

$$\min_{x \geq 0} \ \max_u \ L(x, u);$$

we no longer require $u \geq 0$. The above analysis regarding minimizing L in x is unchanged, but maximizing it in u is now simpler: We need only require that all $\partial L(x, u)/\partial u_i = 0$, which is precisely the same as requiring that all $g_i(x) = 0$. In other words, x must satisfy the constraints.

The final result is the following:

The point x solves the modified programming problem if and only if there exist $v_j \geq 0$ and u_i such that

$$\nabla f(x) + \sum_i u_i \nabla g_i(x) = v$$

and

$$vx = 0.$$

This version of the saddle-point theorem can be justified geometrically (Fig. 9). Letting e_j be the jth coordinate vector, we have

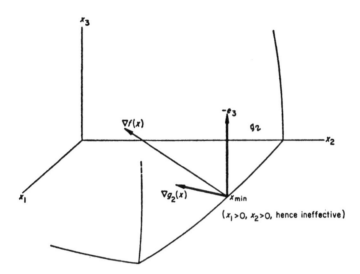

Fig. 9. Geometric representation of saddle-point theorem.

$$\nabla f(x) = \sum_j v_j e_j + \sum_i (-u_i) \nabla g_i(x),$$

with ∇f expressed as a linear combination of normals to the boundaries of the constraint set. We have $v_j \geq 0$ because ∇f must not point outward across the boundaries $x_j \geq 0$. If, however, one of these boundaries is ineffective (i.e., $x_j > 0$), then its normal is also ineffective (i.e., $v_j = 0$), and hence $vx = 0$. This point of view is developed in more detail by Tucker [8].

7. Quadratic Programming

The version of the saddle-point theorem given above yields, rather surprisingly, a walking method [9] for the solution of an important class of nonlinear problems: those in which the constraints are linear and the objective function quadratic. The method is almost exactly the simplex method, although the presence of the multipliers in our formulation of the problem enlarges its size.

The quadratic problem is the following:

$$\text{Minimize } f(x) = px + x^T C x = \sum_j p_j x_j + \sum_{j,k} x_j C_{jk} x_k,$$

$$\text{subject to } x \geq 0 \quad \text{and} \quad Ax = b \left(\text{i.e., } \sum_j a_{ij} x_j = b_i \right).$$

The index i ranges from 1 to m, and j and k range from 1 to n. The superscript T denotes matrix transposition.

Our general requirement that f be a convex function means that the matrix C must be positive semidefinite; that is, that $x^T C x \geq 0$ for all x. The method we shall describe requires further a special sort of "non-degeneracy" condition, namely, that whenever $px = 0$, we have $x^T C x > 0$.

For this problem, $\nabla f(x)$ is just the vector $p + 2Cx$, and each $\nabla g_i(x)$ is the constant vector $(a_{i1}, \cdots, a_{in})$. The saddle-point result thus becomes, in matrix notation:

The point x solves the quadratic programming problem if and only if there exist $v \geq 0$ and u such that $p + 2Cx + uA^T = v$ and $vx = 0$.

The feature that allows us to devise a walking method for quadratic programming is precisely the linearity of ∇f; the only nonlinear condition in the above formulation is $vx = 0$, and it is of a very special form.

The process is begun with an extreme point x^0 of the constraints $x \geq 0$, $Ax = b$. Initially use $v^0 = 0$ and $u^0 = 0$; the condition $vx = 0$ will then be satisfied but, of course, the other condition will not be. We can turn to the device of "artificial variables" to work with this last condition: Let $z_j^0 = p_j + 2(Cx^0)_j$, and let $e_j = \pm 1$ be chosen so that

$$2(Cx^0)_j + e_j z_j^0 = -p_j.$$

We have now chosen an initial feasible simplex basis consisting of part of x^0, u^0, and z^0 for the problem:

Minimize $\sum_j z_j$ under the constraints $x \geq 0$, $v \geq 0$, $z \geq 0$,

$$\sum_j a_{ij} x_j = b_i,$$

$$\sum_k 2C_{jk} x_k + \sum_i u_i a_{ij} - v_j + e_j z_j = -p_j,$$

$$\sum_j v_j x_j = 0.$$

We shall employ the simplex method in this minimization, with one difference in order to handle the last restriction: In considering any of the variables x_j or v_j as candidates for the new basis (i.e., to be made positive), do not allow $x_j > 0$ unless $v_j = 0$, and do not allow $v_j > 0$ unless $x_j = 0$.

It can be shown that this routine will terminate in a finite number of steps, just as in linear programming, with a zero of the objective $\sum_j z_j$. Then the conditions of the theorem above are satisfied, and the

x part of the solution of this problem solves the quadratic programming problem. It is of interest that the simplex solution is usually achieved more quickly for an $m \times n$ quadratic problem than for an $(m + n)$-equation linear problem.

References

1. Dantzig, G. B., A. Orden, and P. Wolfe, "The Generalized Simplex Method for Minimizing a Linear Form under Linear Inequality Restraints," *Pacific J. Math.*, Vol. 5, No. 2, June, 1955, pp. 183–195.
2. Frank, M., and P. Wolfe, "An Algorithm for Quadratic Programming," *Naval Res. Logist. Quart.*, Vol. 3, 1956, pp. 95–110 [Sec. 6].
3. Rosen, J. B., "The Gradient Projection Method. Part I. Linear Constraints," *J. Soc. Indust. Appl. Math.*, Vol. 8, March, 1960, pp. 181–217.
4. Pyne, I. B., "Linear Programming on an Electronic Analogue Computer," *American Institute of Electrical Engineers Transactions Annual*, 1956, pp. 56–147.
5. Kuhn, H. W., and A. W. Tucker, "Nonlinear Programming," *in* J. Neyman (ed.), *Proceedings of the Second Berkeley Symposium on Mathematical Statistics and Problems*, University of California Press, Berkeley and Los Angeles, 1951, pp. 481–492.
6. Arrow, K. J., L. Hurwicz, and H. Uzawa, *Studies in Linear and Non-Linear Programming*, Stanford University Press, Stanford, California, 1960.
7. Kose, T., "Solutions of Saddle Value Problems by Differential Equations," *Econometrica*, Vol. 24, 1956, pp. 59–70.
8. Tucker, A. W., "Linear and Non-Linear Programming," *Operations Res.*, Vol. 5, 1957, pp. 244–257.
9. Wolfe, P., "The Simplex Method for Quadratic Programming," *Econometrica*, Vol. 27, 1959, pp. 382–398.

Chapter 12

The Number of Simplices in a Complex[†]

JOSEPH B. KRUSKAL

1. Introduction

A familiar puzzle for children poses the following question: Given six sticks all the same size, how can you put them together to make four triangles all of the same size? The answer, of course, is a tetrahedron. Similarly we may ask: Given n edges, how many triangles can we make? More generally, suppose that a complex has exactly n r-dimensional simplices. Then we may ask: What is the maximum number of r'-dimensional simplices ($r' > r$) that the complex can have? In this chapter we give an elegant answer to this question.

Since we are concerned here with abstract complexes not embedded in any space, a simplex consists merely of a set of vertices. It will be more convenient for us to label a simplex by the number of its vertices than by its dimension. We speak of an r-set rather than an $(r - 1)$-dimensional simplex. For us a complex is simply a finite set of vertices together with a class of subsets with the subset closure property; that is, if any subset belongs to the complex then all its subsets also belong to the complex.

By $\binom{k}{r}$ we denote the general binomial coefficient. As k increases, this binomial coefficient increases. If n is any nonnegative integer, we define its *r-canonical representation* to be

$$n = \binom{n_r}{r} + \binom{n_{r-1}}{r-1} + \cdots + \binom{n_i}{i},$$

where we first choose n_r to be as large as possible without having the

---

† The problem treated in this chapter was suggested by D. Slepian. The author wishes to thank him and S. Lloyd for many helpful suggestions.

initial binomial coefficient exceed n, and then we choose n_{r-1} as large as possible without having the first two terms exceed n, and so on until we finally obtain equality. We can always obtain equality, for if we do not obtain it before we get to the binomial coefficient with the denominator 1, then n_1 can always be chosen to ensure equality. Furthermore it is clear from our construction that this r-canonical representation is unique. As an illustration we give the 5-canonical representations of several numbers:

$$1 = \binom{5}{5},$$

$$5 = \binom{5}{5} + \binom{4}{4} + \binom{3}{3} + \binom{2}{2} + \binom{1}{1},$$

$$6 = \binom{6}{5},$$

$$62 = \binom{8}{5} + \binom{5}{4} + \binom{3}{3}.$$

It is not difficult to show that a set of integers $n_r, \cdots, n_i$ is associated with the canonical representation of some integer if and only if the following conditions are satisfied:

$$n_r > \cdots > n_i \geq 1.$$

If $r \leq r'$, we define $f(n; r, r')$ to be the *greatest* number of r'-sets that occur in any complex having precisely n r-sets. If $r \geq r'$, we define $f(n; r, r')$ to be the *smallest* number of r'-sets that occur in any complex having precisely n r-sets. The following theorem answers not only the question posed at the beginning of this chapter but a natural dual question as well.

THEOREM 1. *If*

$$n = \binom{n_r}{r} + \cdots + \binom{n_i}{i} \quad canonical,$$

then

$$f(n; r, r') = \binom{n_r}{r'} + \binom{n_{r-1}}{r' - 1} + \cdots + \binom{n_i}{r' - r + 1}.$$

As usual, we take 0 as the value of any binomial coefficients in which either the numerator or the denominator is negative, or in which the numerator is strictly less than the denominator. Contrary to usual practice, however, we also let the binomial coefficient $\binom{0}{0}$ equal 0.

If $n_r, \cdots, n_i$ is any sequence of integers, we let

$$[n_r, \cdots, n_i]_r = \binom{n_r}{r} + \cdots + \binom{n_i}{i}.$$

We call this expression, and also the sequence of integers $(n_r, \cdots, n_i)$, *r-canonical* if the above expression is the r-canonical expression for the integer it equals. We may sometimes omit the qualifying r if its value is clear from context. Note that the void sequence is canonical, and that it corresponds to the representation of 0.

We define a *fractional pseudopower* $n^{(r'/r)}$ as follows: If

$$n = [n_r, \cdots, r_i]_r \text{ canonical,}$$

then we let

$$n^{(r'/r)} = [n_r, \cdots, n_i]_{r'}.$$

To illustrate this concept we have the accompanying table of (4/3)

SOME (4/3) PSEUDOPOWERS

n	3-canonical representation of n	$n^{(4/3)}$
0	[ ]₃	0
1	[3]₃	0
2	[3, 2]₃	0
3	[3, 2, 1]₃	0
4	[4]₃	1
5	[4, 2]₃	1
6	[4, 2, 1]₃	1
7	[4, 3]₃	2
8	[4, 3, 1]₃	2
9	[4, 3, 2]₃	3
10	[5]₃	5
11	[5, 2]₃	5
12	[5, 2, 1]₃	5
13	[5, 3]₃	6
14	[5, 3, 1]₃	6
15	[5, 3, 2]₃	7
16	[5, 4]₃	9
17	[5, 4, 1]₃	9
18	[5, 4, 2]₃	10
19	[5, 4, 3]₃	12
20	[6]₃	15

pseudopowers. In the pseudopower terminology our theorem becomes the following:

THEOREM 1'. *Under the hypothesis of Theorem 1, we have* $f(n; r, r') = n^{(r'/r)}$.

We remark that our fractional pseudopowers depend separately on the numerator and the denominator, that is, the ($\frac{4}{2}$) and ($\frac{8}{2}$) fractional powers are different. Our notation is perhaps justified by the fact that, as n increases, the fractional pseudopower is asymptotically proportional to the ordinary fractional power, and by other properties that we shall demonstrate later.

2. Application to Sequences of 0's and 1's

Let a *binary word* of length n be a sequence of n 0's and 1's. Write

$$a_1 \cdots a_n \leq b_1 \cdots b_n,$$

if $a_i \leq b_i$ for all i. Let a *binary complex* C be a set of binary words, all of the same length, with the property that if b is in C and $a \leq b$, then a is in C. (Such a set C is sometimes called a "lower" set.) We say that a binary word has *weight* r if it has r 1's in it.

Let C be a binary complex of words of length n. Let $v_1, \cdots, v_n$ be abstract vertices. Let v_i correspond to the ith position in a binary word. If $b = b_1 \cdots b_n$ has r 1's, let it be associated with the set R of those r vertices corresponding to the positions with 1's in them. That is, let

$$R(b) = \{v_i \mid b_i = 1\}.$$

Then C corresponds to a class of sets. We easily see that C is a binary complex if and only if this class of sets is an ordinary complex.

Therefore our theorem can be restated thus in terms of binary complexes:

THEOREM 2. *If a binary complex C has exactly n words of weight r, and if $r' \geq r$ [$r' \leq r$], then the maximum [minimum] number of words of weight r' that it can have is $n^{(r'/r)}$.*

It is hoped that this theorem will have an application to the probability-of-error calculations of group codes (see for example [1]); indeed, this was the source of the problem solved here.

3. Some Lemmas and Definitions

We remind the reader that

$$\binom{n}{i} = \binom{n-1}{i} + \binom{n-1}{i-1} \quad \text{if} \quad i \neq 1,$$

$$\binom{n}{1} = \binom{n-1}{1} + \binom{n}{0} \quad \text{if} \quad i = 1.$$

We need to distinguish the case $i = 1$, because for $n = 1$, the top line would then yield $\binom{1}{1} = \binom{0}{1} + \binom{0}{0}$; this is false because of our convention that $\binom{0}{0} = 0$. From the above we see that

$$[n_r, \cdots, n_i]_r$$
$$= \begin{cases} [n_r - 1, \cdots, n_i - 1]_r + [n_r - 1, \cdots, n_i - 1]_{r-1}, & \text{if } i \neq 1. \\ [n_r - 1, \cdots, n_1 - 1]_r + [n_r - 1, \cdots, n_2 - 1, n_1]_{r-1}, & \text{if } i = 1. \end{cases}$$

We call a sequence $(n_r, \cdots, n_i)$ r-semicanonical if either

$$(n_r, \cdots, n_i) \quad \text{or} \quad (n_r, \cdots, n_{i+1}, n_i - 1)$$

is r-canonical. We call a representation $n = [n_r, \cdots, n_i]_r$ r-semicanonical if the $(n_r, \cdots, n_i)$ is r-semicanonical. We note that if $(n_r, \cdots, n_i)$ is r-semicanonical, then it is r-canonical if and only if $n_{i+1} \neq n_i$.

While an integer n has a unique r-canonical representation, it may have several r-semicanonical representations. For example, suppose that

$$n = [n_r, \cdots, n_i]_r \quad \text{canonical}$$

and suppose that $n_i > i$. Then all the semicanonical representations of n consist of the canonical representation and the following:

$$[n_r, \cdots, n_{i+1}, n_i - 1, n_i - 1]_r,$$
$$[n_r, \cdots, n_{i+1}, n_i - 1, n_i - 2, n_i - 2]_r,$$

$$\cdots$$

$$[n_r, \cdots, n_{i+1}, n_i - 1, n_i - 2, \cdots, n_i - i + 1, n_i - i + 1]_r.$$

Suppose we have a semicanonical representation for n, that is,

$$n = [n_r, \cdots, n_i]_r \quad \text{semicanonical.}$$

Then to find its canonical representation we may proceed as follows:

If this representation is already canonical, we have finished; if not, then $n_{i+1} = n_i$, so

$$n = [n_r, \cdots, n_{i+2}, n_{i+1} + 1]_r \text{ semicanonical.}$$

If this is already canonical, we have finished; if not, then $n_{i+2} = n_{i+1}$, so

$$n = [n_r, \cdots, n_{i+3}, n_{i+2} + 1]_r \text{ semicanonical.}$$

If this is canonical, we have finished; if not, we continue in the same manner until we finally reach the canonical representation for n.

Note that if $n = [n_r, \cdots, n_i]_r$ semicanonical, then

$$n^{(r'/r)} = [n_r, \cdots, n_i]_{r'},$$

even though in the definition of pseudopowers we used canonical representations. We leave the proof of this to the reader.

Let us order finite sequences of positive integers lexicographically; that is, one sequence precedes another if and only if it would come earlier than the other in a "dictionary" of such sequences. Using elementary facts about binomial coefficients, we can prove that if $[n_r, \cdots, n_i]_r$ and $[m_r, \cdots, m_j]_r$ are canonical, then

$$[n_r, \cdots, n_i] \begin{Bmatrix} < \\ = \\ > \end{Bmatrix} [m_r, \cdots, m_j]$$

according as

$$(n_r, \cdots, n_i) \begin{Bmatrix} < \\ = \\ > \end{Bmatrix} (m_r, \cdots, m_j)$$

in the lexicographical sense. Also if $[n_r, \cdots, n_i]_r$ and $[m_r, \cdots, m_k]_r$ are semicanonical, then

$$[n_r, \cdots, n_i]_r \begin{Bmatrix} \leq \\ \geq \end{Bmatrix} [m_r, \cdots, m_k]_r$$

according as

$$(n_r, \cdots, n_i) \begin{Bmatrix} \leq \\ \geq \end{Bmatrix} (m_r, \cdots, m_k).$$

It now follows easily that

$$n \leq m \quad \text{implies} \quad n^{(r'/r)} \leq m^{(r'/r)}.$$

LEMMA 1. *The respective relations*

$$(n^{(r'/r)})^{(r/r')} \begin{Bmatrix} \geq \\ = \\ \leq \end{Bmatrix} n$$

hold according as r/r' $\{\gtreqless\}$ 1; *further,*

$$\left[(n^{(r'/r)})^{(r/r')}\right]^{(r'/r)} = n^{(r'/r)}.$$

The second result follows easily from two applications of the first. We prove the first only in the case that $r/r' < 1$. Suppose $n = [n_r, \cdots, n_i]_r$ canonical. Then $n^{(r'/r)} = [n_r, \cdots, n_i]_{r'}$. This expression is canonical unless, for some j,

$$n_j < r' - r + j.$$

If this inequality holds for some j, then it also holds for all smaller values. Now let $j + 1$ be the smallest value for which this does not hold. Then we have

$$n^{(r'/r)} = [n_r, \cdots, n_{j+1}]_{r'} \text{ canonical,}$$

so that

$$(n^{(r'/r)})^{(r/r')} = [n_r, \cdots, n_{j+1}]_r \leq n.$$

This proves Lemma 1.

LEMMA 2. *If* $r \leq r'$, *then* $n^{(r/r')}$ *is the smallest integer* k *such that* $k^{(r'/r)} \geq n$, *and* $n^{(r'/r)}$ *is the largest integer* k *such that* $k^{(r/r')} \leq n$.

PROOF. Consider $n^{(r/r')}$. By Lemma 1, it is a member of the class of integers k such that $k^{(r'/r)} \geq n$. Suppose it were not the smallest member of this class. Then $n^{(r/r')} - 1$ would be in this class also. We would then have

$$(n^{(r/r')} - 1)^{(r'/r)} \geq n,$$

so that

$$((n^{(r/r')} - 1)^{(r'/r)})^{(r/r')} \geq n^{(r/r')}.$$

From the preceding lemma we then would have

$$n^{(r/r')} - 1 \geq n^{(r/r')},$$

which obviously is a contradiction. The other half of Lemma 2 may be proved similarly.

Suppose that $(n_r, \cdots, n_i)$ is a sequence of integers with the property that

$$n_r > \cdots > n_{i+1} \geq n_i \geq 1.$$

Then we define a *cascade* of type $(n_r, \cdots, n_i)$ to consist (see Fig. 1)

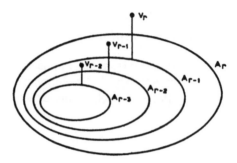

Fig. 1. A cascade of type $(n_r, \cdots, n_i)$.

of a set A_{r+1} of vertices, of distinguished subsets

$$A_r, \cdots, A_i,$$

and of distinguished vertices

$$v_r, \cdots, v_{i+1},$$

with the following properties:

  (a) A_{r+1} contains $n_r + 1$ vertices if $r > i$, and n_r vertices if $r = i$;
  (b) A_j contains n_j vertices for $r \geq j \geq i$;
  (c) $A_{r+1} \supset A_r \supset \cdots \supset A_i$;
  (d) v_j is in $A_{j+1} - A_j$ for $r \geq j \geq i + 1$;
  (e) all the distinguished vertices are distinct.

We say that the following sets are *naturally associated* with the cascade:

  (0) every subset of A_r;
  (1) $v_r +$ each subset of A_{r-1};
  (2) $v_r + v_{r-1} +$ each subset of A_{r-2};
  $\cdots$
  $(r - i)$ $v_r + \cdots + v_{i+1} +$ each subset of A_i.

In other words, suppose R is any subset of A_{r+1}. Let us define its *degree* j, where $r \geq j \geq i$, to be the smallest integer j such that $v_r, \cdots, v_{j+1}$ belong to R. Then v_j does not belong to R. A set R of degree j is naturally associated with the cascade if and only if

$$R - \{v_r, \cdots, v_{j+1}\} \subset A_j.$$

LEMMA 3. *The class of all sets naturally associated with a cascade*

*forms a complex. If the cascade is of type $(n_r, \cdots, n_i)$, then the complex
has $[n_r, \cdots, n_i]_s$ s-sets for any s. Suppose that for some s, $(n_r, \cdots, n_i)$
is s-semicanonical, and let $n = [n_r, \cdots, n_i]_s$. Then for any s' the complex has $n^{(s'/s)}$ s'-sets.*

We leave the proof of this simple lemma to the reader. We remark
that $(n_r, \cdots, n_i)$ is s-semicanonical exactly for those values of s
such that

$$\left.\begin{array}{c} n_i + r - i \\ \text{and} \\ n_{i+1} + r - i - 1 \end{array}\right\} \geq s \geq r - i + 1.$$

We call any complex like that in Lemma 3 a *cascade complex.*
    Using cascade complexes, we easily find an inequality for $f(n; r, r')$.
    LEMMA 4. *If $r' \geq r$, then*

$$f(n; r, r') \geq n^{(r'/r)}.$$

The proof of Lemma 4 is very simple. Let

$$n = [n_r, \cdots, n_i]_r \text{ canonical.}$$

Consider the complex associated with a cascade of type $(n_r, \cdots, n_i)$.
By our lemma this complex has n r-sets and $n^{(r'/r)}$ r'-sets. This proves
the lemma.
    If C is any complex and v is any vertex in C, then by definition the
complement of v consists of all sets in C that do not contain v. It is
easy to see that the complement of v is a subcomplex of C. The *star* of
v consists of all sets in C that do contain v. Of course, the star is not a
subcomplex. The *star boundary* of v consists of all sets obtained by
taking each set in the star and deleting v from it. It is easy to prove
that the star boundary of v is a subcomplex of the complement of v.

4. Proof of the Theorem

We wish to prove that $f(n; r, r') = n^{(r'/r)}$ for all $n \geq 1$, $r \geq 1$, and
$r' \geq 1$.
    LEMMA 5. *If*

$$f(n; r, r') = n^{(r'/r)}$$

for all n and fixed r and r', then

$$f(m; r', r) = m^{(r/r')}$$

for all m and the same r and r'.

COROLLARY. *If the theorem holds for all triples $(n; r, r')$ with $r \leq r'$, then it holds for all triples whatsoever.*

The corollary follows trivially from the lemma. The lemma is trivial if $r = r'$. Suppose next that $r < r'$. By definition, $f(m; r', r)$ is the minimum integer n such that a complex with m r'-sets can have exactly n r-sets. But a complex with n r-sets must have no more than $f(n; r, r') = n^{(r'/r)}$ r'-sets. Therefore, we have $m \leq n^{(r'/r)}$. Furthermore, if it were true that $m \leq (n - 1)^{(r'/r)}$, then there would exist a complex with m r'-sets and only $n - 1$ r-sets, which contradicts the definition of n. Thus we see that n is the least integer such that $n^{(r'/r)} \geq m$. By Lemma 2 we then find that $n = m^{(r/r')}$, so that $f(m; r', r) = m^{(r/r')}$, as desired. The case in which $r > r'$ may be proved similarly.

Henceforth we consider only the case $r \leq r'$. The theorem (in this case) is trivial if $n = 1$, for we find by use of the various definitions that

$$f(1; r, r') = \begin{cases} 0 & \text{if } r < r', \\ 1 & \text{if } r = r', \end{cases}$$

and

$$1 = [r]_r \text{ canonical,}$$

so that

$$1^{(r'/r)} = [r]_{r'} = \begin{cases} 0 & \text{if } r < r', \\ 1 & \text{if } r = r'. \end{cases}$$

Our proof is by induction on n.

Because of the inequality we proved in Section 3 for $f(n; r, r')$, it is sufficient for us to prove the following inductive step: Assume that C is a complex with exactly n r-sets and exactly $f(n; r, r')$ r'-sets; assume that $f(\bar{n}; \bar{r}, \bar{r}') = \bar{n}^{(\bar{r}'/\bar{r})}$ whenever $\bar{n} < n$ and $\bar{r} \leq \bar{r}'$; from these assumptions, prove that C has $\leq n^{(r'/r)}$ r'-sets. We may assume that $n > 1$, for if $n = 1$ we are in the initial case; and we may assume that $r < r'$, for if $r = r'$ then we trivially have

$$f(n; r, r) = n = n^{(r/r)}.$$

Furthermore, we may assume that every vertex in C occurs in at least one r-set in C, for otherwise without changing C in any essential way we could drop all vertices that do not occur in r-sets.

LEMMA 6. *If $n > 1$, then any complex that has exactly n r-sets contains at least one vertex v with the following property: The number j of r-sets in the star of v satisfies the inequality*

$$j \leq (n - j)^{(r-1/r)}.$$

Since the proof of this lemma is long and technical, we defer it until Section 5.

Using Lemma 6, we pick a vertex $\bar{v}$ in C with star having exactly j r-sets, where

$$j \leq (n - j)^{(r-1/r)}.$$

We also have

$$0 < j < n.$$

The left-hand inequality holds because every vertex lies in at least one r-set. The right-hand inequality holds because (since $n > 1$) not every vertex can belong to every r-set.

Let C_0 be the complement of $\bar{v}$, let C_1 be the star of $\bar{v}$, and let C_1^* be the star boundary of $\bar{v}$. Then

$$
\begin{array}{lll}
C_0 & \text{has} & n - j \quad r\text{-sets,} \\
C_1 & \text{has} & j \qquad r\text{-sets, and} \\
C_1^* & \text{has} & j \qquad (r-1)\text{-sets.}
\end{array}
$$

Let d_0 and d_1 be the number of r'-sets in C_0 and C_1, respectively. Then

$$
\begin{array}{lll}
C_0 & \text{has} & d_0 \quad r'\text{-sets,} \\
C_1 & \text{has} & d_1 \quad r'\text{-sets, and} \\
C_1^* & \text{has} & d_1 \quad (r'-1)\text{-sets.}
\end{array}
$$

We see that

$$d_0 + d_1 = f(n; r, r').$$

Using the induction hypothesis on C_0, we find that

$$d_0 \leq f(n - j; r, r') = (n - j)^{(r'/r)}.$$

Using the induction hypothesis on C_1^*, we find that

$$d_1 \leq f(j; r - 1, r' - 1) = j^{(r'-1/r-1)}.$$

Because C_1^* is a subcomplex of C_0, it has no more $(r'-1)$-sets than C_0. Thus by using the induction hypothesis on C_0, we find that

$$d_1 \leq f(n - j; r, r' - 1) = (n - j)^{(r'-1/r)}.$$

Now write

$$n - j = \lfloor p_r, \cdots, p_i \rfloor_r \text{ canonical,}$$

$$j = [q_{r-1}, \cdots, q_k]_{r-1} \text{ canonical.}$$

We easily see that

$$(n - j)^{(r-1/r)} = \begin{cases} [p_r, \cdots, p_i]_{r-1} \text{ canonical,} & \text{if } i > 1, \\ [p_r, \cdots, p_3, p_2 + 1]_{r-1} \text{ semicanonical,} & \text{if } i = 1. \end{cases}$$

Since $j \le (n - j)^{(r-1/r)}$, we have

$$(q_{r-1}, \cdots, q_k) \le \begin{cases} (p_r, \cdots, p_i) & \text{if } i > 1, \\ (p_r, \cdots, p_3, p_2 + 1) & \text{if } i = 1. \end{cases}$$

We now make an important distinction between the following two cases: In the *normal case*, we have

$$(q_{r-1}, \cdots, q_k) < (p_r, \cdots, p_i);$$

in the *special case*, we have

$$(q_{r-1}, \cdots, q_k) \ge (p_r, \cdots, p_i).$$

Let us first consider the special case. Here we have

$$(p_r, \cdots, p_i) \le (q_{r-1}, \cdots, q_k) \le \begin{cases} (p_r, \cdots, p_i), & \text{if } i > 1, \\ (p_r, \cdots, p_3, p_2 + 1), & \text{if } i = 1. \end{cases}$$

If $i > 1$, we clearly have $k = i - 1$ and

$$(q_{r-1}, \cdots, q_{i-1}) = (p_r, \cdots, p_i),$$

so that

$$j = (n - j)^{(r-1/r)}.$$

If $i = 1$, we clearly have

$$(q_{r-1}, \cdots, q_2) = (p_r, \cdots, p_3) \quad \text{and} \quad p_2 < q_1 \le p_2 + 1.$$

(Of course we must have $k = 1$.) Thus $q_1 = p_2 + 1$. It is now easy to see that for $i = 1$ we also have

$$j = (n - j)^{(r-1/r)},$$

so that this equation always holds in the special case.

Now we show that in the special case C has no more than $n^{(r'/r)}$ r'-sets. First we remark that

$$\begin{aligned} n &= (n - j) + j = (n - j) + (n - j)^{(r-1/r)} \\ &= [p_r, \cdots, p_i]_r + [p_r, \cdots, p_i]_{r-1} \\ &= [p_r + 1, \cdots, p_i + 1]_r. \end{aligned}$$

It is easy to see that this representation is canonical, so that we have

$$n^{(r'/r)} = [p_r + 1, \cdots, p_i + 1]_{r'}.$$

Using the inequalities for d_0 and d_1 established above, we see that C has

$$
\begin{aligned}
f(n; r, r') = d_0 + d_1 &\leq (n - j)^{(r'/r)} + (n - j)^{(r'-1/r)} \\
&= [p_r, \cdots, p_i]_{r'} + [p_r, \cdots, p_i]_{r'-1} \\
&= [p_r + 1, \cdots, p_i + 1]_{r'} = n^{(r'/r)}
\end{aligned}
$$

r'-sets. This completes the proof of the inductive step in the special case.

Let us consider the normal case. From the assumption characterizing this case, we see that there is a number μ with $r \geq \mu \geq 1$ such that

$$p_r = q_{r-1}, \cdots, p_{\mu+1} = q_\mu, \qquad p_\mu > q_{\mu-1},$$

where we interpret $q_{\mu-1}$ as 0 if $q_{\mu-1}$ does not exist, that is, if $k = \mu$. We observe that

$$
\begin{aligned}
n = (n - j) + j &= [p_r, \cdots, p_i]_r + [q_{r-1}, \cdots, q_k]_{r-1} \\
&= [p_r + 1, \cdots, p_{\mu+1} + 1, p_\mu]_r \\
&\quad + [p_{\mu-1}, \cdots, p_i]_{\mu-1} + [q_{\mu-1}, \cdots, q_k]_{\mu-1}.
\end{aligned}
$$

Using the inequalities established above for d_0 and d_1, we see that C has

$$
\begin{aligned}
f(n; r, r') = d_0 + d_1 &\leq (n - j)^{(r'/r)} + j^{(r'-1/r-1)} \\
&= [p_r, \cdots, p_i]_{r'} + [q_{r-1}, \cdots, q_k]_{r'-1} \\
&= [p_r + 1, \cdots, p_{\mu+1} + 1, p_\mu]_{r'} \\
&\quad + [p_{\mu-1}, \cdots, p_i]_{r'-r+\mu-1} + [q_{\mu-1}, \cdots, q_k]_{r'-r+\mu-1}
\end{aligned}
$$

r'-sets. To simplify our discussion of these complex expressions, let us make the following definitions. Suppose for the moment that $p = (p_r, \cdots, p_i)$ is any r-canonical sequence, and that $q = (q_{r-1}, \cdots, q_k)$ is any $(r - 1)$-canonical sequence, and suppose that $p > q$. Then let $\mu = \mu(p, q)$ be the integer such that

$$p_r = q_{r-1}, \cdots, p_{\mu+1} = q_\mu, \qquad p_\mu > q_{\mu-1}.$$

For any $s \geq r$, let us define

$$
\begin{aligned}
F_s(p, q) = [p_r + 1, \cdots, p_{\mu+1} + 1, p_\mu]_s &+ [p_{\mu-1}, \cdots, p_i]_{s-r+\mu-1} \\
&+ [q_{\mu-1}, \cdots, q_k]_{s-r+\mu-1}.
\end{aligned}
$$

Then it follows that

$$n = F_r(p, q)$$

and

$$f(n; r, r') \leq F_{r'}(p, q).$$

LEMMA 7. *If p is an r-canonical sequence, if q is an $(r-1)$-canonical sequence, and if we constrain p and q by $p > q$ and by the equation $F_r(p, q) = n$, then $F_{r'}(p, q)$ achieves its maximum when $n = [p_r, \cdots, p_i]_r$ and q is the void sequence. The value of this maximum is $n^{(r'/r)}$.*

As the proof of Lemma 7 is long and technical, we defer it until the end of Section 6. This lemma enables us to complete the proof of our theorem, since we see that

$$f(n; r, r') \leq n^{(r'/r)}.$$

The proof of the inductive step in the normal case, and thereby the proof of the whole theorem, is thus completed.

5. The Number of r-Sets That Contain a Vertex

This entire section leads up to the proof of Lemma 13, which states that under certain circumstances, a complex with n r-sets must contain a vertex v lying in only j r-sets, where j satisfies $j \leq (n-j)^{(r-1/r)}$. This is one of the two results that we have already used but have not proved.

Let $k_r(n)$ be defined as the largest integer k such that

$$k \leq (n-k)^{(r-1/r)}.$$

LEMMA 8. *If*

$$n - 1 = [n_r, \cdots, n_i]_r \text{ semicanonical,}$$

then

$$k_r(n) = \begin{cases} [n_r - 1, \cdots, n_i - 1]_{r-1} & \text{if } i > 1, \\ [n_r - 1, \cdots, n_2 - 1, n_1]_{r-1} & \text{if } i = 1. \end{cases}$$

PROOF. First suppose that the representation given above for $n-1$ is canonical. Let k be the expression above. Let j be the largest integer such that $n_j \leq j$, and let $j = i - 1$ if this never happens. We have $j + 1 \geq i$. Using a formula for square-bracket expressions that we noted earlier, we have

$$n - k = \begin{cases} [n_r, \cdots, n_i]_r + 1 - [n_r - 1, \cdots, n_i - 1]_{r-1} & \text{if } i > 1, \\ [n_r, \cdots, n_i]_r + 1 - [n_r - 1, \cdots, n_2 - 1, n_1]_{r-1} & \text{if } i = 1, \end{cases}$$
$$= [n_r - 1, \cdots, n_i - 1]_r + 1 \text{ for any } i,$$

whence

$$n - k = [n_r - 1, \cdots, n_{j+1} - 1]_r + 1$$

$$= \begin{cases} [n_r - 1, \cdots, n_{j+1} - 1, j]_r \text{ canonical,} & \text{if } j > 0, \\ [n_r - 1, \cdots, n_2 - 1, n_1]_r \text{ semicanonical,} & \text{if } j = 0, \end{cases}$$

so that

$$(n - k)^{(r-1/r)} = \begin{cases} [n_r - 1, \cdots, n_{j+1} - 1, j]_{r-1} & \text{if } j > 0, \\ [n_r - 1, \cdots, n_2 - 1, n_1]_{r-1} & \text{if } j = 0. \end{cases}$$

We see directly that $k \leq (n - k)^{(r-1/r)}$. Now consider $k + 1$. As above, we have

$$n - (k + 1) = [n_r - 1, \cdots, n_{j+1} - 1]_r \text{ canonical.}$$

Thus,

$$(n - (k + 1))^{(r-1/r)} = [n_r - 1, \cdots, n_{j+1} - 1]_{r-1}.$$

We see directly that $(k + 1) \nleq (n - (k + 1))^{(r-1/r)}$. Thus if the representation for $n - 1$ is canonical, the lemma is proved. But it is easy to see that the formula for k_r gives the same value when applied to a semicanonical representation for $n - 1$ as when applied to the canonical representation.

We need the following simple lemma:

LEMMA 9. *Let g be a positive-valued convex function of a real variable. Let the domain of definition of g be any compact subset of the nonnegative real numbers. Then the radial vector from the origin to any point of the graph of g attains its minimum slope at an endpoint of the domain of definition.*

We leave the proof of this simple lemma to the reader. We shall use it only with finite domains.

We say that the r-canonical sequence $(n_r, \cdots, n_i)$ has *degree i*, and that the integer $[n_r, \cdots, n_i]_r$ to which it corresponds has *degree i* or *r-degree i*. We say that two r-canonical sequences [or two integers] with a certain property are *adjacent* with that property if there are no other r-canonical sequences [or integers] between them. By a *run of degree i*, if $i < r$, we mean an increasing sequence of adjacent r-canonical sequences of degree $\geq i$ with interior sequences having degree exactly i and first and last sequences having degree $> i$. For example,

$$(p_r, \cdots, p_{i+1}),$$
$$(p_r, \cdots, p_{i+1}, i),$$
$$(p_r, \cdots, p_{i+1}, i+1),$$
$$(p_r, \cdots, p_{i+1}, i+2),$$

$$\cdots$$

$$(p_r, \cdots, p_{i+1}, p_{i+1} - 1),$$
$$(p_r, \cdots, p_{i+1} + 1)$$

is a run of degree i if the last sequence happens to be canonical. (If the last sequence is only semicanonical, then by substituting in its place the canonical sequence corresponding to the same integer, we obtain a run of degree i.) We also use the word "run" to refer to the run of integers corresponding to the sequences in a run of sequences. For example,

$$[p_r, \cdots, p_{i+1}]_r,$$
$$[p_r, \cdots, p_{i+1}, i]_r,$$
$$[p_r, \cdots, p_{i+1}, i+1]_r,$$
$$[p_r, \cdots, p_{i+1}, i+2]_r,$$

$$\cdots$$

$$[p_r, \cdots, p_{i+1}, p_{i+1} - 1]_r,$$
$$[p_r, \cdots, p_{i+1} + 1]_r$$

is a run of degree i. By a *run of degree* r we mean any finite increasing sequence of adjacent canonical sequences of degree r, or any finite increasing sequence of adjacent integers of degree r. For example,

$$(p_r), \qquad (p_r + 1), \qquad (p_r + 2), \cdots, (p_r + k)$$

and

$$[p_r]_r, \qquad [p_r + 1]_r, \qquad [p_r + 2]_r, \cdots, [p_r + k]_r$$

are runs of degree r. For any canonical sequence of degree i, where $i > 1$, we form the *augmented sequence* by adjoining a new last term that is smaller by 1 than the old last term. Thus the augmented sequence of $(p_r, \cdots, p_i)$ is $(p_r, \cdots, p_i, p_i - 1)$. By an *augmented run* we mean a sequence of sequences each of which comes from the preceding one by augmentation; for example,

$$(p_r, \cdots, p_i),$$

$$(p_r, \cdots, p_i, p_i - 1),$$

$$(p_r, \cdots, p_i, p_i - 1, p_i - 2),$$

$$\cdots$$

$$(p_r, \cdots, p_i, p_i - 1, p_i - 2, \cdots, p_i - i + 1).$$

We also refer to *augmented* integers and *augmented runs* of integers with the obvious meaning. Finally, we define a *mixed* run, either of canonical sequences or of integers, to be a run that can be split into two segments sharing precisely one term (the *middle* term), such that the initial segment is a consecutive interval from some run of some degree i, the middle term has degree i, and the terminal segment is an augmented run; for example,

$$(p_r, \cdots, p_{i+1}, p_i),$$

$$(p_r, \cdots, p_{i+1}, p_i + 1),$$

$$\cdots \qquad\qquad\qquad \}\text{initial segment}$$

$$(p_r, \cdots, p_{i+1}, \bar{p}_i - 1),$$

$$(p_r, \cdots, p_{i+1}, \bar{p}_i), \qquad\qquad\ \}\text{middle term}$$

$$(p_r, \cdots, p_{i+1}, \bar{p}_i, \bar{p}_i - 1),$$

$$\cdots \qquad\qquad\qquad \}\text{terminal segment}$$

$$(p_r, \cdots, p_{i+1}, \bar{p}_i, \bar{p}_i - 1, \cdots, \bar{p}_i - i + 1).$$

LEMMA 10. *If the values of n are restricted so that $n - 1$ runs through all terms of some run of constant degree, or of some augmented run, or of some mixed run, then k_r is convex for this domain of definition.*

PROOF. We remark that if a function is defined on a finite set, then to prove convexity it is sufficient to prove that, for every three adjacent points on the graph of the function, the slope of the line connecting the first two points is greater than or equal to the slope of the line connecting the last two points. What sort of segments of length 3 can the values of $n - 1$ take on? We classify these segments of length 3 as follows: (a) an initial segment from a run of degree i; (b) an interior segment from a run of degree i; (c) a terminal segment from a run of degree i; (d) a segment from an augmented run; (e) the "middle" segment from a mixed run, that is, a segment of which the first two terms are adjacent of degree $\geq i$, the middle term has degree i, and the last term is the augmented sequence of the middle term. In case (a), $n - 1$ takes on the values

$$[n_r, \cdots, n_{i+1}],$$
$$[n_r, \cdots, n_{i+1}, i],$$
$$[n_r, \cdots, n_{i+1}, i+1].$$

The slopes of the two lines are found by direct calculation to be $1/1$ and $(i-1)/i$. In case (b), the values of $n - 1$ are

$$[n_r, \cdots, n_{i+1}, n_i],$$
$$[n_r, \cdots, n_{i+1}, n_i + 1],$$
$$[n_r, \cdots, n_{i+1}, n_i + 2].$$

By direct calculation the slopes of the two lines are found to be $(i-1)/n_i$ and $(i-1)/(n_i+1)$. The calculation for case (b) also covers case (c) if we note that the formula for $k_r(n)$ is also valid when used with a semicanonical representation of n. In case (d), $n - 1$ takes on the following values:

$$[n_r, \cdots, n_i],$$
$$[n_r, \cdots, n_i, n_i - 1],$$
$$[n_r, \cdots, n_i, n_i - 1, n_i - 2].$$

By direct calculation the slopes are found to be $(i-1)/(n_i-1)$ and $(i-2)/(n_i-2)$. Since $n_i \geq i$, the former slope is $\geq$ the latter slope. In case (e), the values of $n - 1$ are

$$[n_r, \cdots, n_{i+1}, n_i],$$
$$[n_r, \cdots, n_{i+1}, n_i + 1],$$
$$[n_r, \cdots, n_{i+1}, n_i + 1, n_i].$$

In this case both slopes by direct calculation are found to be $(i-1)/n_i$. This proves the lemma.

Using the last two lemmas, we obtain the following lemma:

LEMMA 11. *For either of the following two domains of definition, $k_r(n)/n$ achieves its minimum at one or the other endpoint of the domain: (a) n is restricted so that $n - 1$ takes on all integer values between and including two adjacent integers of degree $\geq i$; (b) n is restricted so that $n - 1$ takes on all values between and including $r + 1$ and $\binom{v}{r} - 1$, where $v \geq r + 2$.*

PROOF. We first prove part (a) by an induction on i. The initial case is $i = 2$. In this case the values of $n - 1$ form a run of degree 1 (namely a sequence of consecutive integers), so that, by the preceding Lemma 10, k_r is convex over this domain of definition. Hence, by Lemma 9, the radius vector to the graph of k_r achieves its minimum slope at one end or the other of this domain. The slope of the radius vector, however, is $k_r(n)/n$. Now we proceed to the inductive step. Suppose $n - 1$ takes on all values between two adjacent numbers of degree $\geq i$. If these two numbers of degree $\geq i$ happen to have the forms

$$[n_r, \cdots, n_{i+1}] \quad \text{and} \quad [n_r, \cdots, n_{i+1}, i],$$

then our domain has only two points, so our result is trivial. If not, the two numbers must be of the form

$$[n_r, \cdots, n_{i+1}, n_i] \quad \text{and} \quad [n_r, \cdots, n_{i+1}, n_i + 1].$$

This interval is broken up into subintervals by points of degree exactly $i - 1$. In each of these subintervals, $k_r(n)/n$ achieves its minimum at one or the other of the endpoints of the subinterval. Thus the possible places where the minimum might occur are restricted to values in a run of degree $i - 1$. Now restricting $n - 1$ to these values, we see that k_r becomes convex and thus $k_r(n)/n$ will achieve its minimum at one of the two endpoints. This proves part (a).

We prove part (b), using part (a), by a similar method. First break up the interval of values for $n - 1$ by all numbers of degree r. These numbers are $[r + 1]_r$, $[r + 2]_r$, $\cdots$, $[v - 1]_r$. Call the resulting subintervals, except for the last one, intervals of the *first kind*. Further break up the last interval, which starts with $[v - 1]_r$, by the values of the augmented sequence starting with $[v - 1]_r$, that is, the values

$$[v - 1]_r, \quad [v - 1, v - 2]_r, \cdots, [v - 1, v - 2, \cdots, v - r]_r.$$

Let these resulting subintervals be termed intervals of the *second kind*. Now break up each interval of the second kind as follows: If the interval ends in a number of degree exactly i, then break up the interval by all numbers of degree exactly $i - 1$; call the resulting subintervals intervals of the *third kind*. Each interval of the third kind is the type of interval considered in part (a), so that $k_r(n)/n$ achieves its minimum value on each interval of the third kind at one endpoint of the interval. Within each interval of the second kind, however, the endpoints of the intervals of the third kind form a run of constant degree. Therefore, by the same argument we used earlier, $k_r(n)/n$ must achieve its minimum within each interval of the second kind at one endpoint of the

interval. Each interval of the first kind is the type of interval considered in part (a). Therefore, $k_r(n)/n$ achieves its minimum within
each interval of the first kind at one endpoint of the interval. Thus
within the whole large interval under consideration, $k_r(n)/n$ achieves
its minimum at one of the endpoints of the intervals of either the first
or second kind. These particular endpoints form a mixed run in the
sense of our preceding lemma. Hence, on this domain of definition, k_r
is convex. Therefore, $k_r(n)/n$ achieves its minimum on the whole large
interval when $n - 1$ takes on one of the values at either end of the
large interval. This proves Lemma 11.

LEMMA 12. *If $v > r \geq 1$, $n \geq 1$, if C is a complex with v vertices and
n r-sets, and if j is the minimum, over all vertices, of the number of r-sets
that contain a fixed vertex, then $j \leq k_r(n)$.*

PROOF. *Case I*: $n - 1 \leq r$. In this case we easily calculate that
$k_r(n) = n - 1$. If $n = 1$ then there is only one r-set in the complex.
But as we assumed $v > r$, there is some vertex that is not in this r-set,
so $j = 0$ and our result holds. On the other hand, if $n \geq 2$ we note
that two distinct r-sets cannot have exactly the same vertices in them.
Thus not every vertex can belong to every r-set. Consequently
some one vertex belongs to at most $(n - 1)$ r-sets. Therefore
$j \leq n - 1 = k_r(n)$.

Case II: $n - 1 \geq r + 1$. Recall that $v > r$. If $v = r + 1$, we have

$$n \leq \binom{v}{r} = \binom{r+1}{r} = r + 1,$$

which contradicts the case II assumption. Thus we have $v \geq r + 2$.
Now as $n - 1$ ranges from $r + 1$ to

$$\binom{v}{r} - 1 = [v - 1, v - 2, \cdots, v - r + 1]_r,$$

$k_r(n)/n$ achieves its minimum at one of the endpoints by the preceding
lemma. Thus $k_r(n)/n \geq$ the smaller of

$$\frac{k_r(r+2)}{r+2} \quad \text{and} \quad \frac{k_r\left(\binom{v}{r}\right)}{\binom{v}{r}}.$$

By direct calculation these values are $r/(r+2)$ and r/v; since $v \geq r+2$, we find that $k_r(n)/n \geq r/v$. Thus we have

$$k_r(n) \geq rn/v.$$

On the other hand, rn/v is obviously the average, over all vertices, of the number of r-sets that contain a fixed vertex. Therefore $j \leq rn/v \leq k_r(n)$. This proves Lemma 12.

LEMMA 13. *If $r \geq 1$, $n > 1$, and C is a complex with n r-sets, then C contains a vertex such that the number j of r-sets containing this vertex satisfies*

$$j \leq (n-j)^{(r-1/r)}.$$

The proof of this lemma is now almost trivial. Since $n > 1$, we have $v > r$. Therefore Lemma 12 applies. Suppose that the vertex that lies in the smallest number of r-sets lies in j r-sets. Then we have $j \leq k_r(n)$. By the definition of k_r, it follows that $j \leq (n-j)^{r-1/r}$.

6. The Function $F_s(p, q)$

This section is devoted to proving the final corollary in it. This corollary is the same as Lemma 7 about the function $F_s(p, q)$, which was used earlier but not proved. We repeat here the definition of $\mu(p, q)$ and $F_s(p, q)$. Throughout this section, p will always represent an r-semi-canonical sequence $(p_r, \cdots, p_i)$ and q will always represent an $(r-1)$-semicanonical sequence $(q_{r-1}, \cdots, q_k)$. We shall always assume that $p > q$.

Let us define $\mu = \mu(p, q)$ to be the integer such that $p_r = q_{r-1}, \cdots, p_{\mu+1} = q_\mu$, $p_\mu > q_{\mu-1}$. In this definition we interpret q_j to be 0 if q_j does not exist; this remark is needed only if $\mu = k$. We notice that μ exists because we have $p > q$. Notice that $r \geq \mu$, $\mu \geq i$, and $\mu \geq k$. We define

$$F_s(p, q) = [p_r + 1, \cdots, p_{\mu+1} + 1, p_\mu]_s + [p_{\mu-1}, \cdots, p_i]_{\mu-1+s-r}$$
$$+ [q_{\mu-1}, \cdots, q_k]_{\mu-1+s-r}.$$

We claim that the next larger r-semicanonical sequence than p (in the lexicographical ordering, of course) is given by:

a. $(p_r, \cdots, p_i, i-1)$ if $p_{i+1} \neq p_i$ and $i > 1$,

b. $(p_r, \cdots, p_2, p_1 + 1)$ if $p_{i+1} \neq p_i$ and $i = 1$,

c. $(p_r, \cdots, p_{i+2}, p_{i+1} + 1)$ if $p_{i+1} = p_i$.

We claim that the next smaller $(r-1)$-semicanonical sequence than q is given by:

a'. $(q_{r-1}, \cdots, q_{k+1})$ if $q_k = k$,

b'. $(q_{r-1}, \cdots, q_2, q_1 - 1)$ if $q_k > k$ and $k = 1$,

c'. $(q_{r-1}, \cdots, q_{k+1}, q_k - 1, q_k - 1)$ if $q_k > k$ and $k > 1$.

Notice that if p represents n, then the next larger sequence than p represents $n + 1$ in cases (a) and (b) but represents n in case (c). Similarly, if q represents n, then the next smaller sequence than q represents $n - 1$ in cases (a') and (b') but represents n in case (c'). The act of changing a sequence to the next larger or next smaller sequence will be called a *strong change* in cases (a), (b), (a'), and (b'), and will be called a *weak change* in cases (c) and (c').

LEMMA 14. *Let p' be the next larger sequence than p and let q' be the next smaller sequence than q.*
  I. *If the change from p to p' is weak, then $F_s(p', q) = F_s(p, q)$.*
  II. *If the change from q to q' is weak, then $F_s(p, q') = F_s(p, q)$.*
  III. *If the change from p to p' is strong, then $F_r(p', q) = F_r(p, q) + 1$.*
  IV. *If the change from q to q' is strong, then $F_r(p, q') = F_r(p, q) - 1$.*

PROOF. We prove the four parts of this lemma one by one.
  I. We have $p_{i+1} = p_i$. Let $\mu' = \mu(p', q)$. If $\mu \geq i + 1$, then $\mu' = \mu$, and it is easy to see $F_s(p', q) = F_s(p, q)$. If $\mu = i$, then $\mu' = \mu + 1$ and we have

$$F_s(p', q) = [p_r + 1, \cdots, p_{\mu+2} + 1, p_{\mu+1}]_s$$
$$+ [p_\mu]_{\mu+s-r} + [q_\mu, \cdots, q_k]_{\mu+s-r}.$$

Canceling like terms, we see that

$$F_s(p', q) - F_s(p, q) = \binom{p_{\mu+1}}{\mu + 1 + s - r} - \binom{p_{\mu+1} + 1}{\mu + 1 + s - r}$$
$$+ \binom{q_\mu}{\mu + s - r} = 0,$$

by using $q_\mu = p_{\mu+1}$ and a familiar binomial identity.
  II. We have $q_k > k$ and $k > 1$. Let $\mu' = \mu(p, q')$. If $\mu \geq k + 1$, then $\mu' = \mu$, and it is easy to see that $F_s(p, q') = F_s(p, q)$. If $\mu = k$, then we see that $\mu' = \mu + 1$ and

$$F_s(p, q') = [p_r + 1, \cdots, p_{\mu+2} + 1, p_{\mu+1}]_s$$
$$+ [p_\mu, \cdots, p_i]_{\mu+s-r} + [q_\mu - 1, q_\mu - 1]_{\mu+s-r}.$$

Canceling the like terms, we see that

$$F_s(p, q') - F_s(p, q) = \binom{p_{\mu+1}}{\mu + 1 + s - r} - \binom{p_{\mu+1} + 1}{\mu + 1 + s - r}$$

$$+ \binom{q_\mu - 1}{\mu + s - r} + \binom{q_\mu - 1}{\mu + s - r - 1} = 0$$

by first using the binomial identity on the q_μ terms, then using $q_\mu = p_{\mu+1}$, and finally using the binomial identity again.

III. Let $\mu' = \mu(p', q)$. The sequence p must fall into either case (a) or case (b), above. In each case it is clear that $\mu' = \mu$. In case (a), we have

$$F_r(p', q) = F_r(p, q) + \binom{i - 1}{i - 1} = F_r(p, q) + 1.$$

In case (b), we have

$$F_r(p', q) = F_r(p, q) - \binom{p_1}{1} + \binom{p_1 + 1}{1} = F_r(p, q) + 1.$$

IV. Let $\mu' = \mu(p, q')$. The sequence q must fall into either case (a') or case (b'). In case (a'), if $\mu \geq k + 1$, then we have $\mu' = \mu$ and

$$F_r(p, q') = F_r(p, q) - \binom{q_k}{k} = F_r(p, q) - 1.$$

In case (a'), if $\mu = k$, then $\mu' = \mu + 1$ and we find that

$$F_r(p, q') = [p_r + 1, \cdots, p_{\mu+2} + 1, p_{\mu+1}]_r + [p_\mu, \cdots, p_i]_\mu.$$

Canceling like terms, we find that

$$F_r(p, q') - F_r(p, q) = \binom{p_{\mu+1}}{\mu + 1} - \binom{p_{\mu+1} + 1}{\mu + 1} = -\binom{p_{\mu+1}}{\mu}$$

$$= -\binom{q_\mu}{\mu} = \binom{q_k}{k} = -1.$$

In case (b'), if $\mu \geq 2$, we have $\mu' = \mu$ and

$$F_r(p, q') = F_r(p, q) + \binom{q_1 - 1}{1} - \binom{q_1}{1} = F_r(p, q) - 1.$$

In case (b'), if $\mu = 1$, we find $\mu' = 2$ and

$$F_r(p, q') = [p_r + 1, \cdots, p_3 + 1, p_2]_r + [p_1]_1 + [q_1 - 1]_1.$$

Canceling like terms, we obtain

$$F_r(p, q') - F_r(p, q) = \binom{p_2}{2} - \binom{p_2 + 1}{2} + \binom{q_1 - 1}{1}$$

$$= -p_2 + q_1 - 1 = -1.$$

This proves the lemma.

Suppose that

$$p' = (p_r, \cdots, p_{j+1}, q_j, \cdots, q_k),$$

$$q' = (q_{r-1}, \cdots, q_{j+1}, p_j, \cdots, p_i).$$

We say that p' and q' are obtained from p and q by *interchange of tails* at j. We call this interchange *admissible* if

$$p_{j+2} > p_{j+1} \quad \text{and} \quad \left\{ \begin{array}{c} p_{j+1} > q_j \\ \text{or} \\ p_{j+1} = q_{j+1} = q_j \end{array} \right\} \quad \text{and} \quad q_j > p_j.$$

We agree that $q_j > p_j$ for this purpose if q_j exists and p_j does not.

LEMMA 15. *If p' and q' are formed from p and q by an admissible interchange, then p' is r-semicanonical and q' is $(r - 1)$-semicanonical. Also, we have $\mu - 1 \geq j \geq 1$, $p' > p$, $q' < q$, and $p' > q'$. Finally, we have $F_\bullet(p', q') = F_\bullet(p, q)$.*

PROOF. To prove this lemma, first write $p' = (p'_r, \cdots, p'_k)$, $q' = (q'_{r-1}, \cdots, q'_i)$. It is obvious that $p'_h \geq h$ and $q'_h \geq h$ for all h. Suppose the interchange occurred at j. Since

$$q'_{j+1} = q_{j+1} \geq q_j > p_j = q'_j,$$

we easily see that

$$q'_{r-1} > q'_{r-2} > \cdots > q'_{i+1} \geq q'_i.$$

Thus we see that q' is $(r - 1)$-semicanonical. If $p_{j+1} > q_j$, then we see that

$$p'_{j+1} = p_{j+1} > q_j = p'_j,$$

and also, by admissibility, that

$$p_{j+2} > p_{j+1}$$

(even if p_j should fail to exist), so that

$$p'_r > p'_{r-1} > \cdots > p'_{k+1} \geq p'_k.$$

Thus in this case we see that p' is r-semicanonical. On the other hand, if $p_{j+1} = q_{j+1} = q_j$ then q_{j-1} does not exist, so we see that $j = k$. In this case we have

$$p' = (p_r, \cdots, p_{j+1}, p_{j+1}),$$

so p' is r-semicanonical. (Again we have $p_{j+2} > p_{j+1}$ by admissibility even if p_j does not exist.)

Now suppose for the moment that $j \geq \mu$. By the admissibility condition, we have either

$$p_{j+1} > q_j \quad \text{or} \quad p_{j+2} > p_{j+1} = q_{j+1}.$$

The definition of μ implies both

$$p_{j+1} = q_j \quad \text{and} \quad p_{j+2} = q_{j+1}.$$

This contradicts the preceding statement, and shows that $j \geq \mu$ is false. Thus $\mu - 1 \geq j \geq 1$.

Now it is easy to see that $p' > p$ because we have

$$p'_r = p_r, \cdots, p'_{j+1} = p_{j+1}, \qquad p'_j = q_j > p_j$$

and that $q' < q$ because we have

$$q_{r-1} = q'_{r-1}, \cdots, q_{j+1} = q'_{j+1}, \qquad q_j > p_j = q'_j.$$

Now we have $p' > p > q > q'$ from which $p' > q'$ follows. From the fact that $\mu - 1 \geq j$ it follows that $\mu(p', q') = \mu$, and hence easily that $F_s(p', q') = F_s(p, q)$.

We define the standard transforms of p and q to be p^* and q^*, which are defined as follows:

i. If p satisfies case (c), then p^* is the next larger sequence than p and $q^* = q$;

ii. If (i) does not apply and if q falls into case (c'), then $p^* = p$ and q^* is the next smaller sequence than q;

iii. If neither (i) nor (ii) applies, and if for some j with $\mu - 1 \geq j \geq 1$ we have $p_j < q_j$, then p^* and q^* are formed by interchanging the tails of p and q at the largest such j;

iv. If neither (i) nor (ii) nor (iii) applies, then p^* is the next larger sequence than p and q^* is the next smaller sequence than q.

LEMMA 16. *Let p^* and q^* be the standard transforms of p and q. Let $\mu^* = \mu(p^*, q^*)$. Then the following statements are true:*

I. *If case (iii) of the standard transform definition applies to p and q, then the interchange involved is admissible.*

II. *The transform p^* is r-semicanonical, the transform q^* is $(r-1)$-semicanonical, and $p^* > q^*$.*

III. *The transforms p^* and q^* satisfy $p^* > p$ and $q^* < q$.*

IV. *We have $F_r(p^*, q^*) = F_r(p, q)$.*

V. *For $s \geq r$, we have $F_s(p^*, q^*) \geq F_s(p, q)$.*

COROLLARY. *If $r' > r$, if p is a variable r-semicanonical sequence, if q is a variable $(r-1)$-semicanonical sequence, and if p and q are constrained by the requirements $p > q$ and $F_r(p, q) = n$, then $F_{r'}(p, q)$ reaches its maximum value when $n = [p_r, \cdots, p_i]_r$ and q is the void sequence. The value of this maximum is $n^{(r'/r)}$.*

PROOF. First we prove that the corollary follows from the lemma. If p and q are any pair satisfying the constraints, then according to the lemma the standard transforms of p and q also satisfy the constraints and furthermore yield at least as large a value for $F_{r'}$. Now let us take standard transforms repeatedly. Each time we do so, the value of $F_{r'}$ will not decrease and may perhaps increase. Let p and q now be the final pair of sequences we obtain for which it is not possible to form standard transforms. Clearly q is the void sequence. Then $\mu(p, q) = r$ and

$$n = F_r(p, q) = [p_r]_r + [p_{r-1}, \cdots, p_i]_{r-1} = [p_r, \cdots, p_i]_r.$$

Thus we find that the pair p and q such that

$$[p_r, \cdots, p_i]_r = n \quad \text{and} \quad q \text{ void}$$

yields at least as large a value for $F_{r'}$ as *any* pair of sequences satisfying the constraints. Thus the maximum value for $F_{r'}$ does indeed occur at p and q, and the value of this maximum is

$$F_{r'}(p, q) = [p_r, \cdots, p_i]_{r'} = n^{(r'/r)}.$$

Thus the corollary follows from the lemma.

Now we proceed with the proof of Lemma 16. We prove the five parts of it one by one.

I. Since this part is concerned with case (iii), we know that p does not fall into case (c) nor q into case (c'). Thus we have $p_{i+1} \neq p_i$ and hence $p_r > \cdots > p_{i+1} > p_i$. Now let j be the largest integer with $\mu - 1 \geq j \geq 1$ such that $p_j < q_j$. We know that such an integer exists because we are considering case (iii). We claim that if $\mu > j + 1$ then

$$p_{j+2} > p_{j+1} \geq q_{j+1} \geq q_j > p_j.$$

The first inequality follows from our remark above. The second follows (because $\mu > j + 1$) from the definition of j. The third inequality is trivial and the fourth follows from the definition of j. Clearly we must have either $p_{j+1} > q_j$ or else $p_{j+1} = q_{j+1} = q_j$, and in either case we find that the interchange is admissible. This settles the case $\mu > j + 1$. If $\mu = j + 1$, we claim that

$$p_{j+2} > p_{j+1} > q_j > p_j.$$

The first inequality follows as before. The second inequality simply states that $p_\mu > q_{\mu-1}$, which follows from the definition of μ. The third inequality is trivial. Thus in this case also, the interchange is admissible.

II. In cases (i), (ii), and (iv) the results are all trivial. In the case (iii) the results follow because we now know that the interchange is admissible.

III. This is trivial in all cases.

IV and V. In case (iii) the standard transforms are formed by an admissible interchange. Therefore, by an earlier lemma we know that $F_s(p^*, q^*) = F_s(p, q)$. In cases (i) and (ii), respectively, parts (I) and (II) of Lemma 14 yield that $F_s(p^*, q^*) = F_s(p, q)$. In case (iv), parts (III) and (IV) of that same lemma yield that

$$F_r(p^*, q^*) = F_r(p^*, q) - 1 = F_r(p, q) + 1 - 1.$$

It remains to prove (V) in case (iv). Clearly we may assume that $s > r$. First suppose that q falls into case (b'). Then $q_1 > 1$. Since we are dealing with case (iv), we have $p_1 \geq q_1$, whence $p_1 > 1$. Thus p must fall into either case (b) or case (c). Since this is case (iv), however, p must actually fall into case (b). Therefore, we have

$$p^* = (p_r, \cdots, p_2, p_1 + 1),$$

and

$$q^* = (q_{r-1}, \cdots, q_2, q_1 - 1).$$

As we have $p_2 > p_1 \geq q_1$, we find that $\mu \geq 2$ and hence $\mu^* = \mu$. Then, canceling like terms, we obtain

$$
\begin{aligned}
F_s(p^*, q^*) - F_s(p, q) &= \binom{p_1 + 1}{s - r + 1} - \binom{p_1}{s - r + 1} \\
&\quad + \binom{q_1 - 1}{s - r + 1} - \binom{q_1}{s - r + 1} \\
&= \binom{p_1}{s - r} - \binom{q_1 - 1}{s - r} \geq 0,
\end{aligned}
$$

which gives us the desired inequality.

Next suppose that q falls into case (a'), so that $q_k = k$. We claim that $\mu > k$. Thus, we know that $\mu \geq k$, and if we had $\mu = k$, then we would have $p_{k+1} = q_k = k$, which violates the basic requirement $p_{k+1} \geq k + 1$. From $\mu > k$ it follows that $\mu^* = \mu$, for we have

$$p_r^* = p_r = q_{r-1} = q_{r-1}^*, \cdots, p_{\mu+1}^* = p_{\mu+1} = q_\mu = q_\mu^*,$$

$$p_\mu^* = p_\mu > q_{\mu-1} \geq q_{\mu-1}^*.$$

Now if p falls into case (a), we have

$$F_s(p^*, q^*) - F_s(p, q) = \binom{i-1}{s-r+i-1} - \binom{k}{s-r+k} = 0 - 0.$$

On the other hand, if p falls into case (b), then we have

$$F_s(p^*, q^*) - F_s(p, q) = \binom{p_1+1}{s-r+1} - \binom{p_1}{s-r+1} - \binom{k}{s-r+k}$$

$$= \binom{p_1}{s-r} - 0 \geq 0.$$

This proves Lemma 16.

Reference

1. Slepian, D., "A Class of Binary Signalling Alphabets," *Bell System Tech. J.*, Vol. 35, January, 1956, pp. 203–234.

Chapter 13

Optimization in Structural Design

WILLIAM PRAGER

1. Introduction

To explain some fundamental concepts of the plastic analysis of structures, let us consider a thick-walled circular tube that consists of a ductile material and is subjected to internal pressure. When this pressure is gradually increased, starting from zero, the tube at first behaves in a purely elastic manner; that is, all deformations are reversible and disappear completely when the internal pressure is reduced to zero. The first permanent deformations occur at the interior surface of the tube, when the internal pressure reaches a critical value that depends on the ratio of the interior and exterior radii of the tube. As the pressure continues to increase, the plastic region—the region in which permanent deformations are occurring—grows in diameter; but as long as it remains surrounded by an elastic region in which all deformations are reversible, the permanent deformations in the plastic region remain of the same order of magnitude as the reversible deformations in the elastic region. Plastic flow—that is, the rapid increase of permanent deformations under substantially constant internal pressure—becomes possible only when the interface between plastic and elastic regions reaches the exterior surface of the tube. For practical purposes, the corresponding value of the internal pressure represents the load-carrying capacity of the tube.

These three successive stages of purely elastic deformation, contained elastic-plastic deformation, and plastic flow can be observed in any statically indeterminate structure that is made of a ductile material. Since the load-carrying capacity is reached only well beyond the elastic range, the rational design of such a structure must be based on the theory of plasticity.

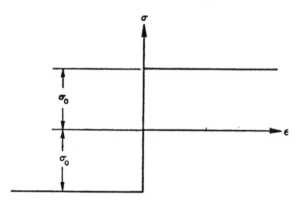

Fig. 1. Stress and strain in a perfectly plastic solid under simple
tension or compression.

2. Rigid, Perfectly Plastic Behavior

Figure 1 shows the stress-strain diagram of a rigid, perfectly plastic
solid in simple tension or compression: the intensity σ of the uniaxial
stress is used as ordinate and the corresponding unit extension ϵ as
abscissa. The absolute value of σ cannot exceed the *yield limit* σ_0. As
long as $|\sigma| < \sigma_0$, the specimen remains rigid; that is, the strain rate $\dot{\epsilon}$ is
zero. When $|\sigma| = \sigma_0$, the specimen is free to assume any rate of ex-
tension $\dot{\epsilon}$ that has the same sign as the stress σ. Since only the sign
and not the magnitude of the strain rate depends on the stress, the
material is inviscid. This absence of viscosity is typical for structural
metals at moderate temperatures and strain rates.

When the specimen flows plastically with the strain rate $\dot{\epsilon}$, the
mechanical energy that is dissipated in a unit volume per unit of time
—that is, the specific power of dissipation—is given by

$$\sigma\dot{\epsilon} = \sigma_0 |\dot{\epsilon}| . \tag{2.1}$$

It is worth noting that this relation contains the complete specification
of the considered mechanical behavior. In fact, it follows from this re-
lation that $\sigma = \sigma_0$ when $\dot{\epsilon} > 0$, and $\sigma = -\sigma_0$ when $\dot{\epsilon} < 0$, and $\dot{\epsilon} = 0$
when $-\sigma_0 < \sigma < \sigma_0$.

In the one-dimensional *stress space* with Cartesian coordinate σ, the
convex domain $-\sigma_0 \leq \sigma \leq \sigma_0$ represents the set of uniaxial states of
stress that can be attained in the considered material. The specific
power of dissipation $D(\dot{\epsilon}) = \sigma_0 |\dot{\epsilon}|$ may be regarded as the supporting
function of this convex domain of attainable states of stress. Note
that this function is homogeneous of the first order and convex.

For an *isotropic* material, the following hypotheses yield an acceptable generalization of this rigid, perfectly plastic behavior under uniaxial stress:

a. The stress tensor and the strain-rate tensor have a common system of principal axes.

b. The specific power of dissipation is a single-valued function of the principal strain rates $\dot{\epsilon}_1$, $\dot{\epsilon}_2$, $\dot{\epsilon}_3$; this *dissipation function* is homogeneous of the first order and convex.

The state of stress specified by the principal stresses σ_1, σ_2, σ_3 will be called *compatible* with the strain rate specified by the principal values $\dot{\epsilon}_1$, $\dot{\epsilon}_2$, $\dot{\epsilon}_3$ if

$$\sigma_1\dot{\epsilon}_1 + \sigma_2\dot{\epsilon}_2 + \sigma_3\dot{\epsilon}_3 = D(\dot{\epsilon}_1, \dot{\epsilon}_2, \dot{\epsilon}_3), \tag{2.2}$$

where the right-hand side is the dissipation function.

In applications, this dissipation function is often piecewise linear. For example, for *plane* states of stress with the principal values σ_1, σ_2, and $\sigma_3 = 0$, the yield condition of Tresca [1] is widely used, corresponding to the specific power of dissipation

$$\sigma_1\dot{\epsilon}_1 + \sigma_2\dot{\epsilon}_2 = \tfrac{1}{2}\sigma_0(|\dot{\epsilon}_1| + |\dot{\epsilon}_2| + |\dot{\epsilon}_1 + \dot{\epsilon}_2|). \tag{2.3}$$

This relation contains the complete specification of the mechanical behavior of the considered material under plane states of stress. If, for instance, $\dot{\epsilon}_1 > 0$ and $\dot{\epsilon}_2 < 0$, a comparison of the coefficients of $\dot{\epsilon}_1$ and $\dot{\epsilon}_2$ on the two sides of (2.3) yields $\sigma_1 = \sigma_0$, $\sigma_2 = 0$. In the two-dimensional stress space with rectangular Cartesian coordinates σ_1 and σ_2, this state of stress is represented by the point A in Figure 2. On the other hand,

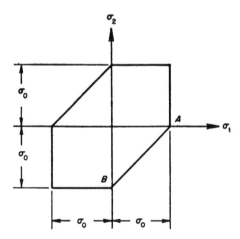

Fig. 2. Two-dimensional stress space.

if $\dot{\epsilon}_1 > 0$, but $\dot{\epsilon}_2 < 0$ and $\dot{\epsilon}_1 + \dot{\epsilon}_2 < 0$, one finds in a similar way that $\sigma_1 = 0$, $\sigma_2 = -\sigma_0$ (point B in Fig. 2). It then follows by continuity that all states of stress represented by the points of the segment AB, that is, the states of stress satisfying $\sigma_1 - \sigma_2 = \sigma_0$ with $\sigma_1 > 0$ and $\sigma_2 < 0$, are compatible with the same type of strain rate satisfying $\dot{\epsilon}_1 > 0$, $\dot{\epsilon}_2 < 0$, and $\dot{\epsilon}_1 + \dot{\epsilon}_2 = 0$. A complete discussion of all possible combinations of the signs of $\dot{\epsilon}_1$, $\dot{\epsilon}_2$, and $\dot{\epsilon}_1 + \dot{\epsilon}_2$ in (2.3) furnishes the hexagon in Figure 2 as the boundary of the region of attainable states of plane stress. We shall say that this hexagon represents Tresca's *yield limit* for plane states of stress.

3. Plastic Analysis

In the fundamental problem of plastic analysis, one considers a body consisting of a perfectly plastic material and assumes that the points of the part S_V of its surface are not allowed to move, that is, have vanishing velocity, while the remainder S_T of the surface of the body is subjected to given surface tractions. One wishes to know whether plastic flow will or will not occur under these conditions.

In the discussion of this problem the following concepts prove useful:

a. A stress field that is defined throughout the volume V of the considered body is called *statically admissible* if the stress components are continuously differentiable functions of position that satisfy the equilibrium conditions in V and on S_T.

b. A velocity field that is defined throughout the volume V is called *kinematically admissible* if the velocity components are continuously differentiable functions of position that vanish on S_V.

As a matter of fact, the condition that these fields should be continuously differentiable can be relaxed to some extent, but a detailed discussion of this possibility would exceed the scope of this chapter.

The load-carrying capacity of the body is characterized by the following results:

THEOREM 1. *The given surface tractions represent the load-carrying capacity of the body if there exists a statically admissible stress field, the stresses of which are at or below the yield limit and compatible with the strain rates of a kinematically admissible velocity field that does not represent a rigid-body motion of the entire body.*

THEOREM 2. *The given surface tractions exceed the load-carrying capacity if there exists a kinematically admissible velocity field with strain* *rates such that the integral of the dissipation function extended over the volume V is smaller than the power of the given surface tractions on the velocities of the points of S_V.*

The proof of these theorems must be omitted for lack of space, but the following immediate consequences should be noted. In the circumstances of Theorem 1, quasi-static plastic flow will occur, in which the power P of the given surface tractions must equal the integral of the dissipation function extended over the volume V:

$$P = \int D \, dV. \tag{3.1}$$

On the other hand, in the circumstances of Theorem 2 we have

$$P > \int D \, dV. \tag{3.2}$$

4. Plastic Design

In the fundamental problem of plastic design, one considers a region V_0 of space, each surface element of which belongs to one of the following three classes: at a *loaded* surface element of dS_T, a nonvanishing surface traction is prescribed; at a *supported* surface element dS_V, the velocity vanishes; and at a *free* surface element dS_0, the surface traction vanishes. A region V is to be determined that is contained in V_0 and satisfies the following conditions:

a. The surface of V consists of the loaded surface S_T, the supported surface S_V, and a third surface S.

b. If V is completely filled with the considered rigid, perfectly plastic material, a body is obtained for which the given surface tractions on S_T represent the load-carrying capacity when it is supported on S_V and free from surface tractions on S.

c. The volume of V has the smallest value that is compatible with the preceding conditions.

We shall distinguish three types of plastic design. A body will be said to represent an *optimal*, an *admissible*, or an *excessive* design according as conditions (a), (b), and (c) are fulfilled, or only conditions (a) and (b), or only condition (a) and a modified form of condition (b) that is obtained by specifying that the given surface tractions should be *below* the load-carrying capacity of the body.

Denoting the dissipation function of the considered material by D, let C be a body that represents an admissible or an optimal design. In either case, quasi-static plastic flow will take place under the given surface tractions. If the volume of C is denoted by V_C, by (3.1) we therefore have

$$P = \int D \, dV_C. \tag{4.1}$$

It will be shown that C represents an optimal design if the velocity field of the considered quasi-static plastic flow can be so continued in $V_0 - V_C$ that the dissipation function has a constant value throughout V_0.

Let the body C^* that occupies the region $V_C^* \in V_0$ represent an excessive design. The extended velocity field of the quasi-static flow of C is a kinematically admissible velocity field. Since the surfaces of both C and C^* contain the entire loaded surface S_T, the power of the given surface tractions on the considered velocity field has the same value for these two bodies. If this power were to exceed the integral of the dissipation function over V_C^*, the second fundamental theorem of plastic analysis would indicate that the given surface tractions exceed the load-carrying capacity of C^*. However, since this body is to represent an excessive design, we must have

$$\int D \, dV_C^* \geq P. \tag{4.2}$$

If, now, the considered velocity field yields a constant value of the dissipation function throughout the region V_0, which contains both V_C and V_C^*, it follows from (4.1) and (4.2) that

$$\int dV_C \leq \int dV_C^*. \tag{4.3}$$

In other words, the volume of C is not greater than the volume of an arbitrary body C^* of excessive design that is contained in V_0, no matter how close this body may be to plastic flow.

5. Example

To illustrate the discussion of Section 4, consider a thin plane disc of variable thickness attached to a rigid foundation along a given arc, the *foundation arc*. Loads that are applied to the edge of the disc and act in its plane produce a plane state of stress for which the dissipation function of the rigid, perfectly plastic material of the disc is supposed to be given by the right-hand side of equation (2.3).

We shall assume that the principal stresses in the plane of the disc are such that $\sigma_1 > 0$ and $\sigma_2 < 0$ throughout the disc. The state of stress at an arbitrary point of the disc is then represented by an interior point of the segment AB in Figure 2. From what has been seen above, the principal strain rates then satisfy the condition

$$\dot{\epsilon}_1 + \dot{\epsilon}_2 = 0, \qquad (5.1)$$

with $\dot{\epsilon}_1 > 0$. In these circumstances, the dissipation function on the right-hand side of equation (2.3) can be written as $D = \sigma_0 \dot{\epsilon}_1$, and the optimality condition of Section 4 reduces to

$$\dot{\epsilon}_1 = \text{const.} \qquad (5.2)$$

We choose rectangular axes x, y in the plane of the disc and denote the velocity components with respect to these axes by u and v, respectively. The condition (5.1) is then equivalent to

$$u_x + v_y = 0, \qquad (5.3)$$

where the subscripts indicate differentiation with respect to the coordinates, and (5.2) yields

$$4u_x v_y - (v_x + u_y)^2 = \dot{\epsilon}_1^2 = \text{const.} \qquad (5.4)$$

To satisfy (5.3), introduce a stream function $\psi(x, y)$ so that

$$u = \psi_y, \qquad v = -\psi_x. \qquad (5.5)$$

According to (5.4), this stream function must satisfy

$$4\psi_{xy}^2 + (\psi_{xx} - \psi_{yy})^2 = -4\dot{\epsilon}_1^2 = \text{const.} \qquad (5.6)$$

On the foundation arc, we have $u = v = 0$; that is, the function ψ is constant and its normal derivative vanishes. Since the velocity field determines the stream function only to within an additive constant, we may set $\psi = 0$ on the foundation arc.

The partial differential equation (5.6) is of hyperbolic type. In the neighborhood of the foundation arc, the function ψ may therefore be determined by the method of characteristics from the initial conditions on this arc. A graphical method for this has been described in an earlier paper [2].

If, in particular, the foundation arc is circular, the characteristics— which incidentally are the lines of principal stress in the disc—are logarithmic spirals which intersect the radii at 45°. Let us assume that

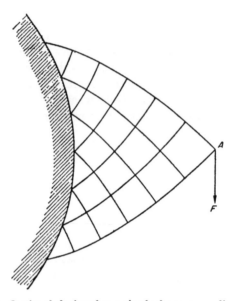

Fig. 3. Optimal design for a single force on a disc with
circular foundation arc.

the single force F shown in Figure 3 acts on the disc. The characteristics through the point of application A of this force then form the
edges of the disc, which must be reinforced by ribs along these edges.
Since each rib follows a line of principal stress, the forces transmitted
by the disc to the rib are normal to the latter. Accordingly, the axial
forces in the ribs have constant intensities, which are readily found by
decomposing the force F in the directions of the ribs at A. These intensities and the curvatures of the ribs determine the forces transmitted by the ribs on the edges of the disc. These edge forces set
boundary conditions of Riemann type for the principal forces N_1 and
N_2 in the disc. These principal forces must satisfy the equilibrium
conditions

$$-\frac{\partial N_1}{\partial s_1} + \frac{N_1 - N_2}{\rho_2} = 0,$$

$$\frac{\partial N_2}{\partial s_2} - \frac{N_2 - N_1}{\rho_1} = 0,$$

(5.7)

where ρ_1, ρ_2 are the radii of curvature and s_1, s_2 the arc lengths of the
characteristics. Finally, the variable thickness h of the disc is obtained
from the equation of the side AB of the hexagon in Figure 2, namely,

$$N_1 - N_2 = \sigma_0 h.$$

(5.8)

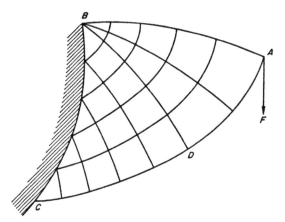

Fig. 4. Modification of optimal design for foundation arc with end point.

Figure 4 shows how the optimal design is modified when the foundation arc ends at B. In the region BCD the characteristics are constructed as before from the conditions on the foundation arc. This set of characteristics is then continued beyond BD by using the already-known conditions on BD and the fact that B must be an intersection of characteristics.

6. Historical Remarks

The mechanical behavior of a rigid, perfectly plastic solid is usually specified by its yield condition and flow rule (see, for instance [3]). The first characterizes the states of stress under which plastic flow is possible, and the second indicates the type of plastic flow that occurs under each of these states of stress. The possibility of specifying the mechanical behavior of a rigid, perfectly plastic solid by its dissipation function was pointed out by Prager [4].

The fundamental theorems of plastic analysis were developed by Greenberg, Drucker, and Prager in a series of papers ([5]–[8]). For typical applications of these theorems see [9]–[11], where numerous further references will be found.

The problem of plastic design for minimum weight was first treated for continuous beams and frames by Heyman [12] and Foulkes [13], [14]. Although isolated problems of optimal plastic design of plates [15], [16] and shells [17]–[19] had been treated before, the general principle presented in Section 4 is due to Drucker and Shield [20]. For a more detailed discussion of the example in Section 5 and related problems, see Te Chiang Hu [21].

Procedures for optimal structural design that are not based on the concepts of plastic analysis have been discussed by Michell [22], Cox [23], [24], and Hemp [25].

References

1. Tresca, H., "Mémoire sur l'écoulement des corps solides," *Mém. prés. par div. savants*, Vol. 18, 1868, pp. 733–799.
2. Prager, W., "On a Problem of Optimal Design," *Proceedings of a Symposium on Non-Homogeneity in Elasticity and Plasticity* (Warsaw, 1958), Pergamon Press, New York, 1959, pp. 125–132.
3. Hill, R., *The Mathematical Theory of Plasticity*, Oxford University Press, New York, 1950.
4. Prager, W., "Théorie génerale des états d'équilibre," *J. Math. Pures Appl.*, Vol. 34, 1955, pp. 395–406.
5. Greenberg, H. J., and W. Prager, "On Limit Design of Beams and Frames," Brown University, *Technical Report No. A18-1*, Providence, R.I., 1949; *Proc. Am. Soc. Civil Engrs.*, Vol. 77, Separate No. 59, 1951; *Trans. Am. Soc. Civil Engrs.*, Vol. 117, 1952, pp. 447–484 (with discussion).
6. Drucker, D. C., H. J. Greenberg, and W. Prager, "The Safety Factor of an Elastic-plastic Body in Plane Stress," *J. Appl. Mech.*, Vol. 18, 1951, pp. 371–378.
7. Drucker, D. C., W. Prager, and H. J. Greenberg, "Extended Limit Design Theorems for Continuous Media," *Quart. Appl. Math.*, Vol. 9, 1952, pp. 381–389.
8. Prager, W., "General Theory of Limit Design," *Proceedings of the 8th International Congress of Applied Mechanics* (Istanbul, 1952), Vol. 2, Istanbul, 1956, pp. 65–72.
9. Baker, J. F., M. R. Horne, and J. Heyman, *The Steel Skeleton*, Vol. 2, Cambridge University Press, London, 1956.
10. Hodge, P. G., Jr., *Plastic Analysis of Structures*, McGraw-Hill Book Company, Inc., New York, 1959.
11. Prager, W., *An Introduction to Plasticity*, Addison-Wesley Publishing Company, Inc., Reading, Mass., 1959.
12. Heyman, J., "Plastic Design of Beams and Frames for Minimum Material Consumption," *Quart. Appl. Math.*, Vol. 8, 1950, pp. 373–381.
13. Foulkes, J. D., "Minimum-weight Design and the Theory of Plastic Collapse," *Quart. Appl. Math.*, Vol. 10, 1953, pp. 347–358.
14. Foulkes, J. D., "Minimum-weight Design of Structural Frames," *Proc. Roy. Soc. (London)*, Ser. A, Vol. 223, No. 1155, 1954, pp. 482–494.
15. Hopkins, H. G., and W. Prager, "Limits of Economy of Material in Plates," *J. Appl. Mech.*, Vol. 22, 1955, pp. 372–374.
16. Prager, W., "Minimum Weight Design of Plates," *De Ingenieur* (Amsterdam), Vol. 67, 1955, pp. 0.141–0.142.
17. Onat, E. T., and W. Prager, "Limits of Economy of Material in Shells," *De Ingenieur* (Amsterdam), Vol. 67, 1955, pp. 0.46–0.49.

18. Freiberger, W., "Minimum Weight Design of Cylindrical Shells,"
    J. Appl. Mech., Vol. 23, 1956, pp. 576–580.
19. Hodge, P. G., Jr., "Discussion of Ref. [18]," *J. Appl. Mech.*, Vol.
    24, 1957, pp. 486–487; for a sequel to this discussion, see W. Freiberger,
    "On the Minimum Weight Design Problem for Cylindrical Sandwich
    Shells," *J. Aero. Sci.*, Vol. 24, 1957, pp. 847–848.
20. Drucker, D. C., and R. T. Shield, "Design for Minimum Weight,"
    Proceedings of the 9th International Congress of Applied Mechanics
    (Brussels, 1956), Vol. 5, Brussels, 1957, pp. 212–222.
21. Hu, Te Chiang, "Optimum Design for Structures of Perfectly Plastic
    Materials," unpublished Ph.D. thesis, Brown University, 1960.
22. Michell, A. G. M., "The Limits of Economy of Material in Frame-
    structures," *Phil. Mag.* (6), Vol. 8, 1904, pp. 589–597.
23. Cox, H. L., *The Theory of Design*, Aeronautical Research Council
    (Great Britain) Report No. 19791, 1958.
24. Cox, H. L., *Structures of Minimum Weight: The Basic Theory of Design
    Applied to the Beam under Pure Bending*, Aeronautical Research
    Council (Great Britain) Report No. 19785, 1958.
25. Hemp, W. S., *Theory of Structural Design*, College of Aeronautics,
    Cranfield, England, Report No. 115, 1958.

Chapter 14

Geometric and Game-Theoretic Methods in Experimental Design†

G. ELFVING

1. Introduction

The purpose of this chapter is to draw attention to certain relatively recent developments in the theory of experimental design. The emphasis here is on ideas and connections rather than on actual techniques.

We shall throughout be concerned with *linear experiments*, that is, experiments composed of independent *observations* of form

$$y_\nu = a_\nu'\alpha + \eta_\nu, \quad \nu = 1, \cdots, n, \tag{1.1}$$

where $\alpha = (\alpha_1, \cdots, \alpha_k)'$ is an unknown parameter vector, the $a_\nu = (a_{\nu 1}, \cdots, a_{\nu k})'$ are known coefficient vectors, and the independent error terms η_ν have mean 0 and a common variance, which we may, for convenience, take to be 1. In matrix notation, the equation (1.1) may be written $y = A\alpha + \eta$. Provided A is of full rank k, it is well known that the least-squares estimator vector $\hat{\alpha}$ is given by

$$\hat{\alpha} = (A'A)^{-1}A'y = \alpha + (A'A)^{-1}A'\eta \tag{1.2}$$

and has covariance matrix cov $\hat{\alpha} = (A'A)^{-1}$. The matrix

$$M = A'A = \sum_{\nu=1}^{n} a_\nu a_\nu' \tag{1.3}$$

is termed the *information matrix* of the experiment; it essentially determines its estimatory properties.

† This work was supported by an Office of Naval Research contract at Stanford University.

Thus far we have been concerned with a *given* experiment. On the design level, of course, the experiment has to be composed of observations chosen from a set of potentially available ones. Since an observation, in our setup, is described by the corresponding coefficient vector a_ν, we shall assume that there is given a bounded and closed set $\mathcal{Q}$ in k-space from which the observations a_ν have to be selected, each one being independently repeatable any number of times. It is no restriction to assume that $\mathcal{Q}$ is symmetric about the origin, since the observation $-y$ is automatically available along with y. The set $\mathcal{Q}$ may be finite (as is usually the case in analysis of variance, the coefficients being 0 or 1), or it may be described, for example, by means of a continuous parameter x (as in polynomial or trigonometric regression models).

An actual experiment, then, is described by its *spectrum* $a = (a_1, \cdots, a_r)$, $a_j \in \mathcal{Q}$, indicating the different observations selected, and its *allocation* $(n_1, \cdots, n_r)$, indicating the number of times that each selected observation is to be repeated. The number r is, of course, part of the design. Denoting by $n = \sum n_j$ the total number of actual observations, and writing $p_j = n_j/n$, we may describe the experiment by the pair $e = (a, p)$, where $p = (p_1, \cdots, p_r)$, $p_j > 0$, $\sum p_j = 1$. The p_j are the relative allocations, to be chosen by the experimental designer, whereas n is usually fixed by cost considerations. In practice, the p_j run over multiples of $1/n$; in a large sample theory, however, we may consider them as continuous.

The information matrix of an experiment $e = (a, p)$ of size n is, by (1.3),

$$M = M(e) = \sum_a n_j a_j a_j' = n \sum_a p_j a_j a_j'. \tag{1.4}$$

Here $M(e)$ determines the value of any particular goodness criterion, such as the variance of the least-squares estimator of a particular parameter, the sum of all such variances, the largest variance, or the like. The solution of the design problem depends, of course, on what criterion we have in mind.

2. Estimating a Single Parametric Form

Consider a particular linear form $\theta = c'\alpha$ in the parameters. The variance of its least-squares estimator $\hat{\theta} = c'\hat{\alpha}$ is

$$\text{var } \hat{\theta} = c'M^{-1}c; \tag{2.1}$$

through M, it depends on the design e. The straightforward minimization of (2.1) with respect to the design is difficult because of the complex character of $e = (a, p)$. Two indirect methods have been devised and will be briefly presented below, their interrelation being pointed out at the same time. Both are based on interchanging extremizations.

Geometric Method

This approach, suggested by Elfving [1], works primarily in the case of two or at most three parameters.

The information matrix $M = n \sum p_j a_j a_j'$ is that of r independent observations $\bar{y}_j$ with means $a_j'\alpha$ and error variances $(np_j)^{-1/2}$ $(j = 1, \cdots, r)$. The least-squares estimator $\hat{\theta}$ is a linear combination

$$\hat{\theta} = \sum_a q_j \bar{y}_j,$$

fulfilling the unbiasedness vector condition

$$Q_c: \qquad \sum_{a_j \in a} q_j a_j = c, \tag{2.2}$$

and the minimum variance condition

$$\mathrm{var}\left(\sum q_j \bar{y}_j\right) = \sum_a \frac{q_i^2}{np_j} = \min_{\substack{q \in Q_c}}. \tag{2.3}$$

In our design problem we are thus faced with the double minimization

$$\min_{(a,p)} \min_{q \in Q_c} \sum_a \frac{q_i^2}{p_j}, \tag{2.4}$$

in which the minimizing $e = (a, p)$ is the desired optimal experiment.

Now we interchange the order of minimization: To an arbitrary spectrum a, and an arbitrary corresponding set of nonvanishing coefficients q_j, we find the minimizing $p = p(a, q)$ and the least variance; this will, in turn, be minimized by a certain choice a^*, q^*, from which finally $p^* = p(a^*, q^*)$ is obtained.

Minimizing the sum in (2.4) with respect to p, under the conditions $p_j > 0$, $\sum p_j = 1$, gives

$$p_j = \frac{|q_j|}{\sum_a |q_h|}, \qquad \min_p \sum_a \frac{q_i^2}{p_j} = \left(\sum_a |q_j|\right)^2. \tag{2.5}$$

To minimize the latter expression with respect to (a, q) under condition

(2.2), we shall write this condition in the form

$$\sum_h |q_h| \sum_j \frac{|q_j|}{\sum |q_h|} \ \text{sgn} \ q_j \cdot a_j = c, \tag{2.6}$$

where all sums are taken over the spectrum a. Call the point represented by the latter sum c_q. For any a and any q, this is a point inside, or on the boundary of, the convex polyhedron spanned by the vectors $\pm \ a_j \in a$, and hence within the convex hull a^* of the set a of available observations. Conversely, choosing the spectrum and the relative size of the $|q_j|$ appropriately, we may make c_q equal to any vector in a^*. The condition (2.6) requires that c_q be proportional to c; if this condition is fulfilled, we have

$$\sum |q_h| \ = \ \frac{\| \ c \ \|}{\| \ c_q \ \|} \ .$$

In order to make the minimum in (2.5) as small as possible, we obviously have to choose for c_q the intersection point c^* between the vector c and the boundary of the convex hull a^*. The vector c^* may be expressed as a convex combination of r points $a_j \in a$, $r \leq k$. The points form the spectrum a of the desired experiment in which the allocations

$$p_j \ = \ \frac{|q_j|}{\sum |q_h|}$$

are the barycentric coordinates of c^* with respect to $a_1, \cdots, a_r$.

The situation is illustrated in Figure 1, where $k = 2$, $c = (0, 1)$;

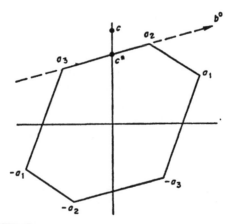

Fig. 1. Geometric method of allocation determination; here $c = (0, 1)$, $k = 2$, and a consists of three pairs of opposite points.

α consists of 6 (pairwise opposite) points, and α^* is a hexagon. The desired spectrum is $a = (a_2, a_3)$, and the allocation $p_2:p_3$ is given by the ratio of the segments $c^*a_3: c^*a_2$.

Game-Theoretic Method

In 1959 Kiefer and Wolfowitz [2] suggested a game-theoretic approach to the allocation problem. Following these authors, we shall show how to find the optimal design for estimating, say, the parameter α_k.

Using a convenient transformation of the parameters $\alpha_1, \cdots, \alpha_{k-1}$, we can easily show that, for any design $e = (a, p)$,

$$\frac{1}{n \text{ var } \hat{\alpha}_k} = \min_b K(e, b), \tag{2.7}$$

where

$$K(e, b) = \sum_a p_j(a_{jk} - b_1 a_{j1} - \cdots - b_{k-1} a_{j,k-1})^2, \tag{2.8}$$

and $b = (b_1, \cdots, b_{k-1})$ denotes a vector in $(k - 1)$-space. Since we wish var $\hat{\alpha}_k$ to be as small as possible, the design problem consists in maximizing the right-hand side of (2.7).

This formulation suggests the use of game theory. Consider the game with pay-off function $K(e, b)$. This function is convex in b. If by the convex combination (q_1, q_2) of two designs e^1 and e^2 we mean the design with spectrum $a = a^1 \cup a^2$ and allocation $p = \{p_j\}$, $p_j = q_1 p_j^{(1)} + q_2 p_j^{(2)}$, it is obvious that $K(e, b)$ is linear in e. As a consequence, the game is completely determined, and we may interchange the extremizations. More precisely, we may adopt the following procedure:

a. For arbitrary b, find $e = e(b)$ so as to maximize $K(e, b)$; let the maximum be $m^2(b) = K(e(b), b)$. The response $e(b)$ will, in certain cases (and notably, in the relevant ones), not be uniquely determined.

b. Find b^0 so as to minimize $m^2(b)$.

c. Find $e^0 = e(b^0)$ and (in case of nonuniqueness) such that, conversely, $b^0 = b(e^0)$; that is, b^0 is a response to e^0 in the sense that $K(e^0, b) = \min$ for $b = b^0$.

Then, e^0, b^0 constitute a pair of equilibrium strategies, and e^0 is minimax, as desired.

Let us see how this program operates in our particular situation.

The function (2.8) is maximized with respect to e if we include in the spectrum a any finite number of observations a_j for which

$$\left| a_{jk} - b_1 a_{j1} - \cdots - b_{k-1} a_{j,k-1} \right| = \max_{a_j} = m(b), \qquad (2.9)$$

and if we provide them with weights p_j adding up to 1. For most b's, the response spectrum will contain one observation. In case $k = 2$, the geometric interpretation of this operation is the following (see Fig. 1): Let a straight line of slope b_1 move parallel to itself, and consider the outermost positions in which it has points in common with the symmetric set $\mathcal{A}$. Let $B(b_1)$ be the set of these points. For most b_1, $B(b_1)$ contains only two (opposite) elements; when b_1 is the slope of a side in the convex hull $\mathcal{A}^*$ of $\mathcal{A}$, there will be four or more.

Next, the maximum in (2.9) has to be minimized by a proper choice of b. It is clear geometrically that this occurs when the hyperplane determined by b is a supporting plane to $\mathcal{A}^*$ at the point c^* where $\mathcal{A}^*$ is cut by the a_k-axis. When $\mathcal{A}$ is finite, $B(b^0)$ will normally contain k points on each side of the origin.

When $B(b^0)$ contains more than one element, as it usually does, it remains to determine $e^0 = (a^0, p^0)$ such that $b^0 = b(e^0)$. For this to hold, we must have

$$-\frac{1}{2} \frac{\partial K}{\partial b_i} = \sum_a p_j a_{j1}(a_{jk} - b_1 a_{ji} - \cdots - b_{k-1} a_{j,k-1}) = 0,$$

$$i = 1, \cdots, k - 1. \qquad (2.10)$$

The expressions in parentheses have the same absolute value $m(b^0)$ for all $a_j \in B(b^0)$; it is no restriction to assume this value to be positive, since otherwise we have merely to replace a_j by $-a_j$. The conditions (2.10) then become

$$\sum_a p_j a_{ji} = 0, \qquad i = 1, \cdots, k - 1. \qquad (2.11)$$

Hence, the spectrum a and the allocation p have to be chosen in such a way that the compound vector $\sum p_j a_j$ coincides with the a_k-axis, that is, with the direction of the coefficient vector $c = (0, \cdots, 0, 1)$ of the parametric form $\theta = \alpha_k$. Thus, we see that the game-theoretic approach leads to the same procedure as the geometric one, whenever this is practicable.

It should be noted, however, that Kiefer and Wolfowitz aimed primarily at the situation in which $\mathcal{A}$ is described by means of a continuous parameter x, say $a = a(x)$, $x \in \mathcal{X}$. The second step of the procedure then consists in minimizing

$$\max_{x} \left| a_k(x) - \sum_{i=1}^{k-1} b_i a_i(x) \right|.$$

Hence, the b_i have to be the Chebyshev coefficients of the function $a_k(x)$ with respect to the basis $a_1(x), \cdots, a_{k-1}(x)$; for their determination, ready-made results may often be available.

3. Minimax Estimation

For another example of the kind of methods we have in mind, assume that we do not know what parameter, or what linear combination of parameters, will eventually be relevant. In such a situation, a reasonable approach seems to be to minimize, with respect to the design e, the largest estimator variance of any *standardized* linear form in the parameters; that is, the quantity

$$\max_{\|c\|=1} \text{var } c'\hat{a} = \max_{\|c\|=1} c'M^{-1}c, \qquad (3.1)$$

where $\|c\|$ denotes the length of the vector c. The quantity (3.1) obviously depends on e through the function M in accordance with (1.4). For this approach to make any sense, one must, of course, have the parameters measured on such scales as to make the desired accuracy the same for all of them.

Consider the game—between the statistician and nature—with pay-off

$$K(e, c) = c'M^{-1}c = c' \left(\sum_a p_j a_j a_j' \right)^{-1} c. \qquad (3.2)$$

The statistician's set of strategies is obviously that of all $e = (a, p)$, $a \in \mathcal{A}$. The function (3.2) is convex in e; that is,

$$K(q_1 e^1 + q_2 e^2, c) \le q_1 K(e^1, c) + q_2 K(e^2, c),$$

where the weighted average of two e's is interpreted as in Section 2. Hence no extension of the e-set is necessary. Nature's set of strategies is the unit sphere in the k-dimensional c-space. On this set, there can be no question of forming linear combinations, so we have to extend it by introducing mixed strategies. A general mixed strategy, say γ^*, would be a probability distribution on the unit sphere S, and the corresponding pay-off would be

$$K^*(e, \gamma^*) = \int_S c'M^{-1}c \, d\gamma^*. \qquad (3.3)$$

The strategy γ^*, however, may always be replaced by a discrete distribution assigning positive probability only to the end points of k orthogonal diameters in the sphere. As a matter of fact, if we factorize M^{-1} into a sum $\sum m_\mu m_\mu'$ of matrices of rank 1, we have

$$K^*(e, \gamma^*) = \sum_\mu \int_S c' m_\mu m_\mu' c \, d\gamma^* = \sum_\mu m_\mu' \left(\int_S cc' \, d\gamma^* \right) m_\mu. \quad (3.4)$$

The integral is the expected value of a nonnegative definite matrix with trace 1 and hence is itself a matrix of the same kind. It can be written as $\sum_\nu \gamma_\nu c_\nu c_\nu'$, where the c_ν ($\nu = 1, \cdots, k$) are the column vectors of an orthogonal matrix, and where $\gamma_\nu \geq 0$, $\sum \gamma_\nu = 1$. Inserting this result in (3.4), and inverting the order of summation, we find that

$$K^*(e, \gamma^*) = \sum_\mu m_\mu' \left(\sum_\nu \gamma_\nu c_\nu c_\nu' \right) m_\mu = \sum_\nu \gamma_\nu \sum_\mu c_\nu' m_\mu m_\mu' c_\nu$$

$$= \sum_\nu \gamma_\nu c_\nu' M^{-1} c_\nu = K^*(e, \gamma), \quad (3.5)$$

which shows that we may restrict our attention to the particular set of "orthogonal mixed strategies" γ. Statistically, the extension of nature's strategies means that we use for loss function not simply the squared error in the estimate of a single parameter but rather a standardized quadratic form in the estimation errors, with nonnegative definite matrix of trace 1.

After this extension, the fundamental theorem in game theory is applicable. Interchanging the extremizations, we are then faced with the following program:

a. For an arbitrary γ, find the response $e = e(\gamma)$ maximizing $K(e, \gamma)$. The response may not be unique.

b. Find γ^0 so as to minimize $K(e(\gamma), \gamma)$.

c. Find $e^0 = e(\gamma^0)$ and (in case of nonuniqueness) such that, conversely, $\gamma^0 = \gamma(e^0)$.

The implementation of this program may be no less complex than that of the original problem. Already the first step involves solving the allocation problem for an arbitrary quadratic loss function. It can be shown (see [3]) that no more than $k(k+1)/2$ different observations will ever have to be included in the optimal design for any estimation problem within the framework of least squares. Generally, then, one will have to test all "promising" spectra with $k(k+1)/2$ or less observations, to solve equally many equations for the optimal p, to pick the best spectrum, and finally to find the minimizing γ. We have no simple technique to offer in the general case. However, there are situations in which short cuts are possible.

First assume that we are able to find an admissible design e^0 such that $M(e^0) = \lambda I$ (e^0 being admissible means that there is no e making the difference $M(e) - M(e^0)$ nonnegative definite). In practice, this can often be achieved by making the design sufficiently symmetric. It can be shown that any admissible design is optimal for some quadratic loss problem, that is, a response to some γ^0. On the other hand, since $M = \lambda I$, we have

$$c'M^{-1}c = c'c/\lambda = \text{const}$$

on the unit sphere. Hence any γ, and in particular γ^0, is a response to e^0. This implies that e^0 is minimax.

Another simple situation is the following, representing in a sense the opposite of the former one. Assume that there is no e such that the smallest eigenvalue $\kappa_{\min}$ of $M(e)$ is multiple. Since

$$\max_{\|c\|=1} c'M^{-1}c = \left[\min_{\|c\|=1} c'Mc \right]^{-1} = [\kappa_{\min}(M)]^{-1},$$

it is seen that the maximum on the left-hand side is always attained in a single pair of points $\pm c$ on the unit sphere. Therefore, in this case we do not have to consider mixed strategies but may apply the one-parameter technique of Section 2 to find $e(c)$. The corresponding payoff $K(e(c), c)$ is the inverted and squared distance from the origin to the boundary of $\mathcal{Q}^*$ in the direction c. The least favorable direction c^0 is that in which the boundary of $\mathcal{Q}^*$ is closest to the origin. Finally, the minimax allocation is obtained as the response to c^0 through the methods of Section 2. It is not trivial, however, to decide whether the minimum eigenvalue will always be simple. For the case $k = 2$, a geometric criterion is indicated in [3], Theorem 3.3.

References

1. Elfving, G., "Optimum Allocation in Linear Regression Theory," *Ann. Math. Stat.*, Vol. 23, 1952, pp. 255–262.
2. Kiefer, J., and J. Wolfowitz, "Optimum Designs in Regression Problems," *Ann. Math. Stat.*, Vol. 30, 1959, pp. 271–294.
3. Elfving, G., "Design of Linear Experiments," in Ulf Grenander (ed.), *The Harald Cramér Volume*, John Wiley & Sons, Inc., New York, 1959, pp. 58–74.

Part Four

MODELS, AUTOMATION, AND CONTROL

Chapter 15

Automation and Control in the Soviet Union

J. P. LaSALLE

1. Introduction

In mathematics the Soviet Union and the United States lead the world and are at approximately the same level [1]. In the mathematical theory of control and stability, Soviet mathematicians have worked longer and harder than we have; moreover, they scrutinize everything we do. More of Bellman's and Lefschetz's books have appeared in the windows of bookstores in Moscow alone than have been displayed by all of the bookstores in the United States, and their sales in Russia are correspondingly higher. Lefschetz's book on *Differential Equations: Geometric Theory* (Interscience, New York, 1957) was recently translated and published in the Soviet Union. A first printing of 10,000 copies sold out there in two days, whereas in four years the American edition has sold only about 2,300 copies.

2. Pure and Applied Mathematics

Pure mathematicians in the Soviet Union often are deeply concerned with applications and find therein many problems of genuine mathematical interest. By contrast, American pure mathematicians frequently know little about applications and have almost no interest in them. Soviet mathematicians make an effort to communicate the latest mathematical findings to as wide an audience as possible, and they have a receptive audience. Soviet engineers hold mathematicians in high regard, are better trained mathematically than our engineers, and work harder at keeping up with developments in the mathematical world. The Soviet Union has a fine group of mathematical engineers who contribute to both mathematics and engineering and who are continually widening the bridge from the one field to the other. In American universities this bridge is excessively narrow; too often the only interchange is what students carry back and forth.

At the level of theory, it is doubtful that Soviet mathematicians know anything that we do not know, but they are making a greater effort to increase and apply basic knowledge. There is little doubt that capable mathematicians are at the top echelons of science and that they are playing an active role.

The high level of Soviet mathematics—and of Soviet science in general—is not surprising. Russia has a long tradition in the field of science. The Soviet Academy of Sciences dates back to the founding of the St. Petersburg Academy of Science in 1724. Russian science was strong in the prerevolutionary period. The Great Soviet Encyclopedia of the year 1957 says of that period:

The brief survey given of the most important achievements in Russian science in the prerevolutionary period shows that in all the realms of natural science, scientists of Russia were making major discoveries and doing research on the principles of science, which have since become an essential part of the world's treasure house of knowledge. . . . From the theoretical point of view Soviet natural science took as its point of departure the achievements of the previous history of science, particularly in questions of principle. Of enormous significance for the further development of science in our country was the fact that the level of science was very high during the period preceding the Great October Socialist Revolution, thanks to such scientists as A. M. Lyapunov, A. N. Krylov, V. A. Steklov. . . .

As early as April, 1918, which was still a critical period for the Soviet republic, Lenin drew up his "Outline of a Plan for Work in Science and Technology." Even at that time there was a constant emphasis in the Soviet Union on the solution of scientific and technological problems "by the united efforts of large groups of scientists consisting of representatives of many specialities and directed by the most eminent scientists and technologists."

3. Automation and Control

The foregoing statement is equally descriptive of the Soviet effort today in the field of automation and control. The Soviets say that their plans for work in science and technology

. . . have a maximal correspondence to the specific features of the modern level of the development of science, which is distinguished on the one hand by extreme differentiation and specialization of the separate branches of knowledge, and on the other by the appearance of multiple fields lying between these sharply differentiated areas. Under these conditions a profound study of any particular object becomes possible only through a combined effort on the part of many specialized sciences.

It is quite natural that Soviet mathematicians operate successfully in the theory of control. This is an area close to differential equations, a field in which Soviet mathematics has profound roots [2]. There are a number of groups working on this theory and led by world authorities at institutes in Moscow, Kiev, Leningrad, Sverdlovsk, and Alma Ata; yet another center is being formed in Novosibirsk.

In the hierarchy of Soviet institutes there are also some dedicated to basic engineering research. The Institute of Automatics and Telemechanics of the Academy of Sciences is "the central academic institution in the field of automatic control." It has the responsibility not only to assemble results in this field of science but also to expound new concepts, principles, and methods. The Institute of Automatics and Telemechanics is expected to supply other agencies with the basic scientific material required for the production and testing of systems and instruments. Some of the institutes using this basic knowledge are the following: (a) The Central Scientific Institute of Complex Automation in Moscow, which appears to be concerned with the control of plants and processes. This institute was founded in 1957. (b) The Central Laboratory for Automation in Moscow, founded in 1940. Its chief task is automation of the steel industry. (c) The Institute of Automation in Kiev, founded in 1957. In 1959 an expansion of this institute was begun, and it is to be a center of automation; by 1963 it is expected to have four branches and to employ from 6,000 to 10,000 people. Its staff works in close coöperation with industry and is concerned with the development, production, and installation of control systems for a large variety of industries—chemical, metallurgical, electric power, mines, machine tools, etc. Additionally, The Institute for Automatics and Electrometry near Novosibirsk, The Institute for Electromechanics in Moscow, and many other smaller institutes and laboratories have some connection with automation.

It seems apparent that Soviet scientists have organized themselves for an all-out scientific effort in the field of automation and control. Their only weakness, if any, may be a lack of practical engineers. A group of American engineers who attended the First Congress of the International Federation of Automatic Control, held in the summer of 1960 in Moscow, visited factories, laboratories, and institutions and were quite unimpressed by what they saw. Their opinion was that if what they had seen was representative of Soviet capabilities, then the

Soviet Union could not have built a jet transport, produced an atomic bomb, or put up an earth satellite. This sounds somewhat like the reports German scientists gave our government about the Soviet Union after they had returned to Germany around 1950. But the American engineers who attended the Congress were of course well aware that the Soviet Union has nevertheless demonstrated precisely these capabilities; this was simply their way of saying that they could form no judgment on the basis of what they had been shown. Visitors to the USSR are evidently not shown the most modern factories; perhaps the Soviets do not care how unimpressed we are by what they choose to show us in their laboratories and factories. However, it may well be true that in techniques, instrumentation, and production, Soviet technology is far behind ours.

In 1959 a British team of experts on automatic control toured the Soviet Union and they concluded that Soviet work on automatic control was, on the average, not so far advanced as in Britain. On the other hand, they admitted that there was considerable activity and that many of the institutes they visited were only two or three years old. They pointed out that the Soviet effort in engineering education was both massive and effective; in 1958, 83,000 students were graduated in engineering in the USSR. They observed that the connection between teaching, research, and industry in the Soviet Union was strong and the research effort on automatic control was much greater there than in Britain.

A considerable amount of talk in the Soviet Union concerns the coming of automation and its fruits in the way of increased goods and shorter working hours. We also see in a report of the Academic Council of the Soviet Academy of Sciences, dated May, 1959, the following pointed statement:

The opinion is sometimes advanced that the new technological revolution in the Soviet Union will be the practical use of atomic energy as the most powerful source of energy. This is not the case. In fact, it is automation moving productive processes under their own power that today plays the same decisive role as was played by power machines in the industrial revolution of the 18th century.

The principal address at the opening session of the First Congress of the International Federation of Automatic Control in Moscow in the summer of 1960 was delivered by Academician V. A. Trapenznikov, the Chairman of the USSR National Committee of Automatic Control and head of the Institute of Automatics and Telemechanics. In his talk, entitled "Basic Theoretical and Engineering Problems of Automatic Control," he made the following point:

... not only will automation raise productivity, but it will also radically change the very nature of labor. . . . The full utilization of the benefits arising out of automation, however, is only possible in a rationally organized society where the manpower made redundant due to automation in one field is easily absorbed in others. Our firm conviction—which we do not, of course, impose on anybody—is that this possibility is offered only by a socialistic state.

The Soviets believe that through automation they can achieve a rapid acceleration in the rate of their technological progress and at the same time they believe that through automation they can demonstrate the superiority of their system—that only a Communist society can carry out automation of industry without a breakdown in its economy. They expect that we cannot in our system achieve automation as rapidly as they can, and that for us it will mean economic chaos. There is every reason to believe they can achieve automation and achieve it rapidly.

4. Implications

The fruitful connections in the USSR between mathematics and its applications cannot help but strengthen Soviet science and Soviet technology, and in turn strengthen mathematics. The narrow, isolated specialist in mathematics does without doubt make a contribution, but he is not to be compared with the great mathematicians of history who have always understood and used the empirical sources of mathematical inspiration. Not every application is of genuine mathematical interest, but this does not justify the pure mathematician's tendency to make a virtue of a lack of interest in mathematical problems arising in other sciences.

Historically, mathematics has always possessed a paradoxical duality. On the one hand mathematics reflects upon itself and is the most original creation of the human mind; yet there has always existed a close relation between mathematics and those empirical sciences that have risen above the level of pure description. Regardless of its purity, mathematics has the habit of proving to be useful. From a pragmatic viewpoint, of course, this justifies mathematics for its own sake and is the reason mathematicians should not be compelled—even if they could be—to confine themselves to the problems that appear to be of immediate practical interest. But in our country today the greater danger is from those who wish mathematics to be completely separated from its applications. Many mathematics departments in our universities have no contact with science and engineering, and many of our mathematicians engaged in teaching are narrow research specialists

intolerant of the interests of others. They and their students are un-
aware of the historical development of their own subject, of its relation
to other sciences, and often even of its relation to other branches of
mathematics. The danger is that a mathematical field isolated too long
from its empirical source may undergo an abstract inbreeding that
finally results in a sort of killing cancerous growth. The discipline splits
rapidly into a multitude of subdisciplines and tends to become a dis-
organized mass of detail and complexity. The orderly growth of mathematics is inspired and directed by the realities of the empirical sciences,
but it remains true that the strength and power of mathematics is its
own creation. We should force no dichotomy on mathematics.

Up to now the Soviet Union has managed both to encourage science
and to use it. They have recently created a new agency, the State Commission for the Coördination of Scientific Research. Its main purposes
are to strengthen basic theoretical research of great "economic" significance, to narrow the gap between research and production, and to
speed up the introduction of new scientific findings into the Soviet
economy. Those who feed our complacency will soon tell us that this
channeling of science will surely weaken and eventually destroy Soviet
science; but in view of both past history and recent events, there is no
reason to expect the Soviet Union to pull the colossal blunder of "killing
the goose." Rather, it is time that we recognize Soviet ability to direct
science to specific ends without destroying it and that we take steps to
meet the challenge by mobilizing more effectively our own scientific
talent.

References

1. LaSalle, J. P., and S. Lefschetz, *Recent Soviet Contributions to Mathe-
    matics*, The Macmillan Company, New York, 1962.
2. LaSalle, J. P., and S. Lefschetz, "Recent Soviet Contributions to Ordi-
    nary Differential Equations and Nonlinear Mathematics," *J. Math.
    Anal. and Appl.*, Vol. 2, June, 1961, pp. 467–499.

Chapter 16

The Theory of Optimal Control and the Calculus of Variations†

R. E. KALMAN

1. Background

"System theory" today connotes a loose collection of problems and methods held together by a central theme: to understand better the complex systems created by modern technology. Aside from certain combinatorial questions, most of present system theory is concerned with problems in automatic control and in statistical estimation and prediction, with emphasis on solutions that are optimal in some sense. These problems are attacked by a variety of *ad hoc* methods.

Recent research has shown how to formulate and resolve these problems in the spirit of the classical calculus of variations. This provides a unifying point of view. Eventually it should be possible to organize system theory as a rigorous and well-defined discipline. One example of this trend is the author's duality principle (see [1], [2], [3]) relating control and estimation. Conversely, problems in system theory are stimulating further research in the calculus of variations.

Let us look first at the historical background of the hamiltonian formulation of the calculus of variations. There is a long stream of scientific thought concerned with wave propagation and variational principles in physics. It begins with Huygens, continues with the work of John Bernoulli, and receives maturity at the hands of the great masters of the nineteenth century: Hamilton, Jacobi, and Lie. The most articulate representative of this tradition in recent times was C. Carathéodory (1873–1950). Beginning with his famous dissertation

† This research was sponsored by the United States Air Force under Contracts AF 49(638)-382 and AF 33(616)-6952.

of 1904, Carathéodory insisted on the hamiltonian point of view in the calculus of variations throughout his lifetime. The evolution of his thinking on this subject is carefully integrated in his last major work [4]—a book that is hard to obtain and difficult to digest.

The theory of optimal control, under the assumption that the equations of motion are known exactly and the state can be measured instantaneously, may be regarded as a generalization of the problem of Lagrange in the calculus of variations: minimization of an integral subject to side conditions, which may be ordinary or differential equations. Carathéodory's work on the problem of Lagrange is incomplete, consisting of only two papers [5], [6]; these papers are discussed briefly in Chapter 18 of [4]. The problem is one of extreme difficulty and it has received little attention until quite recently.

In [7] the present writer gave a new formulation of the problem of optimal control from the hamiltonian point of view. The purpose of this chapter is to extend this approach. We shall see that this formulation—which differs from Carathéodory's in essential details—explains a number of recent results in the theory of control and provides a very general framework for further research.

2. The Variational Problem in the Theory of Control

We assume that the control object is a dynamical system governed by the differential equation

$$\frac{dx}{dt} = \dot{x} = f(x, u(t), t). \tag{2.1}$$

Here x is a real n-vector, called the *state* of the system; $u(t)$ is a real m-vector for each t; f is a real n-vector that is continuously differentiable in all arguments.

To avoid the cumbersome phrase "the state x at time t," we shall refer to the couple (x, t) as a *phase*. The *phase space* is thus the cartesian product of the *state space* X ($= R^n$) with the space T ($= R^1$) of all values of the time.

We call the function $u(t)$ in (2.1) an *admissible control* if (a) it is piecewise continuous in t; (b) for each t, its values belong to a given closed subset $U(t)$ of R^m.

For any admissible control u and any initial phase (x_0, t_0), there exists a unique absolutely continuous function ϕ of t, denoted by

$$\phi(t) \equiv \phi_u(t; x_0, t_0),$$

that satisfies (2.1) almost everywhere and has the property

$$\phi(t_0) = \phi_u(t_0; x_0, t_0) = x_0.$$

We call $\phi_u(t; x_0, t_0)$ the *motion* of (2.1) passing through x_0 at time t_0 under the action of the control u. Sometimes we shall write $x(t) = \phi(t)$ to emphasize the fact that the value of ϕ at some fixed t is the state of the system at that time.

We call x^* an *equilibrium state* if there is some control u^* such that $\phi_u(t; x^*, t_0) = x^*$ for all t, t_0, or, equivalently, if $f(x^*, u^*(t), t) = 0$.

To state the control problem in its simplest form, it is assumed further that physical measurements are available giving the exact numerical value of the state at every instant of time, though of course this is a gross idealization from the engineering point of view. We want to determine $u(t)$ as a function of $x(t)$ so that motions of (2.1) have certain extremal properties. The expression of $u(t)$ as a function of $x(t)$ is commonly called *feedback* in engineering. We denote this functional relationship by

$$u(t) = k(x(t), t), \tag{2.2}$$

and refer to the function k as the *control law*. A control law is *admissible* if $k(x, t) \in U(t)$ for all t.

Let (x_0, t_0) be an arbitrary phase and let S be a surface in the phase space. Consider the following scalar functional of motions of (2.1):

$$V(x_0, t_0, S; u) = \lambda(\phi_u(t_1; x_0, t_0), t_1)$$
$$+ \int_{t_0}^{t_1} L(\phi_u(t; x_0, t_0), u(t), t)\, dt, \tag{2.3}$$

where L, λ are scalar functions and t_1 is the first instant of time after t_0 when the motion enters the set S. Thus it suffices for λ to be defined only on S. We call t_1 the *terminal time* and assume that L, λ are continuously differentiable in all arguments.

In terms of these notations, we can now state the

OPTIMAL CONTROL PROBLEM. *Given any initial phase (x_0, t_0), find a corresponding admissible control u defined in the interval $[t_0, t_1]$ at which the functional (2.3) assumes its infinum (or supremum) with regard to the set of all admissible controls.*

Actually, for technological reasons one usually sets a slightly stronger objective.

OPTIMAL FEEDBACK CONTROL PROBLEM. *Find a control law such that when (2.2) is substituted in (2.1) the functional (2.3) assumes its infinum (or supremum) with regard to the set of all admissible control laws.*

Bellman's principle of optimality shows that we can always define an optimal control law along every optimal motion. Hence the two foregoing problems are abstractly equivalent.

If Equation (2.1) depends on stochastic factors, however, then the infinum of (2.3) with respect to all admissible control laws will usually be *lower* than with respect to all admissible controls that are uniquely determined by the initial phase. This is owing to the fact that the control law takes into account not only the initial state but successive states as well. The added information so obtained may result in a better optimum.

Before embarking on a detailed analysis of the control problem, let us mention a number of typical examples that may be put into this formulation.

Terminal Control

The problem is to bring the state of the system as close as possible to a given terminal state x_1 at a given terminal time t_1. Then $L = 0$, $\lambda(x)$ is the distance of x from x_1, and $S = X \times \{t_1\}$.

Minimal-Time Control

Suppose we want to reach a state x_1 from (x_0, t_0) in the shortest possible time. We then set $L = 1$, $\lambda = 0$, and $S = \{x_1\} \times T$. This problem ordinarily has a solution only if $U(t)$ is a bounded set for all $t \geq t_0$.

Regulator Problem

We assume that the system is in some initial phase (x_0, t_0) and we wish to return to an equilibrium state x^* in such a way that some integral of the motion is minimized. We then usually take L and λ as non-negative. The dependence of L on u is needed because otherwise the problem may not have a solution. The set S is again $X \times \{t_1\}$.

Pursuit Problem

We are given a moving target $\xi(t)$. The problem is to bring the motion to phase $(\xi(t), t)$ as soon as possible. This is a generalization of the minimal-time problem; we take $S = \{(\xi(t), t); t \in T\}$.

Servomechanism Problem

This is a generalization of the regulator problem. We are given a desired state $\xi(t)$, $t \in T$. The problem is to cause the phase of the controlled motion to be as close as possible to $(\xi(t), t)$ on the interval $[t_0, t_1]$. The instantaneous distance between $(x(t), t)$ and $(\xi(t), t)$ is measured by L. The set S is again as in the regulator problem, above.

Minimum Energy Control

We wish to transfer from an initial phase (x_0, t_0) to a final phase (x_1, t_1) with the expenditure of a minimal amount of control energy. In this case we take L to be a nonnegative function of u, independent of ϕ; S is the set consisting of the single point (x_1, t_1); λ is immaterial.

Isoperimetric Problems

Suppose that the optimal motions must satisfy also the so-called *isoperimetric* constraints

$$\int_{t_0}^{t_1} f_{n+k}(\phi_u(t; x_0, t_0), u(t), t) \, dt \leq \alpha_k, \qquad k = 1, \cdots, N - n. \quad (2.4)$$

These problems reduce immediately to the preceding ones when we replace the n-vector x by an N-vector of which the last $N - n$ components satisfy the differential equations

$$\frac{dx_{n+k}}{dt} = f_{n+k}(x, u(t), t), \qquad k = 1, \cdots, N - n; \quad (2.5)$$

the initial values are $x_{n+k}(t_0) = 0$, and the final values $x_{n+k}(t_1)$ are to lie on a surface S for which $x_{n+k} \leq \alpha_k$.

3. Relations with the Calculus of Variations

The classical problem of *Lagrange* in the calculus of variations is concerned with the minimization of the integral

$$\int L(x(t), \dot{x}(t), t) \, dt \quad (3.1)$$

with respect to any smooth curve $x(t)$ that (a) connects a given point (x_0, t_0) with a point (x_1, t_1) lying on a given surface S, and (b) satisfies the constraints

$$g_i(x(t), \dot{x}(t), t) \equiv 0, \qquad i = 1, \cdots, n - m. \quad (3.2)$$

There are two ways in which the optimal control problem discussed above differs from the problem of Lagrange. First, the function L depends on u rather than on $\dot{x}$; second, the constraints are of a mixed type:

$$\dot{x} - f(x, u(t), t) = 0 \quad \text{and} \quad u(t) \in U(t). \qquad (3.3)$$

Neither of these differences is essential; inequality constraints such as $\alpha \geq 0$ can be replaced by equality constraints such as $\beta(\alpha) = 0$, where β is a smooth function that is zero if $\alpha \geq 0$ and positive otherwise. Similarly, one can always express $u(t)$ by (3.3) as a function of x, $\dot{x}$, t by introducing, if necessary, additional equality constraints. Hence the optimal control problem is formally identical with the problem of Lagrange, though the transformations necessary to establish the equivalence will be usually rather complicated. Moreover, because of difficulties arising from an explicit treatment of the constraints (3.2), the theory of the Lagrange problem today is far from adequate.

We therefore prefer to treat *directly* the problem of minimizing (2.3), subject to the constraints (3.3). This treatment includes the ordinary problem of the calculus of variations, obtained by setting $f(x, u, t) = u$ and $U(t) = R^n$, as well as the Lagrange problem after suitable transformation of the type just discussed.

Using the hamiltonian point of view, we do not need to transform the constraints (3.3) but can treat them directly. The principal idea is the following. We define a hamiltonian function not with the aid of the Legendre transformation, as is usual, but in a more general procedure by means of the so-called *minimum principle*. In this way the optimal control problem can be reduced to the solution of the Hamilton–Jacobi partial differential equation. The existence of a solution of the Hamilton–Jacobi equation is a *sufficient* condition for the solution $V^0(x, t)$ of the optimal control problem. If the function $V^0(x, t)$ is smooth, this condition is also *necessary*.

Unfortunately, quite often $V^0(x, t)$ does not have continuous partial derivatives with respect to x. In that case one cannot state necessary and sufficient conditions solely in terms of *differential* equations. But this is hardly the issue. Early in his career, Carathéodory took the following position:

The distinction between necessary and sufficient conditions seems, however, a little artificial; explicit proof that certain conditions are necessary is of interest only in cases in which one cannot resolve a problem at once, and it serves, above all, to limit the scope for future investigations. When, on the other hand, one has a solution possessing all the properties required by the theorem, it suffices to show that this solution is unique in order to have at the same time the proof that *all the conditions that serve to determine the solution are necessary.*†

It has unfortunately become very common in physical and engineering applications to regard the extremals supplied by the Euler equations as the "solution" of a variational problem. There are two long-standing objections to this: (a) the Euler equations may not exist, as when L is not sufficiently smooth; (b) the solution of the Euler equations may cease to define a minimum or maximum after a certain interval of time, as when the extremal contains conjugate points. The hamiltonian point of view, which aims to obtain sufficient conditions, avoids such difficulties at the outset by considering only those initial phases that can be connected by optimal motion with a phase on S, and by regarding the function $V^0 = \min V$ as abstractly defined in advance.

The dynamic programming method of Bellman proceeds from the same fundamental idea, differing only in detail from the hamiltonian methods. For a nonrigorous but highly enlightening discussion of the relations between the two, see the recent paper of Dreyfus [10].

4. The Hamilton–Jacobi Equation; Minimum Principle

Let us first obtain the sufficient condition. The starting point is the following trivial, well-known, but important observation concerning the optimal control problem.

LEMMA OF CARATHÉODORY [4, p. 198]. *Suppose there is a function $k(x, t)$ continuously differentiable in both its arguments and such that, for all (x, t) in some region G of the phase space,*

   i. $k(x, t) \in U(t)$,

   ii. $L(x, k, t) = 0$,   (4.1)

   iii. $L(x, u, t) > 0$ if $u \neq k(x, t)$.

Consider motions of (2.1) with control law defined by (2.2), that is,

$$\frac{dx}{dt} = f(x, k(x, t), t). \qquad (4.2)$$

Let the initial phase (x_0, t_0) belong to G. Let $\lambda(x, t)$ be identically zero on some surface $S \subset G$ of the phase space. Then the following properties hold for any motion ϕ^0 of (4.2) that connects (x_0, t_0) with a phase on S and remains entirely in G:

---

† Author's translation from French; author's italics. See [9], Introduction.

a. *The value of the integral (2.3) is zero.*

b. *The motion ϕ^0 provides the absolute minimum of (2.3) with respect
to any other motion of (2.1) which connects (x_0, t_0) with S and remains
entirely in G.*

In short, the hypotheses of the lemma mean that at every point in
G the integrand L has a *unique, absolute minimum* $u^0 = k(x, t)$ with
respect to all u satisfying the constraint (3.3). Then k is the unique
optimal control law, and the optimal feedback problem is also solved.

PROOF. Conclusion (a) is immediate, since for any motion ϕ^0 of (4.2)
the integral (2.4) is zero by hypothesis (ii). Now let ϕ^1 be any other
motion of (2.1) that connects (x_0, t_0) with S without leaving G, and for
which $V = 0$. Then by hypothesis (iii) and the continuity of L, it is
clear that along ϕ^1 we must have $u^1(t) = k(\phi^1(t; x_0, t_0), t)$ at every
continuity point of $u^1(t)$, since otherwise we would have $V > 0$. We
would obtain the same motion if we let $u^1(t)$ always be defined by this
relation, that is, $\phi_{u^1}(t; x_0, t_0) = \phi^1(t; x_0, t_0)$. But since k is continuously
differentiable in x, (4.2) defines a unique motion, and the proof of (b) is
complete.

It should be noted that there may be phases in G such that the motion
defined by (4.2) going through these phases is *not* optimal. This is
owing to the possibility that a motion may leave G prior to reaching S.

Now we try to construct a Lagrange function L^* and a corresponding
function k satisfying the requirements of the lemma.

Suppose $V^0(x, t)$ is a scalar function that is twice continuously differentiable in both arguments. Then†

$$\int_{t_0}^{t_1} \left[V_t^0(x, t) + f(x, u(t), t) \cdot V_x^0(x, t)\right] dt = V^0(x_1, t_1) - V^0(x_0, t_0) \quad (4.3)$$

along any motion of (2.1) connecting the phase (x_0, t_0) with the phase
(x_1, t_1) on S. If we let

$$V^0(x, t) = \lambda(x, t) \quad (4.4)$$

on S, then the optimal control problem obtained by replacing λ with
$\lambda^* = 0$ and L with

$$L^*(x, u, t) = L(x, u, t) + V_t^0(x, t) + f(x, u, t) \cdot V_x^0(x, t) \quad (4.5)$$

will be equivalent to the original problem, because the values of V and

---

† The dot denotes the inner product; $V_t = \partial V/\partial t$, $V_x = \text{grad } V$.

V^* will differ only by $V^0(x_0, t_0)$, which does not depend on the control u.

Let p be a real n-vector, called the *costate*.

We define a scalar function H by

$$H(x, p, t, u) = L(x, u, t) + f(x, u, t) \cdot p. \tag{4.6}$$

We assume that H has a unique *absolute* minimum for each t with respect to $u(t) \in U(t)$ at the point

$$u^0(t) = c(k, p, t); \tag{4.7}$$

moreover, c is continuously differentiable in all arguments.

The scalar function H^0, defined by

$$
\begin{aligned}
H^0(x, p, t) &= \min_{u(t) \in U(t)} H(x, p, t, u) \\
&= L(x, c(x, p, t), t) + f(x, c(x, p, t), t) \cdot p,
\end{aligned}
\tag{4.8}
$$

is the *hamiltonian* of the problem.

Finally, we assume that $V^0(x, t)$ satisfies the Hamilton–Jacobi partial differential equation

$$V_t^0 + H^0(x, V_x^0, t) = 0 \tag{4.9}$$

with the boundary condition (4.4).

If these assumptions hold, we let

$$p = V^0(x, t). \tag{4.10}$$

Then

$$L^*(x, u, t) = V_t^0(x, t) + H^0(x, V_x^0(x, t), t) \tag{4.11}$$

will clearly satisfy the hypotheses of the lemma of Carathéodory, with k defined by

$$k(x, t) = c(x, V_x^0(x, t), t). \tag{4.12}$$

Moreover, by the lemma, we also have

$$V^0(x_0, t_0) = \lambda(\phi^0(t_1), t_1) + \int_{t_0}^{t_1} L(\phi^0(t), k(\phi^0(t), t), t) \, dt \tag{4.13}$$

along motions ϕ^0 satisfying (4.2). In other words,

$$V^0(x_0, t_0) = \min_u V(x_0, t_0, S; u) \tag{4.14}$$

is the absolute minimum of the integral (2.3) with respect to admissible controls; (4.12) is the optimal control law, and we have also solved the optimal feedback control problem.

Hence we have established:

THEOREM 4.1. SUFFICIENT CONDITION. *Let H^0 be the absolute minimum of $H = L + f \cdot p$ with respect to $u(t) \in U(t)$. Suppose that $u^0 = c$ is unique, that the differentiability hypotheses hold, that V^0 satisfies the Hamilton–Jacobi partial differential equation $V_t^0 + H^0(x, V_x^0, t) = 0$ in a region $G \subset S$, and that furthermore $V = \lambda$ on S. Then the following properties hold:*

a. The function $V^0(x_0, t_0)$ is the absolute minimum of (2.3) with respect to all motions which connect (x_0, t_0) with a phase on S without leaving G.

b. The optimal control law is given by (4.12); with this control law any motion that eventually reaches S without leaving G is optimal.

The introduction of the hamiltonian function H^0 reduces the problem to one of ordinary minimization, which defines the optimal value of $u(t)$ at each moment through (4.7). To achieve this, we bring in the auxiliary variable p. To make sure that the point-by-point optimization based on p is consistent, we eliminate p by (4.10); the construction succeeds whenever V^0 is a solution of the Hamilton–Jacobi equation defined by H^0.

It is easy to prove the converse result—in other words, that the optimal value V^0 of (2.3) must satisfy the Hamilton–Jacobi equation—provided V^0 is a sufficiently smooth function of x.

THEOREM 4.2. NECESSARY CONDITION. *Let G be a region in the phase space possessing the following properties:*

   i. *There is an optimal motion from every phase in G to a phase on S that never leaves G.*

   ii. *The minimum value of (2.3), denoted by $V^0(x, t)$, is twice continuously differentiable in both arguments.*

   iii. *Every point in G that is not also on S has a neighborhood lying entirely in G.*

   iv. † *For every phase in G, $H(x, V_x^0, t, u)$ given by (4.6) has an absolute minimum $H^0(x, V_x^0, t)$ at $u^0 = k(x, t)$ with respect to $u(t) \in U(t)$.*

   v. † *The function k defining the minimum is differentiable in x and continuous in t.*

Then the function $V^0(x, t)$ satisfies the Hamilton–Jacobi equation $V_t^0 + H^0(x, V_x^0, t) = 0$ in the region G.

---

† These conditions may be checked from the given form of $L, f,$ and U, that is, without solving the variational problem.

PROOF. Let (x_0, t_0) be a phase in G for which the theorem is false. There are then two possibilities. We consider first

$$V_t^0(x_0, t_0) + H^0(x_0, V_x^0(x_0, t_0), t_0) > 0. \tag{4.15}$$

Let $N \subset G$ be an open neighborhood of (x_0, t_0) that is small enough that the inequality (4.15) remains true everywhere in N. It is clear that N exists because of (iii) and because the left-hand side of (4.15) is continuous in x and t.

Let $\phi^0(t)$ be an optimal motion originating at (x_0, t_0), and let $u^0(t)$ be the corresponding optimal control. Then, because of the definition of H^0, for all t such that $(\phi^0(t), t) \in N$ we have

$$H(\phi^0(t), V_x^0(\phi^0(t), t), t, u^0(t)) \geq H^0(\phi^0(t), V_x^0(\phi^0(t), t), t). \tag{4.16}$$

Combining (4.15) and (4.16), we obtain

$$-V_t^0(\phi^0(t), t) - V_x^0(\phi^0(t), t) \cdot f(\phi^0(t), u^0(t), t) \leq L(\phi^0(t), u^0(t), t) - \epsilon(t), \tag{4.17}$$

with $\epsilon > 0$ as long as $(\phi^0(t), t)$ remains in N. Let $t_1 > t_0$ such that $(\phi^0(t), t) \in N$ for all $t \in [t_0, t_1]$. Integrating both sides of (4.17), we get

$$V^0(x_0, t_0) - V^0(x_1, t_1) = \int_{t_0}^{t_1} \left[L(\phi^0(t), u^0(t), t) - \epsilon(t) \right] dt$$

$$< \int_{t_0}^{t_1} L(\phi^0(t), u^0(t), t) \, dt$$

or, using the definition of $V^0(x_1, t_1)$ and letting (x_2, t_2) be the phase on S reached by an optimal motion ϕ^0 starting at (x_1, t_1),

$$V^0(x_0, t_0) < \lambda(x_2, t_2) + \int_{t_0}^{t_1} L(\phi^0(t), u^0(t), t) \, dt,$$

which contradicts the assumption that ϕ^0 is optimal.

Now we suppose that N is an open neighborhood of (x_0, t_0) throughout which the inequality (4.15) holds in the opposite sense. Then, by the definition of H^0, we have

$$-V_t^0(x, t) - V_x^0(x, t) \cdot f(x, k(x, t), t) = L(x, k(x, t), t) + \epsilon(x, t),$$

where $\epsilon > 0$ throughout N.

Hence, integrating along the unique motion $\phi_k(t; x_0, t_0)$ defined by

(4.2), we get

$$V^0(x_0, t_0) - V^0(x_1, t_1) = \int_{t_0}^{t_1} \left[L(x, k(x, t), t) + \epsilon(x, t) \right] dt$$

$$> \int_{t_0}^{t_1} L(x, k(x, t), t)\, dt,$$

provided $(\phi_k(t; x_0, t_0), t) \in N$ for all $t \in [t_0, t_1]$. This contradicts the definition of V^0, by the same argument as above, and establishes Theorem 4.2.

The essence of the arguments in this section is to replace the hamiltonian H by the hamiltonian H^0 by eliminating u with the aid of the minimum operation (4.8). We shall call this the *minimum principle*.

5. Canonical Differential Equations; Pontryagin's Theorem

At this stage, the solution of the optimal control problem is reduced to the problem of solving the Hamilton–Jacobi *partial* differential equation. Following Carathéodory's program, one can go a step further and show that the optimal motions must be solutions of the characteristics of the Hamilton–Jacobi equation, which are a set of *ordinary* differential equations of order $2n$. They are the Euler equations in canonical form—or simply the canonical equations—of the problem.

In this way, the determination of optimal motions reduces to the solution of the canonical equations. But in order to show that a given motion is really optimal, one must still construct—abstractly or explicitly—a solution of the Hamilton–Jacobi partial differential equation or, what is the same thing in view of Theorem 4.2, the function $V^0(x, t)$. Moreover, the solution of the canonical equations does not provide the optimal control law for which, by (4.12), knowledge of V^0 is essential.

Let G be a region in the phase space satisfying the hypotheses (i–v) of Theorem 4.2. Let $\phi^0(t)$ be an optimal motion starting at some phase in G and eventually reaching a phase on S without leaving G. We define

$$\psi^0(t) = V_x^0(\phi^0(t), t). \tag{5.1}$$

Differentiating $\psi^0(t)$ with respect to t, we have

$$\frac{d\psi^0(t)}{dt} = V_{xt}^0(\phi^0(t), t) + V_{xx}^0(\phi^0(t), t) \cdot f(\phi^0(t), u^0(t), t). \tag{5.2}$$

Differentiating the Hamilton–Jacobi equation, regarded as an identity in $V^0(x, t)$, with respect to x yields

$$V^0_{xt}(x, t) + H^0_x(x, V^0_x(x, t), t) + H^0_p(x, V^0_x(x, t), t) \cdot V^0_{xx}(x, t) = 0 \qquad (5.3)$$

throughout G. Recalling the definition of H, it follows that

$$H^0_p = f + [L_u + p \cdot f_u]c_p. \qquad (5.4)$$

We wish to show that the bracketed term is zero. For technical reasons, *we assume that the boundary of $U(t)$ is smooth in the space $R^m \times T$.*

Consider a point (x_0, p_0, t) and the corresponding $u^0_0 = c(x_0, p_0, t) \in U(t)$ at which H assumes its unique absolute minimum. Recall that $U(t)$ is closed. The following possibilities arise:

a. The point u^0_0 is interior to $U(t)$. Then the first derivative of H with respect to u must vanish at u^0:

$$L_u(x_0, u^0_0, t_0) + p_0 \cdot f_u(x_0, u^0_0, t_0) = 0. \qquad (5.5)$$

b. The point u^0_0 is on the boundary of $U(t)$. There are now two subcases.

i. There is at least one point in every neighborhood of (x_0, p_0, t_0) such that the corresponding u^0 is in the interior of $U(t)$. Then (5.5) holds also at (x_0, p_0, t_0) since L_u, f_u, c are continuous in all arguments.

ii. There is a neighborhood N of (x_0, p_0, t_0) such that every u^0 corresponding to points in N lies on the boundary of $U(t)$. In this case, throughout N we must have

$$g^i(c(x, p, t), t) = 0, \qquad i = 1, \cdots, q \le m.$$

Since the boundary of $U(t)$ is smooth, we assume that the functions g^i are differentiable in both arguments and also that the determinant

$$\left| \frac{\partial g^i(u, t)}{\partial u_j} \right|$$

has rank q at the point (u^0_0, t_0). Then the well-known Lagrange multiplier rule [4, p. 166] implies that

$$L_u(x_0, u^0_0, t_0) + p_0 \cdot f_u(x_0, u^0_0, t_0) + \sum_{i=1}^{q} \nu_i g^i_u(u^0_0, t_0) = 0, \qquad (5.6)$$

with $\nu_i \not\equiv 0$. On the other hand, differentiating $g^i(c(x, p, t), t) = 0$ with respect to x and p shows that

$$g_u^i(c(x, p, t), t) \cdot c_x(x, p, t) = 0,$$
$$g_u^i(c(x, p, t), t) \cdot c_p(x, p, t) = 0, \qquad i = 1, \cdots, q.$$

Combining the foregoing two equations, we have

$$[L_u(x_0, u_0^0, t_0) + p_0 \cdot f_u(x_0, u_0^0, t_0)] \cdot c_x(x_0, p_0, t_0) = 0,$$
$$[L_u(x_0, u_0^0, t_0) + p_0 \cdot f_u(x_0, u_0^0, t_0)] \cdot c_p(x_0, p_0, t_0) = 0.$$
(5.7)

Hence we conclude that

$$H_x(x, p, t, c(x, p, t)) = H_x^0(x, p, t),$$
$$H_p(x, p, t, c(x, p, t)) = H_p^0(x, p, t) = f(x, p, t);$$
(5.8)

these equations follow immediately from (5.7) in case (ii) and from (5.6) in the other cases.

In view of (5.8), utilizing also (5.2) and (5.3), we obtain the *canonical equations*:

$$\text{a. } \frac{dx}{dt} = H_p^0(x, p, t) = H_p(x, p, t, c(x, p, t)),$$

$$\text{b. } \frac{dp}{dt} = -H_x^0(x, p, t) = -H_x(x, p, t, c(x, p, t)),$$
(5.9)

which could also be written as the identities

$$\frac{d\phi^0(t)}{dt} = H_p(\phi^0(t), \psi^0(t), t, u^0(t)),$$

$$\frac{d\psi^0(t)}{dt} = -H_x(\phi^0(t), \psi^0(t), t, u^0(t)).$$
(5.10)

The last equations constitute a special case of the following result.

PONTRYAGIN'S THEOREM [8]. *If the motion $\phi^0(t)$ is optimal with control $u^0(t)$, then there must exist a function $\psi^0(t)$ such that (5.10) is satisfied and in addition the relation*

$$H(\phi^0(t), \psi^0(t), t, u) \geq H(\phi^0(t), \psi^0(t), t, u^0(t))$$
(5.11)

must hold for all $u \in U(t)$.

Equation (5.11) is Pontryagin's form of the minimum principle. It is proved [8], [11] by constructing a special first variation of the function $u^0(t)$. (In Pontryagin's paper, H is defined, following the standard convention, as the negative of the quantity (4.6). For this reason, Pontryagin speaks of the "maximum" principle. We feel that the present choice of sign, which is motivated by the dynamic programming approach to the definition of V^0, is more natural.)

Note that Pontryagin's theorem is valid [11] without the strong smoothness assumptions concerning V^0. But in that case one cannot identify $\psi^0(t)$ with $V_x^0(\phi^0(t), t)$, and there remains a gap between the necessary condition represented by Pontryagin's form (5.10) of the Euler equations and the Hamilton–Jacobi–Carathéodory theory that we have sketched above.

Nevertheless, our theory can still be used for the effective solution of problems in which $V^0(x, t)$ does not have continuous second derivatives throughout the phase space.

6. Solution of a Minimal-Time Problem

Consider the linear system (harmonic oscillator),

$$\frac{dx_1}{dt} = x_2,$$
$$\frac{dx_2}{dt} = -x_1 + u_1(t),$$

(6.1)

with

$$|u_1(t)| \leq 1.$$ (6.2)

Determine a control law taking the state of the system to the origin in the shortest possible time. See the minimal-time control problem in Section 2.

This celebrated problem seems to have been first mentioned by Doll [12] in 1943 in a U.S. Patent. The first rigorous solution of the problem appeared in 1952 in the doctoral dissertation of Bushaw [13]. Bushaw states that the problem does not fall within the framework of the classical calculus of variations, and he solves it by elementary but highly intricate direct geometric arguments.

The hamiltonian theory developed above can be applied quite simply to give a rigorous proof of Bushaw's theorem.

We rewrite Equations (6.1) in matrix form as

$$\frac{dx}{dt} = Fx + Gu(t),$$ (6.3)

where

$$F = \begin{bmatrix} 0 & 1 \\ -1 & 0 \end{bmatrix}, \qquad g = G = \begin{bmatrix} 0 \\ 1 \end{bmatrix}.$$

The minimum principle shows that the optimal control must satisfy the relation

$$u^0(t) = -\text{sgn}\,[g'\psi^0(t)], \tag{6.4}$$

where sgn is the scalar function of the scalar $g'p$ that takes on the value 1 when $g'p > 0$, -1 when $g'p < 0$, and is undetermined when $g'p = 0$.

Since the problem is invariant under translation in time, we shall drop the arguments referring to the initial time, which can be taken as 0 for convenience. Instead of considering motions in the phase space (x_1, x_2, t), we need to consider them only in the state space (x_1, x_2).

First we determine all possible optimal motions passing through the origin. There are three of these: Either $\phi^0(t)$ is identically zero, which is trivial, or $\phi^0(t)$ is a solution of (6.1) with $u^0(t) = +1$ or -1.

Let $u^0(t) = 1$. Then the motion of (6.1) passing through the origin is a circular arc γ^+ of radius 1 about the point $(1, 0)$ (see Fig. 1). To check whether this motion is really optimal, we must verify first of all that we have $g'\psi^0(t) < 0$ along the entire arc. Now (5.9b) in this case is

$$\frac{dp}{dt} = -F'p, \tag{6.5}$$

which is independent of x and has the solution

$$\psi^0(t) = \begin{bmatrix} \cos(t - t_0) & \sin(t - t_0) \\ -\sin(t - t_0) & \cos(t - t_0) \end{bmatrix} \psi^0(t_0). \tag{6.6}$$

It is clear that $\psi^0(t)$ is periodic with period 2π; therefore the largest interval over which we have $g'\psi^0(t) < 0$ is at most of length $< \pi$. This is actually achieved by choosing $\psi^0(0) = (\epsilon, 1)$, so that

$$g'\psi^0(t) < 0 \quad \text{for all} \quad 0 \le t < \pi - \epsilon. \tag{6.7}$$

Thus the necessary condition provided by the Euler equation (5.10) is satisfied along the arc γ^+ up to the state $(2, 0)$. (The remaining portion of the arc γ^+ is shown by a dashed curve in Fig. 1.)

Now we must establish a sufficiency condition; in other words, we must show that the arc γ^+ is indeed an optimal motion between the states $(0, 0)$ and $(2, 0)$.

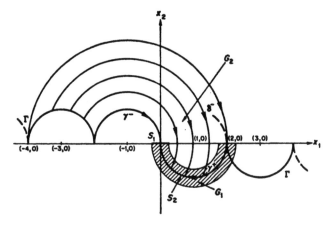

Fig. 1. Solution of minimal-time problem.

Let us define the set S_1 by

$$S_1 = \{x; \; -0.1 < x_1 < 0.1, \quad x_2 = 0\},$$

as shown in Figure 1. We consider the problem of reaching S_1 in minimal time. Since $V_t^0 = 0$, the Hamilton–Jacobi equation for this problem is

$$V_x^0 \cdot (Fx + g) = -1, \tag{6.8}$$

which has the solution

$$V_1^0(x_1, x_2) = \frac{\pi}{2} + \arctan \frac{1 - x_1}{x_2}.$$

The value of V^0 for $x_2 = 0$ is defined by its limit as $x_2 \to 0$ from *negative* values. Then $V^0 = 0$ on S_1, as required. Moreover, the region G_1, where V^0 is to satisfy (6.8), is taken as the semicircular band indicated by the crosshatching in Figure 1.

It follows from the necessary and sufficient conditions given in Section 4 that *if we connect any state on γ^+ with S_1 by means of a motion of (6.1) that is distinct from γ^+ and remains entirely in G_1, then the value of V in (2.3) necessarily is greater than V^0.*

But if it is not possible to reach S_1 from γ^+ faster than by proceeding along γ^+ itself, the same is true a fortiori as concerns reaching the state $(0, 0)$ on S_1. Hence we have proved:

The motion γ^+ is optimal relative to the region G_1.

The same construction establishes the local optimality of the motion γ^- (see Fig. 1).

Now we let $S_2 = \gamma^+$, $\lambda(x) = V_1^0(x)$, and we consider the minimal-time problem relative to S_2. All optimal motions, denoted by δ^- in Figure 1, necessarily correspond to $u^0 = -1$. They are circular arcs of radius 1 about the point $(-1, 0)$. Applying the Euler equations the same way as before, we find that all optimal arcs δ^- must terminate on the semicircle of radius 1 centered at $(-3, 0)$, which is part of the curve Γ in Figure 1. The arcs δ^- therefore fill up a region G_2 bounded by γ^+, γ^-, and the semicircle centered at $(-1, 0)$ that connects $(2, 0)$ with $(-4, 0)$. If we calculate the time needed to reach γ^+ starting from a point to G_2 and proceeding along δ^-, we get a smooth function $V_2^0(x)$ satisfying the Hamilton–Jacobi equation (6.8). Details are left to the reader. This proves that all motions consisting of an arc of δ^- and an arc of δ^+ are optimal.

The construction can be continued in a similar fashion until it covers any point in the plane. The optimal control law is

$$u_1^0(x) = k(x) = \begin{cases} +1 & \begin{array}{l} \text{below the curve } \Gamma \text{ composed of} \\ \text{semicircles of radius 1 and on } \gamma^+; \end{array} \\ -1 & \text{above the curve } \Gamma \text{ and on } \gamma^-. \end{cases} \qquad (6.9)$$

On $\Gamma - (\gamma^+ \cup \gamma^-)$, the value of u^0 is not determined by the minimum principle; it is easily verified that the choice of u^0 on $\Gamma - (\gamma^+ \cup \gamma^-)$ is immaterial as long as $|u_1| \leq 1$.

The control law (6.9) is *Bushaw's theorem*.

It should be noted that the function V^0, which is determined piece-wise as V_1^0, V_2^0, etc., is *not* continuously differentiable at a point P on γ^+. The limit of V_x^0 is infinite if we approach P from below γ^+ along points that lie on the continuation of δ^-; and the same limit is finite if we approach P from above γ^+ along δ^-.

As a result, the Euler equations (5.9) do not have continuous solutions along optimal motions; the conjugate vector p receives an "impulse" on passing through Γ. But the more general proof [11] of Pontryagin's theorem shows that relations (5.10) remain true, so that

$$u^0(t) = -\text{sgn} \, [g'\psi^0(t; p_0, t_0)], \qquad (6.10)$$

where the initial condition p_0 for the adjoint equation (6.5) may be defined by

$$p_0 = k(x_0),$$

in which k is the optimal control law (6.9). Then $\psi^0(t)$ vanishes on Γ, which shows that ψ^0 *cannot be interpreted as V_x^0*.

In using Pontryagin's theorem in the form just mentioned as a necessary condition to determine all possible optimal motions, it is still necessary to carry out the explicit construction given above, for (6.10) can be interpreted as the optimal control only if an optimal motion is known to *exist* connecting x_0 with the origin. In this very special case one can prove sufficiency without the Hamilton–Jacobi theory by noting that for any (x_0, t_0) there is exactly one $u^0(t)$ satisfying (6.10) which transfers x_0 to 0.

7. General Solution of the Linear Optimal Regulator Problem

A class of problems that can be completely solved by the hamiltonian theory is represented by the functional†

$$\left\|\phi_u(t_1; x_0, t_0)\right\|_A^2 + \frac{1}{2}\int_{t_0}^{t_1}[\left\|H(t)\phi_u(t; x_0, t_0)\right\|_{Q(t)}^2 + \left\|u(t)\right\|_{R(t)}^2]\,dt, \quad (7.1)$$

where the motions ϕ_u are defined by the linear differential equation

$$\frac{dx}{dt} = F(t)x + G(t)u(t); \quad (7.2)$$

there are no constraints on u.

This is a slight generalization of the regulator problem given in Section 2.

The matrices $Q(t)$, $R(t)$ are taken to be positive definite for all t. This assumption on R implies that the equation

$$2H(x, p, t, u) = \left\|H(t)x\right\|_{Q(t)}^2 + \left\|u\right\|_{R(t)}^2 + 2p\cdot[F(t)x + G(t)u(t)] \quad (7.3)$$

has a unique absolute minimum for every (x, p, t) at

$$c(x, p, t) = -R^{-1}(t)G'(t)p,$$

so that we have

$$2H^0(x, p, t) = \left\|H(t)x\right\|_{Q(t)}^2 + 2p\cdot F(t)x - \left\|G'(t)p\right\|_{R^{-1}(t)}^2. \quad (7.4)$$

The Hamilton–Jacobi equation corresponding to (7.4) has a unique solution, given any nonnegative definite A and any $t_1 > t_0$. To show this, we assume that (4.9) has a solution of the form

$$2V^0(x, t) = \left\|x\right\|_{P(t)}^2, \quad (7.5)$$

---

† We use the notation $\|x\|_A^2$ for a quadratic form defined by a symmetric nonnegative definite matrix A.

which implies the linear control law

$$k\,(x,\,t)\,=\,-R^{-1}(t)G'(t)P(t)x. \tag{7.6}$$

It is easily checked that (4.9) with the hamiltonian defined by (7.4) has a solution of the type (7.5) if and only if the symmetric matrix $P(t)$ is a solution of the *Riccati equation*

$$-\frac{dP}{dt} = F'(t)P + PF(t) - PG(t)R^{-1}(t)G'(t)P + H'(t)Q(t)H(t). \tag{7.7}$$

Moreover, the boundary condition

$$V^0(x_1,\,t_1) = \|x\|_A^2,$$

which is the concrete form of (4.4), implies that the solution of (7.7) must satisfy the initial condition

$$P(t_1) = 2A. \tag{7.8}$$

Since (7.7) is nonlinear, it is not clear at once that $P(t)$ exists outside of a small neighborhood of t_1. The integral (7.1) may nevertheless be bounded from above by the free motions of (7.2), that is, by setting $u(t) = 0$, which in view of (7.5) is equivalent to a bound on $\|P(t)\|$. Using the a priori bound so obtained in the standard existence theorem for differential equations shows that *solutions of* (7.7) *exist for all* $t \le t_1$. This conclusion is in general no longer valid if A has negative eigenvalues or if $t > t_1$.

Once the existence of solutions of (7.7), and therefore of the Hamilton-Jacobi equation, is ensured, the solutions can be expressed [2], [3], [7] with the aid of solutions of the canonical differential equations

$$\begin{bmatrix} dx/dt \\ dp/dt \end{bmatrix} = \begin{bmatrix} F(t) & -G(t)R^{-1}(t)G'(t) \\ -H'(t)Q(t)H(t) & -F'(t) \end{bmatrix} \begin{bmatrix} x \\ p \end{bmatrix}. \tag{7.9}$$

Further difficulties arise, however, in studying the stability of (7.7) as well as the stability of the optimal motions defined by (7.6). Details of these problems may be found particularly in [7].

8. General Solution of the Linear Optimal Servomechanism Problem

The problem considered in the previous section can be generalized in several ways. We consider here simultaneously two such generalizations.

First, we assume that the motions, in addition to control, are subject to "disturbances" represented by the term $w(t)$ in the equation

$$\frac{dx}{dt} = F(t)x + G(t)u(t) + w(t).$$ (8.1)

Second, we assume that the functional to be minimized is

$$\|\eta(t_1) - H(t_1)\phi_u(t_1)\|_A^2 + \frac{1}{2}\int_{t_0}^{t_1}[\|\eta(t) - H(t)\phi_u(t)\|_{Q(t)}^2$$

$$+ \|u(t)\|_{R(t)}^2]\,dt.$$ (8.2)

We call the p-vector,

$$y(t) = H(t)x(t),$$ (8.3)

the *output* of the system (8.1); by analogy, the vector function $\eta(t)$ is the *desired output*.

This setup is a slight generalization of the servomechanism problem of Section 2. A number of formal solutions have appeared in the engineering literature [14], [15]. The hamiltonian theory provides a simple rigorous proof of the known formulas.

Proceeding exactly as in Section 7, we find that the hamiltonian of the problem is

$$2H^\circ(x, p, t) = \|\eta(t) - H(t)x\|_{Q(t)}^2$$

$$+ 2p\cdot[F(t)x + w(t)] - \|G'(t)p\|_{R^{-1}(t)}^2.$$ (8.4)

To solve the corresponding Hamilton–Jacobi equation (4.9), we assume that

$$2V^\circ(x, t) = \|x\|_{P(t)}^2 - 2z(t)\cdot x + v(t).$$ (8.5)

Substituting, we obtain the following result:

THEOREM. *The function* $V^\circ(x, t)$ *given by (8.5) satisfies the Hamilton–Jacobi equation defined by (8.4), with* $V^\circ(x, t_1) = \|\eta(t_1) - H(t_1)x\|_A$, *if and only if*

a. *the matrix* $P(t)$ *is the solution of the Riccati equation (7.7) with* $P(t_1) = 2A$;

b. *the vector* $z(t)$ *is the solution of*

$$\frac{dz}{dt} = -[F(t) - G(t)R^{-1}(t)G(t)P(t)]'z + P(t)w(t) - H'(t)Q(t)\eta(t),$$ (8.6)

with

$$H'(t_1)A\eta(t_1) = z(t_1);$$ (8.7)

c. *the scalar $\nu(t)$ is the solution of*

$$-\frac{d\nu}{dt} = [\|\eta(t)\|^2_{Q(t)} - \|G'(t)z(t)\|^2_{R^{-1}(t)}] - 2\,z(t)\cdot w(t),\qquad (8.8)$$

with

$$\nu(t_1) = 2\|\eta(t_1)\|^2_A.$$

The control law is *linear*, for it is given by

$$u^0(t) = -R^{-1}(t)G'(t)p(t) = R^{-1}(t)G'(t)[z(t) - P(t)x(t)].\qquad (8.9)$$

This law, however, is *unrealizable*, because it involves $z(t)$ which, according to (8.6) and (8.7), must be computed backward in time and requires the knowledge of $\eta(t)$ and $w(t)$ in the interval $[t_0, t_1]$ and this usually is not known at the time t_0 in practical applications.

It should be noted that the differential equation for $z(t)$, minus the forcing terms, is the *adjoint* of the differential equation of optimal motions given in Section 7.

References

1. Kalman, R. E., "A New Approach to Linear Filtering and Prediction Problems," *Trans. ASME, Ser. D*, Vol. 82D, 1960, pp. 35–45.
2. Kalman, R. E., and R. S. Bucy, "New Results in Linear Filtering and Prediction Theory," *Trans. ASME, Ser. D*, Vol. 83D, 1961, pp. 95–108.
3. Kalman, R. E., "On the General Theory of Control Systems," *Proceedings of the First International Congress of Automatic Control*, held at Moscow, 1960, Butterworth & Co., Ltd., London, 1962.
4. Carathéodory, C., *Variationsrechnung und Partielle Differentialgleichungen erster Ordnung*, B. G. Teubner, Leipzig, 1935.
5. Carathéodory, C., "Die Methode der geodätischen Äquidistanten und das Problem von Lagrange," *Acta Math.*, Vol. 47, 1926, pp. 199–236.
6. Carathéodory, C., "Über die Einteilung der Variationsprobleme von Lagrange nach Klassen," *Comment. Math. Helv.*, Vol. 5, 1933, pp. 1–19.
7. Kalman, R. E., "Contributions to the Theory of Optimal Control," *Bol. Soc. Mat. Mexicana, Second Ser.*, Vol. 5, 1960, pp. 102–119.
8. Pontryagin, L. S., "Optimal Control Processes" (Russian), *Uspehi Mat. Nauk*, Vol. 14, 1959, pp. 3–20.
9. Carathéodory, C., "Sur une méthode directe du calcul des variations," *Rend. Circ. Mat. Palermo*, Vol. 25, 1908, pp. 36–49.
10. Dreyfus, S. E., "Dynamic Programming and the Calculus of Variations," *J. Math. Anal. and Appl.*, Vol. 1, No. 2, 1960, pp. 228–239.
11. Boltyanskii, V. G., R. V. Gamkrelidze, and L. S. Pontryagin, "The

Theory of Optimal Processes. I. The Maximum Principle," *Izv. Akad. Nauk SSSR, Ser. Mat.*, Vol. 24, 1960, pp. 3–42.

12. Doll, H. G., U.S. Patent No. 2,463,362, 1943.
13. Bushaw, D. W., Unpublished Ph.D. dissertation, Department of Mathematics, Princeton University, 1952; also Report No. 463, Experimental Towing Tank, Stevens Institute of Technology, Hoboken, N.J., 1953; "Optimum Discontinuous Forcing Terms," *Contributions to Nonlinear Oscillations*, Vol. 4, Princeton University Press, Princeton, N.J., 1958.
14. Merriam, C. W., III, "A Class of Optimum Control Systems," *J. Franklin Inst.*, Vol. 267, 1959, pp. 267–281.
15. Lee, E. B., "Design of Optimum Multivariable Control Systems," *Trans. ASME, Ser. D*, Vol. 83D, 1961, pp. 85–90.

Chapter 17

Mathematical Model Making as an Adaptive Process

RICHARD BELLMAN

1. Introduction

A great deal of interest now centers on the topic of "intelligent machines." Since there is little agreement as to what the term "human intelligence" means, and less understanding where these islands of unanimity do exist, many feel that hypothetical discussions of capabilities and limitations of machines are not of the immediate essence. The feeling is that the proof is in the pudding—or rather in the program. Pursuant to this operational philosophy, sustained effort has been devoted to the task of duplicating or replicating human achievements in diverse intellectual areas. Numerous groups of mathematicians, logicians, engineers, and philologists are devoting appreciable amounts of time to the construction of devices that will play chess, translate Swahili (for the benefit of those who prefer Swahili literature in translation), compose music, and even produce nonobjective art—activities we associate with cognitive behavior, perhaps charitably in the last-named field of activity.

There is no a priori reason to suppose that a machine accomplishing these tasks performs the intermediate functions along human lines, and certainly there is no need for us to desire this similarity in all applications. It is unquestionably interesting to play this game of duplication of animate activities, and, often, there can be enormous value to this diversion. Thus, for example, if we could riddle the mystery of human memories we could construct computers that would help us attack truly formidable scientific problems. Even where the simulation of human abilities is quite pedestrian, as in the performance of arithmetic by digital computers, the result is still remarkable. Freed of the drudgery of these calculations, the human mind can contemplate problems of real conceptual difficulty, problems that transcend any algorithms we now possess, and thus electronic circuits, now and possibly forever.

As pointed out by N. Wiener in the course of conversation, it is not necessary to suppose that complex tasks will be carried out by machine alone. A man–machine combination, in which the machine performs well-defined tasks quickly and accurately, and the man uses the ill-defined but very real qualities of experience, intuition, and creativity to guide the over-all operation, is clearly superior to either man or machine alone.

One of the reasons most often heard in defense of the allocation of time, effort, and trained personnel to the study of machine simulation of such human attributes as chess-playing or of translation of languages is that the solution of these "simple" problems will provide valuable clues to the solution of the more important difficult problems. This argument has such an appearance of reasonableness that it should be examined with great care.

Probably the principal flaw in this syllogism lies in the adjective "simple." The professional mathematician is well aware, after several hundred years of fruitless effort devoted to the four-color problem, the Goldbach conjecture, and Fermat's last theorem, that a simple verbal statement can conceal unbelievable mathematical difficulty. At the present time, there is an enormous semantic mismatch between mathematical language and the English language. Estimates of the level of difficulty of a problem are seldom reliable, particularly those made before a solution is attained. Consequently, as a pragmatic principle, if one wishes to obtain significant results, it is better to study significant questions from the very beginning. Research in these areas has a higher probability of producing worthwhile by-products. Furthermore, natural problems necessarily possess natural solutions. Despite appearances of perversity and recalcitrance, nature provides many helpful clues. On the other hand, artificial problems, such as chess, poker, and translation, need not possess simple or elegant answers.

In what follows, we wish to sketch a mechanization of the art of mathematical model making. This process can be carried quite far by digital computer because of its many surprisingly routine aspects. The point is that an intellectual activity requiring more training than chess-playing, and presumably on a higher level, is actually easier to program for a computer. This is really not unexpected, since a computer does not possess uniform abilities. Consequently, it can carry out some processes in a superior fashion, and others in a quite inferior manner.

The advantage in carrying out this activity is as discussed above. Freed from the burden of carrying out straightforward and tedious calculations, we can devote our time and energy to grappling with the unknown.

Parenthetically, we would like to remark that there are too many pressing questions, and digital computers are still far too much in demand, for time and energy to be spent upon diversions that are simultaneously of great difficulty, of little intrinsic importance, and the solution of which would actually diminish the pleasures of mankind. Research is more than merely doing things that have not been done before.

2. Mathematical Model Making

It often comes as a bit of a shock to the young scientist when he realizes that the basic problem is more to find the right question than the right answer. For example, we have carloads of data in many fields, but we usually lack the equations that govern the data. Theories are, after all, only mnemonic devices to save us from the impossible task of storing all possible information. Given the basic equations and a small amount of data, we can generate the original information, and more.

In the process of constructing a mathematical image of a physical process, we go through the phases of problem recognition, problem formulation, analytic formulation, numerical calculation, evaluation of results, comparison with observation and prediction of behavior, and finally reformulation, and so on (see [1] for a detailed discussion). The process is further complicated in practice by psychological, esthetic, and practical considerations such as availability of computers and constraints on time. In simplified form, nonetheless, the process can be regarded as a multistage decision process. Since learning is involved, in view of the fact that research automatically implies the unknown, the methodology of adaptive control processes may prove quite useful.

At the moment, we are not concerned with optimal procedures, of the type occurring in medical diagnosis, in the screening of drugs, or in sequential testing in general, but rather in exhibiting the use of digital computers in the performance of certain intellectual activities which appear superficially to be of high level. The point once again is that only a very thorough examination of a process determines whether or not it can be discussed by mathematical techniques with the aid of computer technology.

Rather than discuss the whole field of mathematical model making, in a program which would involve some useful but tedious enumeration of the mathematical schemes available for descriptions of the physical

world, we shall discuss a specific problem, that of constructing a mathematical version of the growth of interacting species of cells.

3. The Biological Problem

Suppose that two types of cells, type A and type B, exist in the same
environment and interact. We are given the size of the respective
groups of cells, $x(t)$ and $y(t)$, as functions of time—results derived from
experiment—and we are asked to provide possible kinetics of growth
and interaction. Abstractly, this is equivalent to the problem concerning two species of fish caught off the Italian coast which attracted the
attention of Volterra and initiated his studies of "la lutte pour la vie."

4. Noninteraction

Let us begin with a deterministic view and assume, initially, that there
is no interaction. The traditional model is then one in which it is assumed that the rate of change of the population at any time depends
only on the size of the population. Hence, we want to determine functions $g(x)$ and $h(y)$ with the respective properties that the solution of

$$
\begin{aligned}
\frac{dx_1}{dt} &= g(x_1), & x_1(0) &= c_1, \\
\frac{dy_1}{dt} &= h(y_1), & y_1(0) &= c_2,
\end{aligned}
\tag{4.1}
$$

fits the observed data, given by the functions $x(t)$ and $y(t)$, as closely
as possible.

There are various ways of tackling this problem, depending on what
we wish to obtain. The standard steps are thus

$$
\begin{aligned}
\frac{dx_1}{dt} &= g_1 x_1, & x_1(0) &= c_1, \\
\frac{dy_1}{dt} &= h_1 y_1, & y_1(0) &= c_2,
\end{aligned}
\tag{4.2}
$$

a *linear* growth model, and then

$$
\begin{aligned}
\frac{dx_1}{dt} &= g_1 x_1 - g_2 x_1^2, & x_1(0) &= c_1, \\
\frac{dy_1}{dt} &= h_1 y_1 - h_2 y_1^2, & y_1(0) &= c_2,
\end{aligned}
\tag{4.3}
$$

a simple self-interaction model.

In both cases, the determination of the constants that yield, say, best quadratic fit over an interval $[0, t_0]$,

$$\int_0^T (x_1 - x)^2 \, dt, \qquad \int_0^T (y_1 - y)^2 \, dt \qquad (4.4)$$

can be carried out in many different ways (see [2]–[5]).

Let us suppose that the fit is rather poor, so that more sophisticated models are to be employed.

5. Block Diagram

So far, a block diagram of the computer program has the form shown in Figure 1. We will now add further stages to take account of the second, unsatisfactory, contingency.

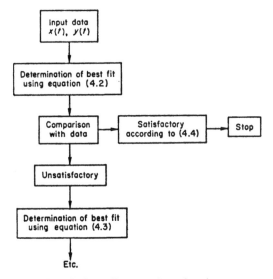

Fig. 1. Block flow diagram for adaptive process.

6. Interactions

Having ascertained the fact that a simple version of independence does not fit the data, we turn to a set of coupled equations of the form

$$\frac{dx_1}{dt} = g(x_1, y_1), \qquad x_1(0) = c_1,$$

$$\frac{dy_1}{dt} = h(x_1, y_1), \qquad y_1(0) = c_2, \qquad (6.1)$$

and a criterion of fit of the form

$$\int_0^T \left[(x - x_1)^2 + (y - y_1)^2 \right] dt. \tag{6.2}$$

There is no reason, of course, to adhere to this simple measure of error. We could use

$$\max_{0 \le t \le T} \left[\,| \, x - x_1 | + | \, y - y_1 | \, \right] \tag{6.3}$$

if we wanted to.

Once again we have a curve-fitting problem, and once again the steps are standard. We begin with a linear model

$$\frac{dx_1}{dt} = g_{11}x_1 + g_{12}y_1, \qquad x_1(0) = c_1,$$

$$\frac{dy_1}{dt} = g_{21}x_1 + g_{22}y_1, \qquad y_1(0) = c_2, \tag{6.4}$$

and use various known methods to determine the coefficients g_{ij} so as to minimize the criterion functions of (6.2) or (6.3). If the error is too great, we introduce nonlinear interaction terms

$$\frac{dx_1}{dt} = g_{11}x_1 + g_{12}y_1 + g_{13}x_1^2 + g_{14}x_1y_1 + g_{15}y_1^2, \qquad x_1(0) = c_1,$$

$$\frac{dy_1}{dt} = g_{21}x_1 + g_{22}y_1 + g_{23}x_1^2 + g_{24}x_1y_1 + g_{25}y_1^2, \qquad y_1(0) = c_2. \tag{6.5}$$

7. Further Adjustments

If the growth curves are unexpectedly obdurate, there are several further standard steps we can take, following the lead of Volterra, Feller, Harris, and others. We can introduce hereditary effects, in the form of time lags or convolution integrals; we can consider a more detailed model based on probabilities of population size and branching processes; we can introduce random forcing terms representing the outside environment; and so on. Each of these models requires some sophisticated analysis, but it is *known* analysis which can be reduced to algorithms and programmed for the computer. We can, without any difficulty, collect the half-dozen or so different mathematical models that have been proposed and apply them sequentially to the particular problem under discussion. The same holds for many classes of processes which have been and are currently being treated in the literature.

8. Discussion

The foregoing procedure can be improved in a number of ways. In the first place, at each step we can arrange to have several alternative models from which to choose, and can make the choice according to the kind of unsatisfactory behavior exhibited by the previous model.

Second, we can apply a man–machine combination, and stop the process from time to time to allow an examination of the results obtained to date. On this basis, we can choose the continuation of sequences of models.

Third, we can introduce adaptive and learning features, in which calls for new information will be made.

In this way, we place ourselves in a position to perform a considerable amount of mathematical experimentation, an activity for which digital computers are ideally suited, and an activity in which they have unfortunately not been prominent to date.

References

1. Bellman, R., and P. Brock, "On the Concepts of a Problem and Problem-solving," *Amer. Math. Monthly*, Vol. 67, 1960, pp. 119–134.
2. Berman, M., E. Shahn, and M. F. Weiss, "The Routine Fitting of Kinetic Data to Models: A Mathematical Formalism for Digital Computers," *Biophys. J.*, Vol. 2, 1962, pp. 275–287.
3. Berman, M., M. F. Weiss, and E. Shahn, "Some Formal Approaches to the Analysis of Kinetic Data in Terms of Linear Compartmental Systems," *Biophys. J.*, Vol. 2, 1962, pp. 289–316.
4. Bellman, R., H. Kagiwada, and R. Kalaba, *Orbit Determination as a Multipoint Boundary Value Problem and Quasilinearization*, The RAND Corporation, RM-3129-PR, May, 1962; in *Proc. Nat. Acad. Sci. USA*, Vol. 48, 1962, pp. 1327–1329.
5. Bellman, R., H. Kagiwada, and R. Kalaba, *A Computational Procedure for Optimal System Design and Utilization*, The RAND Corporation, RM-3174-PR, June, 1962; *Proc. Nat. Acad. Sci. USA*, Vol. 48, 1962, pp. 1524–1528.

For some additional reading on matters of this type, see

6. Bellman, R., C. Clark, C. Craft, D. Malcolm, and F. Ricciardi, "On the Construction of a Multi-person, Multi-stage Business Game," *Operations Res.*, Vol. 5, 1957, pp. 469–503.
7. Bellman, R., "On Heuristic Problem-solving by Newell and Simon," Letter to the Editor, *Operations Res.*, Vol. 6, No. 3, 1958.
8. Bellman, R., "Mathematical Experimentation and Biological Research," *Federation Proc.* (Federation of American Societies for Experimental Biology), Vol. 21, 1962, pp. 109–111.
9. Bellman, R., *Dynamic Programming, Intelligent Machines, and Self-organizing Systems*, The RAND Corporation, RM-3173-PR, June, 1962.

Name Index

Subject Index

Other RAND Books

ARROW, KENNETH J., and MARVIN HOFFENBERG. *A Time Series Analysis of Interindustry Demands.* Amsterdam: North-Holland Publishing Company, 1959.

BAKER, C. L., and F. J. GRUENBERGER. *The First Six Million Prime Numbers.* Madison, Wisc.: The Microcard Foundation, 1959.

BAUM, WARREN C. *The French Economy and the State.* Princeton, N. J.: Princeton University Press, 1958.

BELLMAN, RICHARD. *Adaptive Control Processes: A Guided Tour.* Princeton, N. J.: Princeton University Press, 1961.

BELLMAN, RICHARD. *Dynamic Programming.* Princeton, N. J.: Princeton University Press, 1957.

BELLMAN, RICHARD. *Introduction to Matrix Analysis.* New York: McGraw-Hill Book Company, Inc., 1960.

BELLMAN, RICHARD, and KENNETH L. COOKE. *Differential-Difference Equations.* New York: Academic Press, 1963.

BELLMAN, RICHARD, and STUART E. DREYFUS. *Applied Dynamic Programming.* Princeton, N. J.: Princeton University Press, 1962.

BERGSON, ABRAM. *The Real National Income of Soviet Russia since 1928.* Cambridge, Mass.: Harvard University Press, 1961.

BERGSON, ABRAM, and HANS HEYMANN, JR. *Soviet National Income and Product, 1940–48.* New York: Columbia University Press, 1954.

BRODIE, BERNARD. *Strategy in the Missile Age.* Princeton, N. J.: Princeton University Press, 1959.

BUCHHEIM, ROBERT W., and the Staff of The RAND Corporation. *Space Handbook: Astronautics and Its Applications.* New York: Random House, Inc., 1959.

DAVISON, W. PHILLIPS. *The Berlin Blockade: A Study in Cold War Politics.* Princeton, N. J.: Princeton University Press, 1958.

DINERSTEIN, H. S. *War and the Soviet Union: Nuclear Weapons and the Revolution in Soviet Military and Political Thinking.* New York: Frederick A. Praeger Inc., 1959.

DINERSTEIN, H. S., and LEON GOURÉ. *Two Studies in Soviet Controls: Communism and the Russian Peasant; Moscow in Crisis.* Glencoe, Ill.: The Free Press, 1955.

DORFMAN, ROBERT, PAUL A. SAMUELSON, and ROBERT M. SOLOW. *Linear Programming and Economic Analysis.* New York: McGraw-Hill Book Company, Inc., 1958.

DRESHER, MELVIN. *Games of Strategy: Theory and Applications.* Englewood Cliffs, N. J.: Prentice-Hall, Inc., 1961.

DUBYAGO, A. D. *The Determination of Orbits.* Translated from the Russian by R. D. Burke, G. Gordon, L. N. Rowell, and F. T. Smith. New York: The Macmillan Company, 1961.

EDELEN, DOMINIC G. B. *The Structure of Field Space: An Axiomatic Formulation of Field Physics.* Berkeley and Los Angeles: University of California Press, 1962.

FAINSOD, MERLE. *Smolensk under Soviet Rule.* Cambridge, Mass.: Harvard University Press, 1958.

FORD, L. R., JR., and D. R. FULKERSON. *Flows in Networks.* Princeton, N. J.: Princeton University Press, 1962.

GALE, DAVID. *The Theory of Linear Economic Models.* New York: McGraw-Hill Book Company, Inc., 1960.

GALENSON, WALTER. *Labor Productivity in Soviet and American Industry.* New York: Columbia University Press, 1955.

GARTHOFF, RAYMOND L. *Soviet Military Doctrine.* Glencoe, Ill.: The Free Press, 1953.

GEORGE, ALEXANDER L. *Propaganda Analysis: A Study of Inferences Made from Nazi Propaganda in World War II.* Evanston, Ill.: Row, Peterson and Company, 1959.

GOLDHAMER, HERBERT, and ANDREW W. MARSHALL. *Psychosis and Civilization.* Glencoe, Ill.: The Free Press, 1953.

GOURÉ, LEON. *Civil Defense in the Soviet Union.* Berkeley and Los Angeles: University of California Press. 1962.

GOURÉ, LEON. *The Siege of Leningrad, 1941–1943.* Stanford, Calif.: Stanford University Press, 1962.

HASTINGS, CECIL, JR. *Approximations for Digital Computers.* Princeton, N. J.: Princeton University Press, 1955.

HIRSHLEIFER, JACK, JAMES C. DeHAVEN, and JEROME W. MILLIMAN. *Water Supply: Economics, Technology, and Policy*. Chicago: The University of Chicago Press, 1960.

HITCH, CHARLES J., and ROLAND McKEAN. *The Economics of Defense in the Nuclear Age*. Cambridge, Mass.: Harvard University Press, 1960.

HOEFFDING, OLEG. *Soviet National Income and Product in 1928*. New York: Columbia University Press, 1954.

HSIEH, ALICE L. *Communist China's Strategy in the Nuclear Era*. Englewood Cliffs, N. J.: Prentice-Hall, Inc., 1962.

JANIS, IRVING L. *Air War and Emotional Stress: Psychological Studies of Bombing and Civilian Defense*. New York: McGraw-Hill Book Company, Inc., 1951.

JOHNSON, JOHN J. (ed.) *The Role of the Military in Underdeveloped Countries*. Princeton, N. J.: Princeton University Press, 1962.

JOHNSTONE, WILLIAM C. *Burma's Foreign Policy: A Study in Neutralism*. Cambridge, Mass.: Harvard University Press, 1963.

KECSKEMETI, PAUL. *Strategic Surrender: The Politics of Victory and Defeat*. Stanford, Calif.: Stanford University Press, 1958.

KECSKEMETI, PAUL. *The Unexpected Revolution: Social Forces in the Hungarian Uprising*. Stanford, Calif.: Stanford University Press, 1961.

KERSHAW, JOSEPH A., and ROLAND N. McKEAN. *Teacher Shortages and Salary Schedules*. New York: McGraw-Hill Book Company, Inc., 1962.

KRAMISH, ARNOLD. *Atomic Energy in the Soviet Union*. Stanford, Calif.: Stanford University Press, 1959.

KRIEGER, F. J. *Behind the Sputniks: A Survey of Soviet Space Science*. Washington, D.C.: Public Affairs Press, 1958.

LEITES, NATHAN. *On the Game of Politics in France*. Stanford, Calif.: Stanford University Press, 1959.

LEITES, NATHAN. *The Operational Code of the Politburo*. New York: McGraw-Hill Book Company, Inc., 1951.

LEITES, NATHAN. *A Study of Bolshevism*, Glencoe, Ill.: The Free Press, 1953.

LEITES, NATHAN, and ELSA BERNAUT. *Ritual of Liquidation: The Case of the Moscow Trials*. Glencoe, Ill.: The Free Press, 1954.

McKEAN, ROLAND N. *Efficiency in Government through Systems Analysis: With Emphasis on Water Resource Development*. New York: John Wiley & Sons, Inc., 1958.

McKINSEY, J. C. C. *Introduction to the Theory of Games*. New York: McGraw-Hill Book Company, Inc., 1952.

MEAD, MARGARET. *Soviet Attitudes toward Authority: An Interdisciplinary Approach to Problems of Soviet Character*. New York: McGraw-Hill Book Company, Inc., 1951.

MELNIK, CONSTANTIN, and NATHAN LEITES. *The House without Windows: France Selects a President*. Evanston, Ill.: Row, Peterson and Company, 1958.

MOORSTEEN, RICHARD. *Prices and Production of Machinery in the Soviet Union, 1928–1958*. Cambridge, Mass.: Harvard University Press, 1962.

NEWELL, ALLEN (ed.). *Information Processing Language-V Manual*. Englewood Cliffs, N. J.: Prentice-Hall, Inc., 1961.

O'SULLIVAN, J. J. (ed.). *Protective Construction in a Nuclear Age*. 2 vols. New York: The Macmillan Company, 1961.

THE RAND CORPORATION. *A Million Random Digits with 100,000 Normal Deviates*. Glencoe, Ill.: The Free Press, 1955.

RUSH, MYRON. *The Rise of Khrushchev*. Washington, D.C.: Public Affairs Press, 1958.

SCITOVSKY, TIBOR, EDWARD SHAW, and LORIE TARSHIS. *Mobilizing Resources for War: The Economic Alternatives*. New York: McGraw-Hill Book Company, Inc., 1951.

SELZNICK, PHILIP. *The Organizational Weapon: A Study of Bolshevik Strategy and Tactics*. New York: McGraw-Hill Book Company, Inc., 1952.

SHANLEY, F. R. *Weight-Strength Analysis of Aircraft Structures*. New York: McGraw-Hill Book Company, Inc., 1952.

SMITH, BRUCE LANNES, and CHITRA M. SMITH. *International Communication and Political Opinion: A Guide to the Literature*. Princeton, N. J.: Princeton University Press, 1956.

SPEIER, HANS. *Divided Berlin: The Anatomy of Soviet Political Blackmail*. New York: Frederick A. Praeger Inc., 1961.

SPEIER, HANS. *German Rearmament and Atomic War: The Views of German Military and Political Leaders*. Evanston, Ill.: Row, Peterson and Company, 1957.

SPEIER, HANS, and W. PHILLIPS DAVISON (eds.). *West German Leadership and Foreign Policy*. Evanston, Ill.: Row, Peterson and Company, 1957.

TANHAM, G. K. *Communist Revolutionary Warfare: The Viet Minh in Indochina*. New York: Frederick A. Praeger Inc., 1961.

TRAGER, FRANK N. (ed.). *Marxism in Southeast Asia: A Study of Four Countries*. Stanford, Calif.: Stanford University Press, 1959.

WHITING, ALLEN S. *China Crosses the Yalu: The Decision To Enter the Korean War*. New York: The Macmillan Company, 1960.

WILLIAMS, J. D. *The Compleat Strategyst: Being a Primer on the Theory of Games of Strategy*. New York: McGraw-Hill Book Company, Inc., 1954.

WOLF, CHARLES, JR. *Foreign Aid: Theory and Practice in Southern Asia*. Princeton, N. J.: Princeton University Press, 1960.